Business and Corporate Aviation Management

About the Author

John J. Sheehan is President of Professional Aviation, Inc., which provides management, flight operations, and maintenance consulting services for the civil aviation community and conducts corporate air transportation analyses. He is also an experienced corporate, charter, and military pilot and manager. The International Business Aviation Council is a retainer client. Mr. Sheehan resides in Wilmington, North Carolina.

Business and Corporate Aviation Management

Second Edition

John J. Sheehan

New York   Chicago   San Francisco   Lisbon   London   Madrid
Mexico City   Milan   New Delhi   San Juan   Seoul
Singapore   Sydney   Toronto

Library of Congress Cataloging-in-Publication Data

Sheehan, John J.
    Business and corporate aviation management / John J. Sheehan.—2nd ed.
        p.   cm.
    Includes index.
    ISBN 978-0-07-180190-4 (alk. paper)
    1. Local service airlines—Management.   2. Airlines—Management.
  3. Airports—Management.   4. Aeronautics—Flights—Management.
  5. Aeronautics, Commercial—Management.   6. Aeronautics, Commercial—
  Charters.   7. Airplanes, Company.   8. Private flying.   9. Business travel.
  I. Title.
  HE9785.S54   2013
  387.7′428′068—dc23   2013004871

McGraw-Hill Education books are available at special quantity discounts to use as premiums and sales promotions, or for use in corporate training programs. To contact a representative, please e-mail us at bulksales@mcgraw-hill.com.

Business and Corporate Aviation Management, Second Edition

1 2 3 4 5 6 7 8 9 0 DOC/DOC 1 9 8 7 6 5 4 3

ISBN 9781265942793
MHID 1265942793

Sponsoring Editor Larry S. Hager	**Project Manager** Nancy Dimitry, D&P Editorial Services	**Proofreaders** Don Dimitry, Joseph Cavanagh, D&P Editorial Services
Editing Supervisor Stephen M. Smith	**Composition** D&P Editorial Services	
Production Supervisor Pamela A. Pelton	**Copy Editor** Joseph Cavanagh, D&P Editorial Services	**Indexer** WordCo Indexing Services
Acquisitions Coordinator Bridget L. Thoreson		**Art Director, Cover** Jeff Weeks

To Peggy, my True Course

Contents

Foreword to the Second Edition

When the first edition of John Sheehan's *Business and Corporate Aviation Management* was published in June 2003, it was heralded as one of the best resources for people seeking guidance on how to establish and run a company flight department. For nearly a decade, the book has served as a standard reference for flight departments and the National Business Aviation Association's Corporate Aviation Management program, as well as the primary text for business aviation courses offered by numerous colleges and universities.

However, as everyone in our fast-paced industry can attest, business aviation continues to evolve. Procedures and best practices are changing, not only in order to keep business aviation the most efficient form of transportation possible, but also to ensure a continuing high level of safety, while meeting the international regulatory requirements of what has become a truly global industry.

This second edition of *Business and Corporate Aviation Management* includes descriptions of the characteristics and performance of the latest generation of business aircraft. It also has been revised to underscore the growing importance of safety management systems, including the International Standard for Business Aircraft Operations (IS-BAO). Consequently, the operations and safety chapters have been completely reworked, and substantive changes have been made to the chapters on running and managing a flight department. In addition, the data used throughout the book has been updated.

As business aviation continues to move forward, it is only fitting that the book that has served as the template for so many successful flight departments moves forward, too. I offer my personal congratulations to John Sheehan for helping aviation managers stay on top of what is one of the world's most dynamic industries.

Ed Bolen
President and CEO
National Business Aviation Association
December 2012

Foreword to the
First Edition

Transportation is a necessary component of economic growth and improved quality of life. In Biblical times, King Solomon's power was tied to his system of roads that enabled the peoples of his domain to be effective traders. Nations with the most capable sailing fleets dominated European trade in the centuries before trains and planes. The first U.S. president, George Washington, urged his newly formed Congress to build roads and canals, for to do so would reap rewards. President Lincoln pledged if elected in 1860 to complete the transcontinental railroad to bind the nation and advance commerce. In the mid-twentieth century, President Eisenhower championed legislation that resulted in the interstate highway system, which proved to facilitate significant advances in the nation's economy and the well-being of its citizens.

Today, aviation is the principal form of transportation for business. No current means of communication supports the fast pace of commerce so well. Fax machines, cell phones, videoconferencing, and all the marvels of the communications revolution have not negated the need to travel. Rather, they have simply quickened the pace of business and necessitated the need to be face-to-face with a company's customers and enterprise partners before someone from the competition gets there first.

Think about your own business life. Has your cell phone caused you to slow down or travel less? Has your fax machine taken the place of a visit to a customer or potential customer? Are you comfortable building a bond of trust using the Internet? Do you want to launch your development plans using the telephone or e-mail?

Nothing takes the place of face-to-face contact in keeping clients and growing your business.

While there is universal acceptance that transportation is a necessity for economic development, many business leaders think only of scheduled airlines as a means of satisfying their travel needs. They have yet to understand and use a particularly advantageous form of travel known as *business aviation*.

Scheduled airlines provide safe, secure, and frequent connections between major cities throughout the world, albeit lacking the flexibility to serve many less populated cities and ill-structured to provide time-efficient travel itineraries involving multiple stops. Limited by the strictures of their self-imposed hub and spoke system, scheduled airlines provide frequent and timely service to about 10 percent of the 429 commercial airports that have scheduled operations by air carriers. In fact, nearly three-quarters of all airline passenger enplanements occur at fewer than 50 locations.

Business aviation, the use of general aviation aircraft for business transportation, provides safe and secure access to about 5000 locations within the United States and many more internationally. With the ability to transport employees in a timely fashion, unencumbered by limited airline service and inefficient schedules, companies are able to enhance the productivity of a firm's two most important assets—people and time.

Increasingly, business leaders and planners are looking to business aviation as an effective tool for enhanced productivity and growth. No longer misunderstood or maligned by the specter of an uninformed press, business aviation has emerged as an important travel resource that should be part of a company's travel equation. Over 10,000 U.S. companies own business aircraft, and about 90 percent of the public companies that return the highest dividends and capital gains to shareholders employ some form of business aviation. Worldwide, nearly 14,000 companies own company aircraft. Knowledgeable travel specialists, however, suggest that the number of companies that could employ some form of business aviation advantageously exceeds 100,000.

Business aviation complements rather than competes with scheduled airlines. Members of the National Business Aviation Association (NBAA) are the most active users of business aviation in the world, yet they purchase over $10 billion in airline tickets annually. Companies need to travel, and the most enlightened firms use the airlines when it is most efficient to do so and employ business aviation when that form of transportation is most productive. Thus, understanding business aviation and determining how it can be applied profitably to serve a company's travel needs is a challenging yet beneficial management task.

Also, business aviation comes in several forms, ranging from chartering an aircraft for occasional use, through owning a fractional share of an aircraft managed by a fractional ownership provider, to full ownership by a company and the establishment of a corporate flight department. The enlightened user has access to all aspects of business aviation, selecting the form that best satisfies the company's varying needs. Thus the company with an established in-house flight department also should know when and how to select supplemental lift using charter, for example. The availability of business aviation options reflects the sophistication of this form of business transportation.

Business and Corporate Aviation Management by John Sheehan is a most appropriate and authoritative reference for the company or individual seeking a full understanding of business aviation and its capabilities. Authored by an aviator, educator, and lecturer with 40 years of relevant experience, this handbook encompasses John Sheehan's considerable knowledge of the business aviation community gleaned from years of consulting with the world's safest and most successful flight departments. In his role as a safety auditor over the past 20 years, John has earned the respect of business aviation's leading practitioners. He has prepared and presented seminars on flight department management for the NBAA for more than a dozen years and is scheduled for continuing activity for NBAA members in the areas of business aviation operations.

Good management concepts and techniques are essential to capitalizing on the benefits of business transportation. I recommend *Business and Corporate Aviation Management* by John Sheehan for seasoned flight department managers as well as for company travel department personnel who are exploring how business aviation can benefit their company. This handbook also should be required reading for students seeking to grasp the breadth of transportation capabilities inherent in business aviation. Furthermore, John Sheehan covers the elements of business aviation as a transportation resource in a comprehensive and insightful manner that adds to the handbook's value.

John W. Olcott
President
National Business Aviation Association
November 2002

Acknowledgments for the Second Edition

This revision would not have been possible without the assistance and counsel of the following individuals: on the NBAA staff, Ed Bolen, Steve Brown, Bill Stine, Mike Nichols, Doug Carr, Eli Cotti, and Jay Evans; on the IBAC staff, Ray Rohr, Peter Ingleton, Paul Lessard, Jim Cannon, Sonnie Bates, and Larry Fletcher; on the ICAO staff, Raymond Benjamin, Nancy Graham, Mitch Fox, Richard MacFarlane, and John Illston. David Wyndham of Conklin & de Decker supplied both ideas and data for many of the tables. Tara Harl and her Business and Corporate Aviation Management classes at St. Cloud State University were my insightful supporters. Finally, to the scores of readers who provided suggestions for the new edition, my sincere thanks.

As with the first edition, my wife Peggy was a continuing source of inspiration, encouragement, editing, and moral support.

John J. Sheehan

Acknowledgments for the First Edition

The following people made material contributions to this book through their ideas, experience, and encouragement: Roger Phaneuf, of PAI; Jack Olcott, David Almy, Bob Blouin, and Greg Jackson, of NBAA; Jim Cannon, Rich Messina, Steve Nielsen, Jim West, Len Beauchemin, Pat Cunningham, and Steve Hawkes, flight department managers *par excellence*; Walter Kraujalis, Bloomer DeVere, and Mark Twombly, of WestWord; Frank Hofmann, of IAOPA; Dennis Wright, of FlightTime; Steve Quilty, of BGSU; and Shelley Carr, of McGraw-Hill.

In a larger sense, all my clients and workshop participants through the years actually wrote this book; I just recorded their experiences and observations.

A special thanks to my wife, Peggy, without whose ideas, encouragement, editing, and moral support, this book would never have seen the light of day.

John J. Sheehan

Introduction to the Second Edition

I was surprised when McGraw-Hill asked if I would be interested in writing an updated version of this book. Surprised because the original was 10 years old, and while it had sold moderately well, I thought the work had reached its retirement age and would begin gracing dusty bookshelves. Yet the book has been used by a number of universities, flight departments, corporate executives, and the NBAA Corporate Aviation Management program as a principal reference. So thank you all who purchased the original and have returned for another look. A special thanks to all who provided comments and recommendations regarding the original edition, many of which have been incorporated into this edition.

Some parts of the book remain essentially the same, especially the eternal verities regarding the value of business aviation, selecting the right type of on-demand air transportation, running the business, and maintenance. However, all of these have been brought up-to-date with the latest numbers and techniques.

The major changes to the book involve the complete revisions of the operations and safety chapters. These were the primary revision targets of the new edition because of their importance and impact on the flight department. Substantive changes were made also to the running the business and management chapters. Numerous items were updated, especially the data shown throughout the book.

The new edition contains fewer endnotes and more embedded web addresses, since the Internet can provide current information much more rapidly and in greater depth than non-electronic sources. In fact, I encourage the reader to make more use of his or her computer's search engine whenever a subject of interest presents itself.

While this may sound obvious, especially to the younger generation, I am still puzzled when asked the most elementary questions by clients, questions for which a few searching key strokes will yield an abundance of information.

As before, I incorporate many references to NBAA materials and strongly recommend this organization's excellent website, www.nbaa.org, as a primary source of information. Better still, an NBAA corporate membership will not only yield more personalized information, it will also serve to strengthen the business aviation community.

The operational and safety concepts promulgated in the International Standard for Business Aircraft Operations (IS-BAO) and its attendant Safety Management System are mentioned several times in the book because I strongly believe that this system of best practices and ICAO standards form the best discipline for *any* flight department or charter operation worldwide. In fact, I believe in this system so much that I joined the staff of the International Business Aviation Council (IBAC) a few years ago to help administer the IS-BAO program.

I hope that you will be able to put the new edition to good use and will continue to improve the worldwide business aviation community.

John J. Sheehan
Wilmington, North Carolina
February 2013

Introduction to the First Edition

Why This Book Was Written

This book was written as a consequence of 13 years spent helping individuals and companies to choose the types of aircraft they needed to fulfill their on-demand air transportation needs and to evaluate their flight departments. Certain central themes and patterns of success in business aviation operations have emerged over the years, ideas that I felt should be shared with those desiring to engage in on-demand air transportation.

Finding the Right Aircraft and Method of Delivery

The advantages of having one's own aircraft are evident—creating your own travel schedule, time saved, security, and the ability to go many places that the airlines cannot are all compelling reasons to use on-demand air transportation. Yet few people are sufficiently familiar with the breadth and detail of this form of aviation to choose efficiently among the options available.

Deciding to use on-demand air transportation consists of two tasks: selecting an aircraft and the method of delivering the service. Choosing the aircraft is not as simple as buying an automobile; there are many different types of aircraft, each incorporating a variety of features not apparent or well appreciated by the novice buyer. Seating capacity, runway performance, range versus payload trade-offs, and interior appointments and amenities all make a difference in how the service is delivered. The days of the boss asking the board of directors for $30 million to buy a Gulfstream aircraft without justification are a fond memory in most companies; the board and the boss want to know what they are getting for their money and what the downstream consequences of their purchase are. Likewise, an individual selecting an aircraft and delivery method is faced with a wide variety of choices and combinations much too complex to absorb at a single inquiry.

The common issue left out of the aircraft acquisition equation is, "What are my/our on-demand air transportation needs?" Many acquisitions are made on the recommendation of a trusted acquaintance, senior corporate officer, or aircraft broker, none of whom have a good idea of your total transportation needs or those of the company or what value the aircraft will bring to the equation. An air transportation analysis must be done to determine the types and number of aircraft needed and how the service should be delivered.

While thousands of individuals and companies own and operate aircraft worldwide, other options exist: charter, aircraft management companies, fractional ownership, joint ownership, and combinations of these. Service levels and both capital and operating costs vary widely depending on the type of transportation delivery method chosen. Unfortunately, many of these essential points are lost in the rush to acquire an aircraft or sign up for a fractional share.

The first three chapters of this book address these important issues, attempting to provide insights that will be useful in creating a substantial understanding of how people and companies use their aircraft and decide how the service is to be delivered.

Operating the Flight Department

Running a flight department effectively and efficiently seems to be easy for some but difficult for others. My evaluations originally consisted of examining flight departments for compliance with Federal Aviation Administration (FAA) regulations and the operator's policies to ensure that they were safe. While this was a major responsibility (and the principal factor in causing the evaluation to take place), my work increasingly looked at the business side of the operation. Management, business skills, handling people, administration, reports, planning—all the elements necessary for the effective operation of a flight department—became increasingly important in my evaluations. Not that the safety and compliance aspects were any less important, but it soon become evident that there were some clear distinctions between the good and the not-so-good departments I visited. The single element that made the difference between the two types of departments boiled down to *management*. The desire to take the department's operations beyond tomorrow's flight schedule and the ability to systemize processes, to create control mechanisms, and to handle people well—all these elements of management became evident either by their presence or absence shortly after my arrival at the departments.

At first, it seemed so simple—to detect the differences, catalog them, prescribe remedies, and rapidly leave town. However, it slowly came to me that many of the people with whom I left my remedies did not *understand the management process.* They had never been trained, never been expected to create order out of a complex operation, did not appreciate the value of control and feedback, and never worked in a corporate environment. Many of the flight department managers, chief pilots, and directors of maintenance were *winging it!* These people were fish out of water, out of their natural environment. These good people were used to thinking in terms of flight plans, weather forecasts, air traffic control clearances, maintenance inspection schedules, life-limited parts, and recordkeeping—not budgets, reports, human resources, and long-range plans.

With very few exceptions, these people were bright, well-motivated, sincere people; I knew they could be helped. How best to help them? Some had had elementary management training at a point in their past, some had business degrees, and some had no training at all. There was little common ground from which to start. Thus I fell on the idea of writing a book that would link the theoretical tenets of management with the practical needs of day-to-day flight department operations.

Over the past 11 years, I helped to develop and present workshops for the National Business Aviation Association (NBAA) on developing flight department maintenance and operations manuals and managing small flight departments. These workshops provided me with a wealth of information as a consequence of facilitating thousands of participant interactions about how flight departments should and should not be operated.

The result of my observations and interactions, which you hold, represents my learning experiences from working closely with people in hundreds of aviation organizations over the past 13 years. This is a composite of large and small aviation organizations, low and high usage operations, companies with large jet transports and those with a single piston-powered twin, and one-person departments and those with more than 100 people. I have attempted to draw on the best practices from each segment of the aviation community, but in doing so, I must issue a note of caution: *There is no single right way to make any organization work effectively and efficiently.* Rather, there are as many ways as the human imagination and creativity can conjure; there are literally thousands of *right* ways.

Flight department management must constantly evaluate the organizational culture, economic environment, personnel assigned, mission, and resources to adapt their operations to meet user needs. It is this ability to adapt, to reorient, and to intelligently change the organization and its procedures that makes the difference between the good and not-so-good departments. However, adherence to core value systems and cultural norms is necessary to provide substance and integrity to any organization. Therefore, it is the balance between maintaining values and adjusting to the demands of the environment that can mean the difference between success or failure.

The Audience

This book was written with two basic types of people in mind:

- Individuals seeking information about how to get into the on-demand air transportation business, whether for business or personal reasons
- Flight department managers, their bosses, and those who would become flight department managers

While all these types of people are closely related, they are not necessarily close in outlook, immediate and long-term needs, or background. Yet they all need varying amounts of practical, real-world information to do their job; this is the factor that links them in their common quest for providing safe, reliable, and efficient on-demand air transportation.

Individuals

Whether visiting grandchildren, traveling to a vacation home, or talking with an advisor, individuals have found that going via aircraft according to their own schedule makes life simpler and enjoyable. The aircraft may work out to be a modest single-engine, piston-powered four-seater or a medium-sized turbojet, yet selecting and operating the aircraft sometimes can be a confusing and lengthy task. Where does one begin to understand both the aircraft options and the methods of delivery? And once the decision is made, how is the service to be managed? Similarly, those who work with and advise individuals must understand these features if they are to properly assist their clients and friends.

Business Executives

The harried entrepreneur, chief executive officer (CEO), and senior manager are all victims of the same deficiency: *not enough time.* This is usually coupled with the need to visit with customers, deal makers, and company personnel who will make a difference for their future. They slowly come to realize that they must *make more time for themselves.* Doing so can be difficult, but if they travel frequently, cutting down on their travel time will enable them to get out and back more quickly, spend fewer days (and nights) on the road, and be more productive en route. All these features are the reasons why businesspeople choose on-demand air transportation.

The chairman, CEO, and senior staff are prime candidates for this book, but others within the company will benefit as well. Foremost is the staffer charged with either seeking the right answers to the air transportation questions or attempting to justify the decision made regarding an aircraft type or delivery method made somewhere above his or her level. Such a staffer probably has not thought of all the potential uses of the service nor of the different means of providing it. In doing so, he or she will feel much better about his or her choice and confident about implementing it.

Human resources, finance and accounting, risk management, planning, and legal personnel also need to understand why the company either has an aircraft or will soon get one. Theirs is a need to justify or comprehend the process but, more importantly, to assist the flight operation in succeeding in its mission. By understanding the operation and needs of the flight department, they can become partners in its success.

Finally, for middle managers or salespeople who are desperately attempting to satisfy their pressing travel needs, this book will provide ideas about how to get management's attention and justify their own on-demand air transportation needs. Business aircraft are not just for the few executives at the top of the company; they are business tools that should be used where they make the most sense.

Flight Department Managers

Flight department managers, including the department manager, chief pilot, head scheduler, director of maintenance, and chief inspector, are the front-line, buck-stops-here people who must produce reliable, safe, and on-time on-demand air transportation. Yet their one- or two-aircraft airline does not have the depth of personnel or virtually any other resource a major airline has. They must keep a number of balls in the air constantly; they are master jugglers, with few assistants to relieve them of their many tasks. Before we become overwhelmed with pity for these poor souls, it is well to realize that they also have some of the most rewarding jobs in aviation.

They are highly autonomous, operating in a location remote from the corridors of power and often with a financial independence that is the envy of their peers within the company. In the great majority of cases, they are "aviation junkies," having risen through the ranks of pilot, scheduler, or aviation maintenance technician; they love airplanes and get paid to work with them. What could be better?

This question takes on a new relevance when the CEO is waiting for a mechanical problem to be fixed on the only available aircraft, or the budget proposal is 20 percent over what the boss wants, or word just arrived that the Gulfstream's left engine melted down and it is in Melbourne (Australia, not Florida)! As the old adage goes, this is why they pay the manager the big bucks.

With privilege comes responsibility. With responsibility comes the need to organize, plan, foresee events, and grasp the big picture of where the organization should be in 1, 3, or 5 years. The pilot or technician fresh from the ranks, testing the heady mantle of boss, manager, or leader, may not have had the opportunity to practice the types of skills necessary to assume the responsibilities that accompany the title. He or she rapidly learns that the skills needed to run the show are considerably different from those required of one of the performers.

The Boss

This is the person to whom the flight department manager reports either in a company or in a private operation. This is the person to whom the CEO or principal casually assigned management and oversight of the aviation department, not appreciating that this person's total aviation experience comes from enjoying the perquisites of the first class section of an airliner. In a significant number of cases, the addition of the aviation department to this manager's already impressive list of tasks was not something for which he or she would have volunteered.

These people are typically an executive vice president, senior vice president of administration, vice president of finance, or director of facilities. As far as they are concerned, taking on the aviation department is like assuming the job of elephant master at the circus; airplanes, pilots, and mechanics are strange and sometimes messy animals. However, the fact that they have been put in charge brings pressure to bear—how can they manage it if they do not understand it? Certainly, they must accept the word of their subordinates within the flight department, but this sort of trust only goes so far when the safety of lives and a $2 million annual budget are involved. What are they to do?

The advice of getting smart fast about corporate aviation may be good but largely unattainable for most. The new reporting senior can talk to peers in other companies, talk to his or her friends on the golf course, read everything he or she can lay hands on, or hire a consultant. If only there were a single source of information about everything that goes on out at the airport. This book obviously does not have all the answers, but it will provide a sturdy foundation.

Would-be Managers

An axiom in the airline business states that if you take the most senior line captain and make him or her the chief pilot, you lose the best pilot and gain a lousy manager. This presumes that the captain never had any management training or experience—a reasonable assumption, given the normal career progression of a professional pilot. The same situation is probably not true for an aviation maintenance technician, since there is a logical progression of positions within the maintenance world that requires supervisory and management skills. All this spawns close parallels within the corporate flight department business.

The existing flight department manager may receive a better offer, finally decides to retire, or cannot pass the aviation medical exam—this sets the stage for the senior pilot or technician to succeed to the throne. The call comes from the company reporting senior for a meeting, the offer is made and accepted, and yesterday's happy-go-lucky pilot or technician has just landed in the fire. It is likely, too, that this new-found manager has not been privy to the inner workings and hidden mechanisms that make the flight department run. In essence, this person's new boss has unceremoniously thrown him or her into the deep end and told him or her to start swimming.

The old adage that the best time to know an emergency procedure and the worst time to learn it is in an emergency applies here. If the fresh-caught manager has been preparing for this eventuality for more than a few months, he or she has a head start. With no preparation, the new kid on the block is at a real disadvantage. This book was not specifically designed as a cookbook to be used in the heat of the kitchen, but it will enable the new manager to keep the departmental potatoes from burning while the peas and chicken are tended to.

Finally, the student who is taking a course in aviation management will gain valuable insights regarding the real world of business aviation by reading the book. Note that I said *reading*, not *studying*; this material may enhance and bring reality to a more basic text. Therefore, this book should be considered supplementary reading and not a primary text, since I do not cover all the basic elements associated with planning, finance, personnel management, and the like. With luck, I may convince someone starting in an aviation career to gravitate toward business aviation instead of an alternative.

Satisfying the Needs

Each of these people has a significant need to know how a flight department should be run and what tools are required to make that happen. Much of what they use to accomplish this task comes from prior experience in aviation or outside it. In most cases, this experience is neither appropriate nor fitting to the job at hand. Thus, this book attempts to describe a range of flight department operations and present alternative management methods that will enable the preceding classes of individuals to cope and perhaps thrive.

However, this book alone will not solve all the needs or problems of those interested in business aviation and how it operates; this is just a beginning, a guide, a starting point for those who would manage. It takes curiosity, tenacity, sensitivity, resoluteness, and vision—a wide range of characteristics and traits that mark both leader and manager. These are the skills that draw primarily from life experience, character, and values. The principal characteristic that makes it all work is the individual's *will* to manage or *motivation* to do the job; without these, this book is of little value.

The Book

This book was written to be a practical guide for those who would undertake to manage an on-demand aviation operation, large or small. While the book contains some theory, the majority of its contents are very practical, based on a variety of observations and insights gained from my work with on-demand flight operations. Experiences gained from working with airlines, repair stations, and air taxi operations are used liberally throughout as well.

I realize that there are many different ways to operate a flight department, some of which may be better or more appropriate to a specific situation than a suggestion made within this book. However, it is my intent to provide the uninitiated and those seeking solutions to problems with a starting point for many of the common situations faced by companies and individual operators. Some of the more significant situations described herein contain several alternatives in an attempt to provide flexibility and options, but many offer just a starting point.

While the book may be read cover to cover as a conventional text, the real value should come from its use as a reference. Regardless of how the book is used, however, Chapters 4, "Running the Business," and 5, "Flight Department Management," should be read by all. These chapters serve as starting points for the neophyte and as a refresher for the old hand.

Some sections of this book are more detailed than others, recognizing the fact that some aspects of the manager's job have proved to be either more important or more interesting to managers. Sections as diverse as small flight departments, budgets, safety, working with people in the parent company, and marketing the flight department have been chosen for special, extended treatment. Choosing these aspects of running the flight department over others was difficult, but it is these subjects that have given flight department managers the most trouble or provided the greatest advantages.

The Index should lead readers to the desired topic, if not the general subject area. Readers are encouraged to seek answers to questions in several different topic areas rather than just a single one. The Glossary is designed primarily for those not experienced in aviation issues, attempting to give them a head start in the arcane and voluminous jargon endemic to the world of aviation.

For those wishing to cut to the chase, to get to the real answers of how to run a flight department successfully, see Chapter 9, "Putting It All Together." There I try to sum up the superior performance and wisdom I have witnessed over the years to portray the best methods of serving your company and ensuring flight department survival.

The Chapters

Chapter 1, "Setting the Scene," provides a background to on-demand air transportation, relating reasons for its existence, its definitions and characteristics, and a means for justifying it. An attempt is made to predict the future of the genre, done with more enthusiasm than precision. Each method of delivering this service—employee/owner-flown, in-house flight department, charter, fractional ownership, etc.—is further defined and explored to provide a brief idea of how each one works and might be used to satisfy a company's air transportation needs.

The big issue of determining the shape and size of the air transportation need is discussed in Chapter 2. The air transportation analysis process, evaluating the options for service delivery and evaluating the costs of the venture, is examined. Most importantly, advice on how to select from among the alternatives is presented in a straightforward manner.

"Getting Started" is the title of Chapter 3, which shows in brief form how to begin one of the several types of aviation operations. Emphasis and detail are given to owner-flown and in-house methods, since these are the most common types of operations. Practical advice on finding the right people, setting up a schedule method, and arranging for maintenance is provided.

Chapter 4, "Running the Business," examines the planning, administrative, human resources, and financial details of operating a flight department. A special section is devoted to the small flight department, those operating two or fewer aircraft. This is the most common form of flight department and the type with the greatest burden because of their many tasks and few people available to complete them.

"Management 101" may be a better title for Chapter 5, "Flight Department Management." This covers the background and elements of the practice of management, attempting to show how they can best be used in the context of running a flight department. The chapter examines the flight department manager as a business executive, putting this person into the larger corporate frame of reference.

How the flight operations details of the department work is the subject of Chapter 6, "Operations." It is in this chapter that the details of the flight department are explored: scheduling, standard operating procedures, security, training, and safety. Particular emphasis is placed on the subject of standards and limits to be used in a flight department.

"Maintenance," Chapter 7, is a short but meaty chapter. While much of the glamour associated with flying goes to the operations end of the hangar, maintenance is what really makes it go. The somewhat obscure but essential issues regarding maintenance are covered to provide perspective for technician, pilot, and nonaviation manager alike.

Safety is given special treatment in a chapter of its own, Chapter 8. The enviable safety record of business aviation does not happen by accident; dedicated staff and a conscious adherence to safe operating procedures make it happen. The chapter provides a number of insights and samples of safety programs to use.

Chapter 9, "Putting It All Together," attempts to take all the information imparted in the previous chapters and put it into perspective, attempting to achieve the maximum utility from the more detailed sections of the book. Skim this chapter first, since it contains many of the valuable insights my clients have taught me.

Finally, a Glossary and five Appendices provide useful reference sections. And then, for research, the Index permits fast access to the variety of information provided in the body of this book.

No one book can possibly contain sufficient material to cover a subject in its entirety. This book is no exception, primarily because of the great diversity of operations represented within on-demand aviation. However, this book also contains references that will lead readers to a variety of other, more comprehensive material on the subject. Again, no one work can be truly comprehensive; this is just a beginning.

Being of a mind-set that I should continue to improve my work until I get it right, I would sincerely appreciate reader comments about this book. I would especially be grateful for comments concerning the mix of information and appropriateness of material included for all types of flight departments. You can reach me through the publisher—the second edition will be better because of your comments.

John J. Sheehan
Wilmington, North Carolina
November 2002

CHAPTER 1

Setting the Scene

Four years ago we had four plants and today we have eleven and the only way we have been able to expand our company and to get these locations has been in our own aircraft. —CORPORATE CEO

On-Demand Air Transportation

Within all forms of transportation there are two basic divisions, scheduled and nonscheduled. *Scheduled* is self-explanatory and well understood. Airlines, railroads, buses, and ferry boats operate according to set schedules. *Nonscheduled* requires further explanation; the term expresses a sense of randomness or irregularity, concepts not useful when attempting to define the process. However, *on-demand transportation* is a more descriptive term, signifying that the transportation should be available when requested or needed. Therefore, the purpose of this book is to describe the various types of on-demand air transportation and to tell readers how they can best choose and use the methods comprising it.

Many different people use on-demand air transportation: Individuals, corporate executives, technical troubleshooters, sports teams, entrepreneurs, and families all benefit from the advantages brought by this form of flying. The object for all is to travel to some distant point as safely and comfortably as possible. Some wish to do so in high style, others more economically; some rapidly, others at a more leisurely pace; some doing the flying themselves, others leaving those tasks to professional flight crews. However, all do so because they want to create and maintain their own schedule and control their lives more fully.

This book will examine this form of aviation from two basic perspectives: personal and business aviation. This distinction is made because of the fundamentally different motivation for each, one to serve a personal lifestyle and one to create efficiency in the workplace.

On-Demand Air Transportation Defined

The International Civil Aviation Organization (ICAO) states that all civil aviation operations are divided into three categories: commercial air transportation, general aviation, and aerial work. They are defined as follows:

- *Commercial air transport*—An aircraft operation involving the transport of passengers, cargo, or mail for remuneration or hire.
- *Aerial work*—An aircraft operation in which an aircraft is used for specialized services such as agriculture, construction, photography, surveying, observation and patrol, search and rescue, or aerial advertisement.
- *General aviation*—An aircraft operation other than a commercial air transport operation or an aerial work operation.

For the purposes of this book, we will consider only the first and last types, since they both provide on-demand air transportation, with the fundamental difference being whether the operation is conducted for compensation or for hire (Fig. 1.1).

On-Demand Air Transportation—The transportation of passengers and cargo by aircraft from one point to another in a manner and at a time designated by the person exercising operational control.

Personal aviation may be provided by the owner-operator's own aircraft or by a commercial venture supplying air transportation on demand. Therefore, personal aviation may be a subset of either commercial or general aviation.

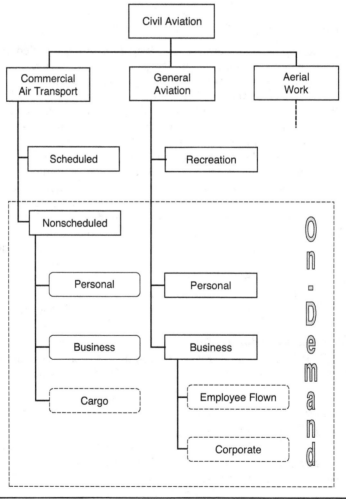

FIGURE 1.1 Civil aviation elements.

Business aviation covers all types of aircraft operations flown in pursuit of business matters. Within this broad classification are employee-flown and corporate operations. Employee-flown operations are predominately owner-flown aircraft, yet these flights may be piloted by any employee, carrying other employees or customers. The key factors involved in employee-flown operations are that the aircraft is operated by a company using a pilot or pilots not specifically employed or compensated to fly the company aircraft. Corporate aircraft operations are distinct from employee-flown operations in that they are operated by a company for its business purposes, using professional pilots (normally two) compensated specifically for their piloting duties. The National Business Aviation Association (NBAA) defines *corporate aviation* as "aircraft owned or leased and operated by a corporation or business firm for the transportation of personnel or cargo in furtherance of the corporation's or firm's business and which are flown by professional pilots receiving a direct salary or compensation for piloting." Note that in either employee-flown or corporate methods, the aircraft does not necessarily have to be owned or leased by the company; it can be rented or even borrowed (some restrictions apply to these operations, however).

Aircraft charter and fractional ownership also may be used to satisfy business requirements for on-demand air transportation.

Personal Aviation

The best means of explaining this segment of air transportation may be via an analogy. Both buses and automobiles are available to people wishing to travel, yet the automobile dominates our ground transportation for obvious reasons: comfort, convenience, and schedule. The automobile provides ground-based on-demand transportation to individuals and groups willing to pay for the service. This is preferred to public transportation for the reasons listed. Perhaps the most important reason for our preference for the automobile is that we wish to create our own schedule and change it at will. Buses (and airliners) seldom permit us this luxury.

There are approximately 350,000 general aviation and aerial work aircraft worldwide, of which an estimated 150,000 are used at least in part for personal transportation. Perhaps one-third of these are used for true on-demand purposes, possessing the ability to fly in weather and to carry two or more passengers. Therefore, personal air transportation constitutes a significant worldwide activity, complementing other forms of transportation that support the needs of individuals. Whether it be a family seeking the fastest and most hassle-free transportation to the ski slopes, an individual flying to a mountain retreat, or a person seeking the best shopping venue in a major city, personal aviation has come of age. These are just a few of the reasons individuals, families, and friends use aircraft.

Some people use an aircraft for recreational, sightseeing, or sport purposes, but these do not constitute on-demand air transportation, which generally refers to flying people or things from point A to point B.

Personal air transportation uses a wide range of aircraft, from small, four-seat, single-engine, piston-powered aircraft to large, multiengine turbojets carrying scores of people over intercontinental distances. However, in this book I will concentrate on three broad classes of personal air transportation: owner-flown operations, those provided by a company as a service, and in-house flight departments using professional flight crews.

Personal and business aviation are separated by a single factor: *purpose*. The organization, regulations, and mechanics of the two operations are essentially the same; only the motivation for the on-demand air transportation services is different. At the lower end of aircraft types used, single-engine aircraft capable of carrying more than two passengers during instrument meteorological conditions (IMC), personal aviation generates an estimated 80 percent of usage worldwide; at the upper end of the spectrum, turbine-powered aircraft are used an estimated 90 percent of the time for business purposes.

Yet both types are organized, regulated, and operated in a similar fashion—the principles are the same. Therefore, within this book, business aviation will be used as the predominant example because it often requires the highest level of organization and standards, therefore making it the form more inclusive of the factors to be considered in many aspects of on-demand air transportation. Where differences exist, they will be discussed.

Our aircraft has enabled my family to be together more frequently and to more fully enjoy those times.
—AIRCRAFT OWNER

Business Aviation

In an average year, hundreds of millions of people travel for business purposes worldwide. This is done in an effort to establish firm connections between businesses, to communicate ideas, to act on opportunities, and to establish a face-to-face connection. While many inter-business communications can be made by telephone, e-mail, fax, and mail, the deals are made, the problems solved, and new frontiers forged by people meeting in person.

Thirty years ago, technology mavens were forecasting a dramatic decrease in business travel due to the advent of cell phones, the Internet, and teleconferencing. Their theory stated that face-to-face meetings were no longer necessary due to the superlative new communications devices; teleconferencing was the next best thing to being there. While these new communications devices did relieve the more routine travel needs, a strange thing happened: Both teleconferencing and business travel increased. Productivity gains provided by communications technology generated the opportunity for more business, which, in turn, generated more business travel. The two connectivity methods worked together to help the productivity of such countries as the United States, Germany, and China become the highest in the world.

The need to be face-to-face in the business world never seems to decrease, only increase. New and important deals are made in person, seldom over the phone or Internet. Customers are cultivated, pampered, listened to, and helped in-person, and new products and services are introduced in-person; business travel is a long-term growth industry. However, there are other means of getting business travelers to their destinations via air.

Most business travel, perhaps two-thirds, is accomplished via the airlines. Every year the world's airlines carry an average of 2.8 billion passengers on 38 million flights worldwide, with roughly one-third of these people traveling for business purposes. The airlines serve thousands of airports, providing frequent, safe, and fairly reliable service to the world's business travelers, yet there are alternatives.

With frequent airline service available to the far corners of the earth, why do we need other aircraft to transport a few people to many of the same destinations? Isn't this much more expensive than the airlines? And how safe is it? These are just a few of the questions asked by both the curious and the critical when considering business aviation.

As this book is being written in 2012, more than 17,000 operators fly in excess of 30,000 turbine (turbojet and turboprop) aircraft worldwide in support of business (Figs. 1.2 and 1.3). Three-quarters of the operators and aircraft are located in North America. In the United

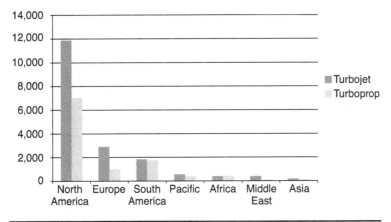

FIGURE 1.2 World turbine-powered business aircraft distribution. (*Flight International.*)

States, each of the approximately 18,000 turbine aircraft flew an average of 450 hours, transporting an estimated 23 million business travelers in 2011.[1] Interestingly, three-quarters of U.S. business flight operations have just one aircraft.

Of the Fortune 500 companies, only one-quarter choose not to operate at least one corporate aircraft. The productivity, net income, and sales of companies operating corporate aircraft exceed those of comparable nonaircraft operators. These companies use their aircraft as productivity tools, as a means of controlling their busy travel schedules, and as a

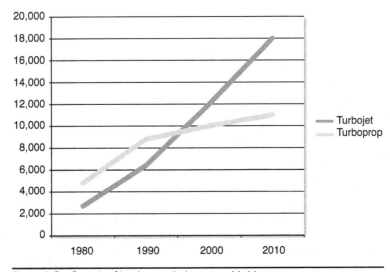

FIGURE 1.3 Growth of business airplanes worldwide.

The Teal Group, a noted forecasting firm, forecast production of 13,879 business aircraft worth $310.3 billion (in 2012 dollars) over the next ten years (2012–2021). This includes 10,249 traditional business jets worth $249.5 billion, 568 corporate versions of jetliners and regional jets worth a combined total of $42.3 billion, and 3062 business turboprops worth a total of $18.6 billion.

convenience for their executives who need to pack every ounce of productive work into each day. In essence, the aircraft becomes a time machine, a modern magic carpet to safely and rapidly transport executives to the next business opportunity.

On-demand air transportation has become a part of corporate culture and promises to be a fixture in the business world for the foreseeable future. It did not take long for companies to discover that having an aircraft of one's own created advantages that the competition did not have. The time savings, flexibility, efficiency, privacy, and security that corporate aviation offered were convincing enough to justify aircraft devoted to a company's exclusive use. However, softer, more intangible benefits accrued to its use too. Increased time at home, the ability to reach even the most remote locations directly, comfort, and fewer airport hassles became sufficient reasons for most to gravitate toward this type of transportation.

The real clincher, the reason that keeps the board of directors and stockholders happy, is the ability of corporate on-demand air transportation to increase productivity for its executives. The corporate aircraft becomes a time multiplier, offering convenient access to as many as 5000 airports in the United States (compared with just 500 that airlines serve) with turbojet speed and an ease not possible on the airlines. It makes little sense to consume 2 days in getting star corporate performers to and from a remote location when the corporate aircraft can get them there, provide a half-day on site for business, and have them home for dinner, *all in the same day.* It is this measurable productivity multiplier effect that keeps selling corporate aviation every day.

Even the accounting department realizes that business flying makes sense. Studies show that companies using their own aircraft have better sales growth, earnings per share, long-term return to investors, and productivity (sales per employee) than companies that do not use business aircraft. The evidence is compelling: Business aircraft are good for the bottom line. The airlines continue to drive executives to corporate aviation in increasing numbers. As scheduled air transportation is forced to restrict and segment its schedules and routes to be more competitive, as airlines overbook and bump passengers in increasing numbers, and as airline terminals become more unmanageable, corporate aviation can only increase in importance; the airlines may be the best reason for using corporate aircraft.

As the global village increasingly becomes a nation's marketplace, the use of corporate aircraft will facilitate this transition. Intercontinental corporate aviation has been a reality for some years and increasingly will serve our interests abroad. Corporate turbojets having nonstop New York to Tokyo and Los Angeles to Paris range are a reality and the hottest-selling aircraft in the business fleet.

Yet corporate aviation is not immune to the business cycle. The health and well-being of the company flight department may be directly tied to the state of the economy, with boom times signaling increased aviation activity and recessions creating not only less activity but also fewer aircraft, hours, and people as well. There will continue to be good times and bad as the economy pursues its mood swings. Despite the bad times, corporate aviation will persist and become stronger as the concept of on-demand air transportation becomes more deeply entrenched in many corporate cultures.

The Beginnings

Personal aviation began with the Wright brothers in 1903. Their objective was to provide a new form of transportation that would enable people to get from point A to point B as rapidly as possible. Their initial 129-ft trip signaled the beginning of personal air transportation. Before the end of the first decade of the twentieth century, individuals were using the airplane as a legitimate, although limited, form of transportation. It was the initial desire for individuals to travel by air rather than groups in larger public aircraft that drove early aviation forward. While interrupted by World War I, personal aviation continued to grow and grow steadily until Lindbergh's transatlantic crossing in 1927, a watershed event for all aviation.

Business aviation began shortly after World War I. Plentiful war-surplus open-cockpit, two-seat biplanes were used to promote various companies' wares via barnstorming trips and advertising courtesy of product logos painted on the fuselage. Hearty entrepreneurs flew or rode in these aircraft to sales opportunities around the country. Yet the rigors of open-cockpit flying and less-than-reliable engines kept these risky trips to a minimum through the 1920s. The world's oil companies, Standard Oil, Texaco, Continental, and Royal Dutch Shell, became the pioneers of true business and corporate flying in the 1920s, using the first multiengine, usually trimotor transports to service their far-flung drill sites.

Standard Oil's chairman, Robert W. Steward, concluded that the advantages of airplane versus train travel "…more than compensate for the expense and make the airplane an economical method of travel…as safe and comfortable as transcontinental train."[2] Despite this glowing endorsement, relatively few companies took advantage of business aviation until the late 1920s. It was the simultaneous availability of closed-cabin monoplanes (with more reliable engines) and Charles Lindbergh's solo transatlantic flight in 1927 that served as the basis for corporate aviation as we know it today.

The Lindbergh flight, combined with national authorities regulating civil aviation, gave travelers new confidence in this fledgling form of transportation. Airlines suddenly sprouted, proving to the world that air travel was safe, efficient, and mostly reliable. Business executives and entrepreneurs, ever seeking a competitive advantage, immediately saw company operated aircraft as a means to facilitate their business activities. The 1930s saw the availability of aircraft specifically designed for business applications, including small single-engine and medium-sized multiengine aircraft. Oil companies again took the lead, using fleets of aircraft, including airliners, to service their international business activities. The growth of business aviation during the 1930s is remarkable in that the great depression curtailed much other business expansion.

Similarly, personal aviation benefited greatly from Lindbergh's transatlantic flight, turning all eyes to the skies. During the 1930s, scores of new aircraft were introduced with the thought of providing personal transportation for the masses. Despite the constraints of the poor economic conditions, light aviation flourished and grew. Business aviation and the airlines provided the new developments and the infrastructure that benefited personal flight.

World War II called a temporary halt to civil aviation development, but in a very real sense, it prepared the world for the aviation boom that followed the cessation of hostilities. Hundreds of thousands of people learned about and became acclimated to the concept of air transportation during their wartime service, leading them to transfer their skills and knowledge of aviation to the business environment. Postwar boom times and the availability of surplus military transports fueled a rapid growth in both personal and business aviation.

Significantly, the rapid advances of wartime aviation technology made civil flying much safer, more reliable, and even practical. Improved communications and navigation equipment, more reliable engines and aircraft systems, and the beginnings of a real air traffic control system facilitated the move toward business and personal aviation.

While business and personal (general) aviation grew substantially during the 1950s, it was the introduction of the business jet at the end of the decade that made the industry take off. In 1957, the Lockheed Jetstar made its first flight and ushered in a new era of speed and convenience for corporate travelers. Fast, comfortable, and practical, "airborne board rooms" became a reality with the jet age. Followed shortly by the Sabreliner and Hawker/ deHaviland 125, the jets provided a quantum increase in business travel effectiveness and efficiency. Then, in 1964, the first Learjet brought jet travel within economical reach of most corporations. The Learjet cost $550,000 and provided speeds of up to 485 mph over distances as great as 1800 mi, accommodating up to seven passengers. Business aviation had arrived.

Commercial helicopters were available as early as 1946, but it was the advent of the Sue-Est Alouette turbine-powered helicopter in 1958 that brought a practical rotary-wing aircraft to business applications. While thousands of business helicopters are used worldwide for business aviation, their numbers remain small relative to the number of fixed-wing aircraft working for business. Yet their importance for specialized applications make them valuable and permanent assets in the corporate transportation inventory.

Personal aviation benefited from these developments, taking aerodynamics, radio, and infrastructure improvements for its own. During the 1960s and 1970s, personal aviation grew rapidly, attempting to reach the dream of practical and safe flying for all. The late 1970s and early 1980s saw a peak in the number of pilots and aircraft produced, essentially realizing the dream within the bounds of economic practicality. Finally, small general aviation aircraft could be used as a practical form of personal transportation through most weather conditions and at speeds and distances that made sense to the consumer.

Through the mid-1970s, business aviation aircraft growth was modest, with fewer than 100 turbine-powered aircraft being produced for the business community in 1972. Good economic times in the late 1970s and early 1980s produced a miniature boom in business turbine aircraft, with more than 400 such aircraft produced at the peak in 1982. During the same year, more than 2000 piston-powered light general aviation aircraft were produced, a significant number destined for business uses. In 2001, more than 1200 turbine-powered business aircraft were shipped worldwide, approaching the 1800 piston-powered airplanes produced in the same year; the emphasis is clearly turning to turbine-powered business aircraft.[3]

In 1986, some 14,000 turbojets and turboprops flew for business around the world; that number had grown to more than 30,000 by the end of 2012 (18,000 of these are turbojets), 100 percent growth in just 26 years. Importantly, 17,000 operators worldwide sponsored all these aircraft. The rate of increase is growing, perhaps as rapidly as 4 percent per year. Business aviation has earned itself a permanent place in the world's corporations.[4]

Personal aviation has taken its place as a practical transportation alternative throughout the world with an estimated 150,000 aircraft used for safe, reliable, transportation.

Travel Is Important for Business

Corporations of all sizes increasingly focus on their travel expenditures, which often represent the third largest expenditure within many companies. As such, corporate stakeholders traditionally have viewed business travel from the narrow perspective of costs as opposed to its relationship to corporate productivity. Understanding the need to better define the corporate perception of travel and the role of travel management, the Institute of Business

Travel Management (IBTM) commissioned an environmental scan to assess the value senior corporate executives place on business travel and travel management in meeting corporate goals and objectives.

With participants ranging from chief executive officers (CEOs), senior vice presidents, vice presidents, and chief financial officers (CFOs) of Fortune 500 companies including Boeing, EDS Corporation, Johnson & Johnson, Microsoft, and The Limited, Deloitte & Touche conducted the first research project of the IBTM.[5]

Key findings included the following:

- *Travel is important to meeting company goals and objectives, and travel management programs contribute significantly to companies.* As a result of the combination of an increasingly global economy and increasing company growth fueled by the economic boom of the last few years, companies have strategically channeled more funds into the cost of conducting business. Like other business units, travel is expected to increase the corporation's return on investment and increase process efficiencies. The findings of this study clearly indicate that senior executives value the role of travel and effective travel management. Further, the study found that senior executives believe that travel management programs make significant contributions to the corporation and support the achievement of corporate missions and goals.

- *Few companies use sophisticated models to calculate the contribution of travel management to company bottom lines and shareholder value.* While senior executives value the contributions of the travel management programs, the study found that few companies use sophisticated means to calculate the true value of their travel programs. Although executives implicitly understand that travel affects other areas of the company such as sales, marketing, and operations, most travel management programs do not employ measurements that effectively calculate travel's bottom-line impact on the corporation. As a result, travel's impact on the company is strongly recognized, but a mechanism to quantify its impact has not been fully developed.

- *Senior management's highest travel priority is to optimize time while being very sensitive to the balance between time and cost.* While minimizing costs for any company is critical, the study found that those interviewed also understand that "time is money." Thus, in crafting travel policy, entering supplier relationships, and developing performance measurements, travel managers should develop methods that integrate the time and cost elements of travel.

The Reasons[6]

Deciding how to travel—via the airlines, via company or charter aircraft, or even by driving or taking a train—involves many considerations. While several of the benefits of business aircraft are tangible and measurable, some are challenging to quantify precisely. Progressive managers routinely consider all the costs and realistically evaluate all the benefits of every travel option before deciding how to go. According to the NBAA these benefits include the following:

Saving Employee Time. Efficient employee scheduling and employee time saved are key advantages of business aircraft use. Because business aircraft have the ability to fly nonstop between any of the 3500 small, close-in airports—ten times the number of locations served by scheduled airlines in the United States—highly efficient employee time management becomes a very real benefit. Additionally, the value of employee time often exceeds its cost to the company by substantial margins, further increasing the importance of employee time savings. Simply stated, business aviation helps a company obtain maximum productivity from its two most important assets—people and time.

Increasing Productivity En Route.   High levels of employee productivity en route to a business destination—in a secure office environment that is free from interruptions, distractions, or eavesdropping—can have substantial value to an employer. A Louis Harris & Associates survey showed that executives felt that they were 20 percent more productive in the company jet than they were in the office. Conversely, they felt that they were 40 percent less productive in an airliner due to distractions and lack of privacy.

Group productivity, maximized due to the common availability of club seating and tables, often is unique to business aircraft. Strategizing before meetings and debriefing afterwards are common practices facilitated and encouraged by business aircraft cabin configurations (Figs. 1.4 and 1.5).

Minimizing Nonbusiness Hours Away from Home.   Family time before and after traditional business hours is critical to most employees. Because a stable, supportive family can have an acute effect on employee morale and productivity, scheduling that minimizes time away from home can be a key benefit.

Ensuring Industrial Security.   For many companies, the protection of personnel from uncontrolled public exposure alone is justification for business aircraft use. Avoiding eavesdropping, reducing travel visibility, and eliminating unwanted and unnecessary conversations and interruptions all support the use of business aircraft to safeguard company employees and the sensitive information they carry.

Maximizing Personal Safety and Peace of Mind.   Turbine-powered aircraft flown by two-person professional crews have a safety record comparable with or better than that of

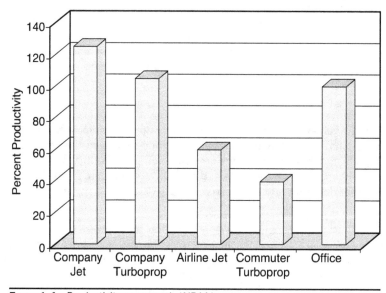

FIGURE **1.4**   Productivity compared. (*NBAA.*)

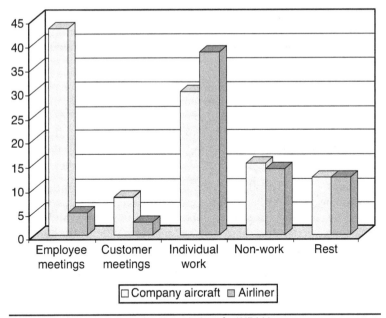

Figure 1.5 Passenger activity aboard aircraft. (*NBAA.*)

scheduled airlines. The peace of mind that results from complete company control over the aircraft flown, passenger and baggage manifests, pilot quality and training, aircraft maintenance, and operational safety standards is substantial. This benefit also can include the rescheduling of flights if weather, mechanical, or other considerations suggest that this is the appropriate course.

Exercising Management Control over Efficient, Reliable Scheduling. The near-total scheduling flexibility inherent in business aircraft—even changing itineraries en route—can be a powerful asset. Since aircraft can arrive and depart on the passenger's schedule, typically waiting for them in the ordinary course of business, meetings can be moved up, back, or extended without penalty, risk, or unnecessary scheduling pressures. Overnight trips also can be avoided. If managed proactively, this benefit can improve business results.

Projecting a Positive Corporate Image. For customers in particular and often for vendors, the arrival and departure of company employees via business aircraft are the sign of a well-run company, signaling the progressive nature of an organization with a keen interest in efficient time management and high levels of productivity. If used for charitable purposes, significant public service contributions, as well as possible public relations benefits, also can be realized.

Attracting and Retaining Key People (Customers Included). The right person in the right place at the right time can change everything. Finding and keeping such people can hinge on many factors, including the ability to maintain reasonable travel schedules, maximizing personal productivity and ensuring family time. Holding on to valuable employees also can prevent companies from spending time and resources on training replacement employees.

Reducing Posttrip Fatigue and Increasing Posttrip Productivity.   Schedules that require late-night travel or longer-than-necessary trips often result in posttrip fatigue, damaging productivity in the day(s) after the trip. Because they can facilitate more efficient scheduling, business aircraft can minimize this loss.

Optimizing Payroll.   Under "rightsizing" initiatives, many organizations have rediscovered the need to maximize the productivity of the same or fewer employees to accomplish equal or greater amounts of work and ensure their competitive position and long-term success. As business aircraft improve employee time management and efficiency, they can help eliminate the need for additional personnel, reduce payroll costs, and help to maximize a company's competitive market advantage.

Truncating Cycle Times.   The compound effect of increased productivity and saved travel time is that more can be accomplished in less time. Consequently, many companies attribute reductions in *cycle times*—when facilities are brought online sooner and projects finished faster—to business aircraft use. Although it is challenging to quantify or attribute entirely to business aircraft use, this benefit often can be substantial.

Charging the Entrepreneurial Spirit.   By minimizing or eliminating many of the barriers to travel, business aircraft allow business opportunities to be considered and acted on more readily. Business cultures and their strategies change as markets, facilities, and customers in rural areas of the country—once practically unreachable and unconsidered—are newly accessible.

How Companies Use Aircraft[7]

There are a nearly infinite variety of ways to use business aircraft. From the obvious, getting the executives out and back rapidly, to transporting teams of specialists, to going out and getting customers to witness your excellent manufacturing or service organization, companies have learned that these and other applications make sense.

The NBAA has studied how companies use their aircraft. Here are just a few ways that organizations have discovered to use their aircraft (Table 1.1):

Key Employee Travel.   This is by far the most frequent use of business aircraft. Getting to essential meetings, conducting site visits, and keeping customers happy are all reasons for the right people to get to the right place in a *timely manner*. The timeliness of these trips means choosing the right time for the trip and then getting out and back as quickly as possible. We all know intuitively that the business adage "time is money" is true, so we try to maximize our time where it counts most. Spending time waiting at airports in security lines does not meet this criterion.

Key employees can be anyone of any rank who is indispensable to a task, not just the CEO. Financial experts, attorneys, technicians, and quality control specialists are all candidates for using the company aircraft when hours count. Not surprisingly, an NBAA study revealed that top-level executives were not the most frequent flyers in company aircraft; middle managers were.

Business aircraft are used commonly by key employees to

- Extend management control and bring operational areas to a manageable size
- Facilitate company, supplier, and/or customer meetings in multiple cities per day
- Take practical and routine daily on-site supervision of facilities in different cities

Key employee travel.   Getting the right person in the right place at the right time.

Customer visits.   Visit customers on their turf.

Customer trips.   Bring customers to you.

Scheduled customer service.   Routine trips to service customer accounts.

Emergency customer service.   Rapid response trips to fix what's broken and "put out fires."

Humanitarian and charitable flights.   Being a good corporate citizen; helping employees.

Sales and marketing blitzes.   Multiday/multicity sales trips covering a region or sales area.

Charter revenue flights.   Offering your aircraft for use by a charter operator.

International flying.   Regularly outside the United States.

Helicopters.   Used to go directly to specific destinations; not just between airports.

Management teams.   Transporting management teams to organization sites.

Engineering teams.   Transporting production or engineering teams to critical work sites.

Corporate shuttles.   Regularly scheduled flights between organization facilities or customer sites.

Make airline connections.   Making airline connections, particularly international flights.

Carry priority cargo, parts, or mail.

Special projects.   Such as advertising shoots.

For goodwill/lobbying.   Transporting elected officials or candidates; going to lawmakers.

Utilitarian purposes.   Mapping, aerial surveys or inspections, etc.

Market expansion.   Evaluating new markets/sites.

The airborne office.   Working/conferring en route.

Personal travel.   Employees and their families.

Attract and retain key people.   A tool to facilitate work or get people home more nights.

Maximize employee safety and industrial security.   Better than the airlines.

TABLE 1.1   Business Aircraft Utilization Strategies (*NBAA*)

- Reduce, sometimes dramatically, travel times to multiple locations versus public transportation
- Facilitate emergency meetings, including those involving the board, partners, and customers
- Efficiently reach remote locations
- Strengthen relationships with customers through shared private time en route to a destination
- Facilitate rapid action on mergers and acquisitions, particularly for companies in remote locations

Customer Visits.   Most businesses place customer contact at the top of their priority list. An integral part of such contacts involves face-to-face visits—visits to introduce new products, discuss existing product lines, explore new customer needs, and in general, be nice to the people who ultimately pay the bills. However, getting to customer locations, which frequently are poorly served by the airlines, can be a several-day chore. Since business aircraft have access to many more airports than do the airlines, a number of customer visits can be made in one day with the company "magic carpet."

Saving Time and More

A control-valve manufacturer in Milwaukee was in the final stages of signing a major contract with a potential customer located in Asheville, North Carolina. A major sticking point arose over a number of issues, including final design, production rate, and costs. The situation was deemed "critical" by the company executive vice president. Clearly, a meeting of the minds was in order. A visit to the Asheville customer was planned.

As is often the case, finding a 3-day window for the meeting for all concerned (at both locations) was difficult. Why 3 days for a meeting that should require at most 6 hours? Airline service from Milwaukee to Asheville was such that the five travelers had to depart the day before and return the day after the meeting. Therefore, the executive vice president chose the only day available to all and scheduled the company aircraft for the trip.

The team rose early on the day of the trip, went to a conveniently located general aviation airport in suburban Milwaukee, boarded the aircraft on arrival (no waiting or processing necessary), and 1.5 hours later arrived in Asheville. Exactly 7 hours later (a little longer than they expected), they reboarded their aircraft and arrived back in Milwaukee in time for a stylishly late dinner.

Not only did team members save an aggregate 60 productive hours, they also saved more than 120 hours of their personal time in the process. Moreover, the 3-hour roundtrip was used profitably in strategy and implementation meetings in the company aircraft. The best part: The cost of the company aircraft was 20 percent less than the combined fares would have been on an airline.

Visits via business aircraft can be used to

- Attend customer-sponsored events, such as a grand opening.
- Service more than one destination in a day.
- Efficiently increase face-to-face contact with the customer, relationship building, interaction between people. "They think more highly of us when they see more of us."
- Facilitate a lean senior staff. "We have a lean management, so it has to cover a wide area quickly." Consequently, business aircraft can facilitate senior management participation in selling.
- Enable a team approach rather than a one-person attempt. "If there is a new business opportunity that requires a team visit, we dispatch them out on that call."
- Demonstrate capability; establish or reinforce an image.
- Support a customer in trouble. "Being there sometimes helps."

Market Expansion.   The attractiveness and potential of new geographic markets may be limited by a lack of physical access to those markets. Business aircraft can open potential new markets, including international ones, and provide dramatically improved access especially to rural areas. Market expansion is facilitated as much by attitude as by access. Business aircraft, because they can facilitate access, can lessen or remove perceived barriers to the management of new markets.

Prospecting within new markets is the first step, often for potential local business partners. Business aircraft sometimes are used to facilitate meetings that take place at a halfway point into these new markets. A new class of business aircraft with nonstop ranges in excess of 7000 miles is making access to global markets practical and common. Almost every flight on business aircraft has an element of market expansion to it.

Management Team Travel.   Management team travel is the most common use of business aircraft. These teams provide a critical mass of expertise, talent, and experience that can be applied to problems and challenges not found at the home base. Getting these individuals out to the problem area and back expeditiously makes good business sense.

Management teams have a substantial aggregate hourly value to their employers. The combined costs of air services, employee travel time (door to door), and other trip expenses (such as hotel, meals, rental cars, etc.) often compare favorably to airline travel between second-tier cities or rural locations. When the value of employee time is considered along with the value of productive time en route and nonbusiness hours away from home (family time), the comparison of business aircraft travel versus public transportation often becomes problematic.

Business aviation is playing a key role in helping with the economic recovery in Europe, a new study finds, saying that each passenger flown on a business aviation flight generates the same gross domestic product as nine business passengers on a scheduled flight. It has been shown that two-thirds of executives declare face-to-face contact to be crucial in deal-making," says Brian Humphries, EBAA president. "If you consider (the study's) finding that 96% of city pairs served by business aviation in 2011 had no scheduled connection, it is little wonder that business aircraft passengers place a value on business aviation flights that (are) between 8 to 15 times higher than those made on scheduled airlines.

Some progressive companies have or are installing conference facilities in corporate hangars to facilitate off-site meetings. Corporate hangars also are being equipped with transient manager offices that can be used for private conversations, study, or sending and receiving e-mail.

Transporting Customers.   A corollary to customer trips is the practice of bringing customers to your facilities to witness your excellence and to impress them with your knowledge and insight. Moreover, fetching customers in the company airplane makes a substantial impression on them; this means that they are very important people to you.

The practice of transporting customers can

- Be used to pick them up, bring them to company facilities, meetings, cultural, entertainment, or sporting events, and then return them home as quickly as is practical, often on the same day
- Provide an opportunity to build relationships and ease communications
- Help those in need to restore service by moving their personnel and equipment
- Improve customer access in both directions
- Be used as a courtesy to correct for company-induced delays
- Engender goodwill
- Facilitate approvals for the delivery of products or services
- Facilitate customer contract signing

Letting Them See Your Excellence

A major office furniture manufacturer provides its products to a worldwide audience; however, most of its sales occur in the United States. Early in its life it saw the need for bringing customers to its Midwest headquarters, first by car and train and then by company aircraft. The investment in customer transportation has paid off, making the company a member of the Fortune 500 and a world-class provider of office furniture.

"We can show them what we make in catalogs and showrooms," says a member of the company's marketing team, "but getting them on our turf and showing them how all the pieces fit together and how dedicated we are to quality is a powerful marketing tool. We couldn't do this without the company airplanes."

International Trips.   The allure of emerging markets—both as potential suppliers and as customers—in China, India, and Russia is of increasing interest. Hundreds of business aircraft fly every day between North America and Europe, Asia, Africa, or South America. A surprising number of the aircraft used for this purpose are small to midsized jets.

Many companies transport management teams internationally for multicity visits, some covering dozens of cities over multiple weeks. Given the limited airline service in some regions, this can be a highly efficient practice. Importantly, as the teams travel, they have an increased opportunity to confer and plan during lengthy en route legs because of the uniquely private and communicative environment provided by the business aircraft cabin.

Canada, Mexico, Central America, and the Caribbean are the most frequent destinations for U.S. companies; many of these markets become closer via business aircraft than cross-country U.S. markets. As major markets and producers move offshore, the need to get to those sources becomes increasingly important; business aviation provides ready access to these areas at greatly reduced cycle times.

Specialty Teams.   Teams of engineering, troubleshooting, survey, and financial employees are commonly dispatched to rural or remote areas to

- Rapidly restore service at "down" facilities or sites
- Monitor, inspect, and review construction progress
- Install, modify, or dismantle equipment
- Deliver and install emergency parts
- Evaluate potential construction sites
- Attend or facilitate critical meetings
- Visit suppliers

Not only do you get the teams there faster, but by using a business aircraft, you also send a message to the passengers and customers alike: This is an important job that deserves attention.

Sales/Marketing Campaigns.   Sales and marketing blitzes can be of nearly any scope and duration, depending on the stamina of the participants and their families. Such blitzes can be highly efficient and concentrated direct selling opportunities. Because they are time-efficient, participants are able to spend more time in the office and less time in airports.

Team Efficiency

A major U.S. catalog clothing retailer regularly ensures the quality of its products by sending quality control teams to its contractors' manufacturing plants to inspect the operation. When the manufacturing took place within the United States, normally in the southeastern part of the country, these visits were relatively simple and quick, involving a week on the road, via airline, at most to cover four to six plants using a team of five to seven inspectors. However, when most of the manufacturing activity went offshore, primarily to Central America and the Caribbean, the inspection teams began to spend more than 3 weeks to accomplish their task.

A chance meeting between the company flight department manager and head of quality control spawned a scheme to make the company more efficient and to get the inspection team home more rapidly. By using the company Learjet, up to seven team members could be taken to the offshore plants, sometimes two per day, in about half the time required by international airline travel. This scenario is now played out several times a year, allowing the company to visit its suppliers more frequently while reducing wear and tear on its inspectors. Importantly, using the company jet saves tens of thousands of dollars on travel and trip expenses.

In some companies, business aviation is the prevailing method used to get senior officers into a specific marketing territory to visit customers and vendors. Because time is so limited, multiday trips facilitate maximum exposure in a limited time.

Sales and marketing blitzes can intensify seasonally, during new product introduction cycles, for initial public offerings (IPOs) or investment/investor swings, during downturns in the economy, or following a restructuring. Potential customers and distributors also can be the target of multiday trips.

Attract and Retain Key People

Attraction:

- Companies commonly use business aircraft as recruiting tools to facilitate the interview and negotiation process for key prospective employees or business partners and their families.
- The greater the distance and the more rural the recruiting base, the greater is the apparent benefit of a company aircraft.
- Anything that helps motivate key employee recruitment is critical.
- The value of the first impression left on a prospective employee is very important.
- The use of business aircraft rapidly establishes credibility.
- Companies make regular flights to universities to recruit graduating students.

Retention:

- Companies can keep personnel longer by making their days shorter.
- The use of business aircraft for commuting or other personal travel can be negotiated as part of a personal services contract.

- Extra "no cost" passengers can fill empty seats as a reward to high-performance employees and their families on certain trips scheduled for business reasons, although this type of travel may be taxable as personal income to those passengers.

- A company philosophy that stresses family-friendly scheduling—reducing non-business hours away from home through the use of business aircraft for efficient day trips—can be very attractive.

- A surprising number of goodwill flights seem to be flown, typically very quietly, to facilitate employee medical treatments.

Corporate Shuttles.   Some companies with multiple operating bases have found it advantageous to institute a scheduled flight operation between those bases to save both time and money relative to airline service. Airline schedules frequently do not fit the daily commute nature of many work requirements, driving companies to establish what amounts to their own private airline. Time savings in airport processing delays with the airlines greatly reduces an employee's availability to accomplish work productively; a private shuttle may reduce this wasted time by as much as 80 percent. Moreover, with just a modest number of travelers, significant cost savings may be realized with a shuttle.

Shuttles are most appropriate for larger companies with decentralized operations and facilities or for companies with repetitive tasks, such as safety checks, management reviews, ongoing employee training, or a need to make airline connections from rural locations. Shuttles are not always permanent, often operating for short periods for specific projects or start-ups. High-priority personnel can be transported regularly via helicopter to several company locations in a major metropolitan area, saving many hours of tedious ground travel.

Priority Cargo, Parts, or Mail.   This subject has immense potential. Many companies use aircraft to carry high-value cargo, replacement or spare parts, or interoffice mail. This practice is common and often unquantified but of substantial commercial value in displaced overnight delivery charges. Any material that cannot be moved practically by conventional shippers due to size, weight, susceptibility to damage, or other physical limitations, and particularly those which need to be delivered within hours, often are best transported via business aircraft.

Projects.   Usually business aircraft are used for infrequent or "one shot" projects, such as multilocation advertising photography "shoots" involving a multidisciplined team of experts. The combination of both personnel and an unusually large amount of project-specific equipment can trigger business aircraft use as the most efficient and cost-effective travel alternative, specifically for transporting delicate equipment of extreme value to the project.

> *It's just essential for us to get out in the field.... Aviation is the most time-efficient way for us to do that.*   —MARKETING MANAGER

Utilitarian.   Business aircraft are used commonly as a convertible platform for mapping; aerial surveys, inspections, and photography; cattle ranching and herding; flying laboratories for airborne experiments or other research and development; educational laboratories; surveillance; and power-line and pipeline patrols, in addition to other uses.

Public, Press, and Investor Relations.   Progressive companies often manage opportunities to project a positive corporate image using business aircraft either as a symbol of efficiency or as simple transportation to facilities or events to further this purpose. Publicity generated by the use of aircraft for noble purposes, such as the rapid delivery of relief supplies after a natural disaster, can be of substantial public relations benefit.

Personal Travel.   Just because the aircraft is operated by a company does not preclude its use for personal transportation. However, the company must define personal-use policies and provide methods for employees to pay for these services.

Emergency Evacuation/Response.   Business aircraft also are used to remove company employees from harm's way, usually as a result of a medical emergency, natural disaster, civil strife, or other security threat, particularly in areas of limited or uncertain scheduled airline service. Similarly, the ability to place key personnel at the scene of an oil spill, manufacturing plant fire, or tornado-damaged office site rapidly makes good business sense. Getting to the scene early with the right people may save additional damage and costly downtime.

Goodwill.   Business aircraft also are used for personal customer or employee needs, such as travel associated with family emergencies, or as a favor. There is not always an immediate, tangible business benefit from the use of business aircraft for customers or employees with special needs or for public officials. A longer view, however, may suggest that the company's strategic interests can be best served by this practice depending on individual circumstances.

Lobbying.   Business aircraft can be used to transport elected or appointed officials. The use of business aircraft for the carriage of public officials is common at all levels of government. Specific Internal Revenue Service (IRS) regulations govern this practice. Legal restrictions also can limit the availability of sponsored travel for certain government employees. All travel of this type is routinely and publicly reported. In rare instances, business aircraft also are used to carry government officials to company-sponsored events. These can be unusual and invaluable opportunities for information exchange.

Other.   Any need to travel can be accomplished safely, reliably, and efficiently with the company aircraft. After beginning to use their aircraft, companies rapidly discover many collateral uses. The only limit on these uses lies in the minds of the users.

Seven Key Enterprise Value Drivers Resulting from Business Aircraft Utilization

Financial	Nonfinancial
Revenue growth	Customer satisfaction
Profit growth	Employee productivity, motivation, and satisfaction
Asset efficiency	Innovation
	Risk management and compliance

Studies show that there is a clear correlation between business aircraft utilization, the associated benefits of use, and the key financial and nonfinancial drivers of enterprise value.
(*Source: NEXA Advisors, 2010.*)

Justifying Business Aviation

Most businesspeople identify quite well with the advantages of on-demand air transportation, yet the inevitable question becomes, "Isn't all of this rather expensive?" The answer is a qualified "Yes."

Much of what a company does requires large quantities of investment capital and operating funds to achieve desired goals and objectives. Yet there is a vision behind every capital and operating expense that permits the company to move ahead and compete effectively in the marketplace. Granted, the company always must be cognizant of bottom-line and rate-of-return concerns, but it is the vision of an individual or small group of individuals that carries a winning project forward to its destiny; such is the case in most companies when the concept of safe, reliable, and efficient transportation is being considered.

In evaluating transportation alternatives, tangible and intangible benefits must be defined to begin to realize the value of on-demand air transportation.

Tangible and Intangible Benefits.   Business aviation creates a number of benefits for individuals and the company. Some can be measured, and others elude quantification. Nonetheless, all these factors are very real and should be considered when evaluating on-demand air transportation options.

The most obvious and quantifiable advantage of on-demand air transportation is the amount of time saved for the passengers. Access to 10 times the number of airline airports, fewer check-in formalities, direct routing, landing near the business destination, and no waiting for luggage save major portions of days for single trips and full days for multiple-leg trips. The time saved means more time on task for employees, greater productivity en route, the ability to confer with fellow travelers en route, and fewer distractions during the entire travel process that detract from focus on the issues at hand.

Operational flexibility, or the ability to react intelligently to changing business situations, often serves companies well. Being locked into an inflexible airline schedule and lack of alternative flights at the last minute serve as sources of frustration for business travelers. Schedule changes, even in flight, can be arranged easily for on-demand transportation.

While often perceived to be more important to the traveler than to the company, the number of days and nights spent on the road are important to both. The company needs its key personnel to be operating at maximum productivity as much as possible; nights in motels and days in airline terminals seldom make it possible to achieve this goal.

The preceding items are quite important features of business life and are measurable. Yet some of the most important aspects of business aviation are immeasurable.

- Security concerns have become very important for all travelers. The very personal, private, and well-controlled nature of business aviation makes this a very secure form of transportation, one that allays the concerns of most travelers.

- Fatigue is a major negative component of travel, especially when public transportation modes are used. Business aviation provides a secure, hassle-free environment that promotes a relaxed travel experience. Further, it is a lot easier to rest in the peace and quiet of an intimate personal space offered by business aircraft.

- "Road warriors," those who spend a major portion of their time traveling, may become discouraged or even depressed at the thought of suffering the indignities associated with airline travel, innumerable nights on the road, and the stress associated with the entire process. On-demand air transportation provides a morale boost for weary travelers, permitting them to regain a better lifestyle and become more productive.

- While the concepts of prestige and image are not normally considered in the employee business equation, the ability to use on-demand air transportation creates a very positive reaction in business travelers. Companies promote additional image and prestige items in many aspects of an executive's life. Why not in the way they travel? This feature will permit the company to attract and retain the best and brightest performers.

- Companies rely on privacy and confidentiality in their daily operations to gain advantage over the competition. These features are also important in ensuring the safety and security of business travelers. It is difficult to work on confidential information in airport terminals and airliners because of prying eyes and ears. On-demand air transportation offers complete privacy, with just a few trusted operational personnel aware of destinations, schedules, and manifests.

- Finally, comfort and convenience may be overlooked or discounted as significant factors in business travelers. One has but to ask busy executives why they value the company aircraft, and these features usually will be among the most frequently mentioned.

The airplane is an extension of our office; we probably have our most productive meetings there.
—CORPORATE CEO

Value of Time. *Value* in this case refers not to the intrinsic monetary benefit of on-demand air transportation but to the worth in terms of usefulness, utility, and merit. The following set of statements will illustrate this: (1) "A coach airline fare to Ft. Wayne will cost X, whereas the company airplane (or charter) will cost $X+$" and (2) "Getting the vice president of marketing to Ft. Wayne tomorrow morning will save the Acme contract."

There are several issues at work here:

- How important is the Acme contract?
- What *value* does the contract hold for the company?
  - Projected revenues
  - Future business
  - Criticality to other products/services
- Can anyone get to Ft. Wayne via the airlines by tomorrow morning?
- Will such a flight require an overnight stay?
- What else could the vice president of marketing be doing with his or her time?

Before we look at any of these variables, it is important to examine the *value* of an employee's time. Many companies mistakenly equate the value of an employee's time with his or her compensation level. For instance, if the vice president of marketing receives $250,000 in total compensation, some would say that this person is worth $125 per hour ($250,000 divided by 2000 hours of work time per year). Yet this would be short-sighted and would sell the *value* of the employee short, as well.

Human resources accounting recognizes employee skills, training, and experience as a capital asset. This field is based on the fact that employees are vitally important to the success of a company. Studies indicate that profits are due in large part to the actions and expertise of management personnel. Human resources accounting places a value on people rather than a cost. These values have been recognized and quantified by the insurance industry through key-person insurance policies for business. The theory behind this insurance is the potential loss to a company as a consequence of the injury or death of a key person; life insurance is issued in the name of the key employee, with the company listed as the beneficiary.

Easier to understand is an inherent belief that each employee must have the power to bring in sufficient revenues beyond his or her compensation to achieve profitability. If this were not true, the company would only pass through revenues to the employees as compensation and have nothing left over for profit. The assumption of an employee's ability to make a profit for the company is understandably higher for upper-level executives, but the expectation is nonetheless levied on each employee within the company. In recognition of this expectation, key-person insurance is written as multiples of an employee's compensation, generally in the range of 2.5 to 15 times salaries and other forms of compensation.

If this concept is to be believed, then the inherent *value* of an employee becomes some multiple of his or her compensation. By extension, this value should extend to all aspects of the employee's work life, including examination of low-level administrative tasks, supervisory duties, and the use of travel time. Therefore, if, in our example, the marketing vice president can be expected to bring revenues of 4 to 6 times compensation, then every aspect of his or her time carries the same value, including travel time. At the higher multiple, this executive is worth $750 per hour, making the value of travel time a pricey commodity. If this executive has to leave the office 3 hours before a scheduled airline departure, spend an overnight at the destination due to inadequate airline schedules, drive a rental car 1 hour from the destination airport to the real destination, and invest another day returning to base, the waste of time, talent, and ultimately, dollars adds up rapidly.

All the travel variables involved in computing the time-value equation have proved too complex in the past to permit rapid or real-time calculation. However, the NBAA has devised a sophisticated software tool to compute just such values. *Travel$ense* is a tool designed to help management better analyze and understand business travel. With surprising accuracy and flexibility, this tool automatically calculates the actual business hours, productivity, trip expenses, and family time saved by traveling via business aircraft versus airline alternatives. For each trip, the tool accounts for and calculates

- *Trip purpose and outcome.*   The benefit in cost-benefit terms.
- *Cost of air service.*   On two business aircraft *and* comparable airline flights using real-time data.
- *Cost of employee travel time.*   Door to door for each passenger, leg, and trip on three aircraft.
- *Other travel expenses.*   Including overnight costs.
- *Value of employee travel time.*
- *Value of productive time en route.*
- *Nonbusiness hours away from home.*
- *Productive hours saved and other trip statistics.*

Executives and flight department personnel have found *Travel$ense* to be a powerful tool to illustrate the extraordinary business advantages of business aircraft.

I gain an additional productive month each year with the time I save by using the company airplane. Love those 13-month years!   —CEO OF A MAJOR SERVICES COMPANY

Creating Shareholder Value. Once the value of time is appreciated, it is relatively easy to calculate the per-trip cost alternatives versus value to the company. Ultimately, some person or persons within the company must fully appreciate the potential benefit of on-demand air transportation to the company and act as a catalyst and spearhead to make this happen. However, others have studied the process from a macro viewpoint and reached some interesting conclusions.

Anderson Consulting assessed the potential financial benefits of operating business aircraft to companies and their shareholders by examining peer groups of companies distinguished by their use or nonuse of business aircraft.[8] The study looked at the companies comprising the Standard and Poor's (S&P) 500 list, which represents the largest U.S. publicly held companies according to their market capitalization. The S&P 500 were classified into 24 industry groups, evaluating each of them as to the number of aircraft operators and nonoperators. Financial performance and share price information were examined for the period 1992 through 1999.

The results showed that aircraft operators earned 146 percent more in cumulative returns than nonoperators from 1992 through 1999 (609 versus 463 percent). Increased shareholder returns primarily were due to improved business performance. Executive interviews revealed that business aircraft were essential "to conduct due diligence for a number of potential targets" and to "get our financial teams into and out of hard-to-reach locations." Aircraft operators achieved sales and earnings growth nearly double that of nonoperators. Key drivers of sales growth include the ability to identify and execute strategic transactions and alliances.

The aircraft helps us increase and protect revenues. —CORPORATE CFO

Other key areas of business performance yielded margins for aircraft operators of up to 100 percent above their nonaircraft peers.

- Average cumulative earnings before interest, taxes, depreciation, and amortization (EBITDA) growth
- Increased productivity
- Asset efficiency ratio (sales to total assets) decline

Quality improvements emerged as a theme that supported several drivers of shareholder value. As one chief financial officer (CFO) stated, the chairman's methodical oversight of the company's network made possible by using business aircraft "sets the tone" across the United States and around the world, helping to ensure a single level of quality. Another executive said that her company used the fleet to "sell time to our executives by executing trips in hours instead of days." As another CFO of a financial services company explained, the key profitability benefits of business aircraft derived from increased productivity and efficiency for top executives—benefits that are "obviously there, but hard to quantify."

Corporate performance relative to the use of company aircraft was revealed in four factors:

- *Strategic transaction orientation.* Being able to facilitate critical transactions was associated most regularly with direct shareholder value creation. Rapid transaction value is essential in several industry sectors, especially those in consolidation.
- *Customer service orientation.* Time-sensitive requirements, such as emergency customer services, supported sales retention and sales growth and could be met most efficiently by some companies using business aircraft (no ready substitute).

- *Process and quality improvement orientation.* Being better able to manage and execute far-flung operations was found to be the most extensively cited trait. Business aircraft enabled executives to visit multiple locations, sometimes more than once a year, by customizing schedules not possible on commercial airlines. Executives were able to review operations, efficiency, quality, and customer service.

- *Meritocracy orientation.* When a company uses aircraft to treat all employees as an important asset, it achieves uncommon results. Because the workday could be lengthened without sacrificing employee family time, shuttling employees between company facilities offered significant productivity gains. Enhanced employee safety and security and the security of intellectual property were characteristics of this orientation.

Our executives often use the aircraft to reach meetings they couldn't otherwise attend, closing transactions that would fund the flight department for years to come. —CFO

The study concluded that there was a strong correlation between business aircraft benefits and shareholder value creation. Its central finding was that business aircraft can make a substantial difference in how a company performs its mission, in many cases generating significant gains in the drivers of shareholder value. Increased mobility was at the core of these gains—satisfying management's need for greater organizational agility, knowledge integration, and transaction speed.

Significantly, in the study's interviews with CFOs, more than 75 percent confirmed that disposing of their business aircraft could, for the same reasons, potentially harm their company's value. In many cases it was found that there often were no ready substitutes for business aircraft without diminishing company performance or losing new business opportunities.

Hands-On Helps

A manufacturer of emergency-response municipal telephone systems introduced a new line of equipment and attempted to market it via conventional means using sales personnel to call on clients in the field. Unfortunately, the innovative features of the equipment were difficult to describe to potential customers. Simulator boxes, call-in demonstrations, and videotapes were employed in an attempt to demonstrate the features of the equipment, with little result.

The vice president of marketing convinced the CEO to use the company's Citation aircraft to bring representatives from two municipalities to company headquarters to attend a hands-on demonstration of the equipment. The first visit achieved a sale to one community and a promise to send municipal officials back for further evaluation. Based on this and subsequent successes, the company aircraft now brings potential customers to headquarters for demonstration once a week. The passengers are impressed with the fact that they travel in a corporate jet and are back home the same day. Best of all, they are able to fully assess the benefits of the product by observing it in operation.

Business Aviation: The Sign of a Well-Managed Company

According to a 2009 NEXA[9] study released by NBAA, using nonfinancial measures, the highest-performing companies appearing on several "Best of" lists reveal a remarkable correlation with business aircraft use among the following categories:

- *Business Week's 2009 "50 Most Innovative Companies."* 95 percent of the S&P 500 companies on that list were users.

- *Fortune's 2009 "100 Best Places to Work."* 86 percent of the S&P 500 companies on that list were users.

- *Business Week's 2009 "25 Best Customer Service Companies."* 90 percent of the S&P 500 companies on that list were users.

- *Business Week/Interbrand's 2008 "100 Best Brands."* 98 percent of the S&P 500 companies on that list were users.

- *Fortune's 2009 "50 World's Most Admired Companies."* 95 percent of the S&P 500 companies on that list were users.

- *CRO Magazine's (now titled Corporate Responsibility (CR) Magazine) 2009 "100 Best Corporate Citizens."* 90 percent of the S&P 500 companies on that list were users.

Safety

Companies engaged in business aviation activities choose to operate or contract for their own aircraft as an alternative to the airlines and charter operators. In doing so, they expect that their operations will be as safe as or safer than the public transportation option. Therefore, company aviation operations work under safety programs that provide the desired level of safety.

Corporate aviation operators are justifiably proud of their excellent safety record, equaling or exceeding the airline accident record for their turbine-powered aircraft operations. For instance, the airlines experienced over a recent five-year period 0.37 accidents per 100,000 departures whereas corporate aviation experienced 0.42 accident on the same basis. (See Chapter 8 for more safety information.)[10] Even though this in an enviable record, business aviation operators strive for a perfect record of no accidents; while this goal has not been achieved, it remains a credible goal for the industry.

Most corporate aviation flight departments voluntarily exceed the minimum operating standards imposed by their national regulatory agencies (such as the Federal Aviation Administration) for their type of operation. In fact, corporate operators' standards and limits closely resemble those used by the airlines and charter operators. Two experienced pilots, certified to airline standards, fly company aircraft and undergo recurrent training in airline quality simulators; maintenance technicians use state-of-the-art techniques and materials to keep the aircraft in first-class condition; and all personnel are trained in ground and flight emergency procedures.

Owner- or employee-flown business aircraft enjoy a considerably better safety record than the whole of general aviation but lag slightly behind that of their corporate aviation brethren. This is understandable because these pilots are probably less experienced and fly aircraft that may be less capable than a corporate turbojet. Yet, while comparisons are difficult, traveling in an employee-flown business aircraft is safer than traveling an equivalent distance in an automobile.

Companies and their pilots, technicians, and support personnel demonstrate their pride in safety through the NBAA Flying Safety Awards. Each year at the NBAA convention, hundreds of individuals and companies receive recognition for their many years of safe operations. A number of companies have achieved the 50 Year Safe Flying Achievement Award, indicating that they have had a corporate flight operation for that long without having had an aircraft accident. Similarly, at a recent convention, more than 900 pilots received recognition that they had flown at least 1500 accident-free hours. These awards are cumulative, often honoring pilots who have achieved tens of thousands of accident-free hours and technicians who have amassed scores of accident-free years of service.

Business aviation operators take safety seriously, since their jobs and reputations depend on each and every flight. All operations are planned carefully and executed in accordance with mature standards and procedures stemming from industry best practices.

Bombardier Aerospace predicts that North America will receive the greatest number of new business jet deliveries between 2012 and 2031 with 9500 aircraft, followed by Europe, with 3920 aircraft. Notably, China will become the third largest market for business jet deliveries, with 2420 deliveries from 2012 to 2031. Bombardier also expects key growth markets, including Brazil, India, Russia and the Commonwealth of Independent States (CIS), Indonesia, Mexico, South Korea, and Turkey, to receive a significant share of business jet deliveries during the next 20 years.

Aircraft

Aircraft come in two broad categories: airplanes and helicopters. Airplanes are what we normally see at the airport—an engine-driven, heavier-than-air device supported by the dynamic action of the air on wing surfaces. Helicopters are marvelous devices that move through the air supported by a rotor revolving above the fuselage; really, the helicopter moves its "wing" before the fuselage moves, as is the case with a conventional airplane.

Each of these devices, in turn, is classified by the type of powerplant used, either piston- or turbine-powered. The piston engine is usually installed in smaller single- and twin-engine versions used in employee-flown aircraft. Similarly, employee-flown helicopters may be powered by a piston engine more often than the more sophisticated turbine engine used for corporate applications.

Turbine engines come in two varieties: turbojets and turboprops. While their propulsive force relies on a gas turbine in both cases, the turbojet relies entirely on its ability to accelerate air through its interior to move it forward. The turboprop uses its gas generator to drive a propeller through reduction gears, which pulls it through the air.

Any aircraft (including both airplanes and helicopters) may use one or more engines to propel it through the air. Smaller aircraft tend to have one engine. The larger the aircraft, the more likely it is to have more than one engine.

Aircraft Characteristics

A four- to six-seat, single-engine, piston-powered airplane is considered ideal for some owner- and employee-flown operations. Its simplicity and relatively low operating cost make it ideal for an entry-level aircraft. This aircraft is ideal for trips of less than 500 miles carrying not more than three passengers. Importantly, in ideal conditions, it can operate safely on runways just 2500 ft long, whereas airliners normally require at least 6000-ft runways[11] (Table 1.2 and Figs. 1.6 through 1.14).

Type of aircraft	Example	Passenger seats	Normal cruise speed (mph)	Range with half seats full (mi)	Acquisition cost new ($) (million)	Direct operating cost ($/hr)
			Airplanes			
Single-engine piston	Cirrus SR22T*	3	200	900	0.5	160
Single-engine turboprop	TBM 850*	5	330	1400	3	600
Multiengine turboprop	Hawker Beechcraft King Air B250	8	340	1600	6	1000
Small twin-engine turbojet	Embraer Phenom 300	7	430	2200	9	1200
Medium twin-engine turbojet	Cessna Sovereign	9	460	3200	18	2200
Large twin-engine turbojet	Gulfstream G650	16	550	8000	65	2900
			Helicopters			
Single-engine turbine	Robinson R44	3	130	325	1	315
Twin-engine turbine medium	Bell 429	6	160	500	5	700
Twin-engine turbine large	Eurocopter Dauphin EC 155	8	160	500	11	700

*Primarily owner- or employee-flown.

TABLE 1.2 Business Aircraft Characteristics

Figure 1.6   Cirrus SR22. (*Courtesy Cirrus Aircraft.*)

Figure 1.7   SOCATA TBM 700. (*Courtesy SOCATA Aircraft.*)

FIGURE 1.8 Hawker Beechcraft King Air 250. (*Courtesy Hawker Beechcraft Corp.*)

FIGURE 1.9 Embraer Phenom 300. (*Courtesy Embraer SA.*)

FIGURE 1.10   Cessna Sovereign. (*Courtesy Cessna Aircraft.*)

FIGURE 1.11   Gulfstream 650. (*Courtesy Gulfstream Aerospace.*)

FIGURE 1.12   Robinson R44. (*Courtesy Robinson.*)

FIGURE 1.13   Bell 409. (*Courtesy Bell Helicopter.*)

Figure 1.14   Eurocopter EC 155. (*Courtesy Eurocopter.*)

The next step up is a four- to eight-seat, twin-engine, piston-powered airplane, also normally used in owner- or employee-flown operations. The additional engine provides increased load carrying and speed capability along with increased safety. Incremental increases in both speed and range are realized over the single-engine aircraft.

A single-engine turboprop is the next level in the upward progression of both capability and costs. The powerplants used in single-engine turboprops are smooth-running and quite reliable. These aircraft also normally are employee-flown, although many companies use either a full-time professional pilot as the principal pilot or as a part-time copilot for the owner/pilot. Additional experience is required for operators of turbine aircraft to obtain reasonably priced hull and liability insurance.

The twin-turboprop is the entry-level professionally flown aircraft, yet owners may pilot this type as well. This versatile aircraft is ideal for trips of less than 700 miles while carrying up to four passengers, although most can fly up to 1500 miles nonstop with up to six passengers. In general, this type of aircraft can operate out of smaller airports than can pure turbojets; a 3500-ft runway is considered adequate under ideal conditions.

While a number of single-engine turbojets are under development, entry-level turbojets currently all have two engines. This class of airplane generally will accommodate four to six passengers with a flight crew of two pilots. They will carry two to four passengers 2000 miles nonstop at speeds approaching 400 mph.

The next step up is a midsized-cabin twin-turbojet featuring up to eight passenger seats and capable of flying nonstop coast to coast, west to east (nonstop westbound transcontinental flights are seldom possible in many of these aircraft due to headwinds, especially during the winter). The range of airports that will support this type becomes smaller because they require at least 5000-ft runways to operate safely under normal conditions.

The large-cabin aircraft accommodate up to 14 passengers and have substantial range capability. Although not possible with all seats full, some of these aircraft can easily make it nonstop from New York to Paris.

Finally, the true intercontinental turbojet can carry 15 passengers in comfort for distances of up to 8000 miles. These aircraft require airline-sized runways and more extensive support facilities.

On the helicopter side, the entry-level single-turbine-engine aircraft will carry three passengers 325 miles at 120 mph. The more capable twin-turbine helicopter will carry four passengers 400 miles at 130 mph. The high-end twin-turbine helicopter accommodates up to seven passengers on journeys of 350 miles at 130 mph.

On-Demand Aviation Methods

One of the great advantages of on-demand air transportation is the number of methods of delivering this service. Companies or individuals do not have to own or even lease their own aircraft, although this is the most common form of operation. Importantly, more than one type of delivery method is often employed to meet emerging user requirements. For example, a company owning an airplane may contract for charter services on occasion to meet unusual requirements or heavy demand. Similarly, an individual may purchase a share of an aircraft from either a company or private owner to form a joint ownership agreement or from a fractional provider to augment existing aircraft.

In each of the methods listed below, the relative advantages and disadvantages will be discussed for each form of delivery. A full appreciation and understanding of these features is not available until the unique requirements of an individual company are known. Therefore, those contemplating on-demand air transportation should seek a wide variety of information sources to be fully informed prior to making a decision.

On-demand aviation includes the following methods:

- Owner/employee-flown
- In-house flight department
- Management company
- Joint ownership
- Interchange
- Time share
- Charter
- Fractional ownership

Owner/Employee-Flown

This is the entry level of on-demand air transportation; in many cases, a businessperson or individual discovers the advantages of on-demand air transportation through others and either learns to fly or uses an already qualified employee to fly an aircraft and manage its operation. The aircraft is purchased and used primarily for either business or personal purposes, depending on the situation.

An aircraft may be purchased, leased, or rented, depending on personal or company preferences. However, most aircraft are wholly owned. In most countries, all administrative and operational costs may be fully charged as business expenses when the aircraft is being flown for a business purpose.

The majority of these operations fall into the single- or twin-engine piston or turboprop airplane category, although an increasing number of these applications use helicopters and small turbojets. While it is difficult to determine the actual number of aircraft used for this type of flying, literally tens of thousands of these applications exist worldwide. In the United States, GAMA statistics showed that during the year 2010, more than 17,000 aircraft were used primarily for employee-flown business operations.

National aviation authorities governing this type of flying impose only the most basic standards governing civil aviation. Holders of a private pilot license may carry passengers anywhere, even internationally. However, insurance requirements and good operating practices dictate that more than minimum standards be employed for this type of flying.

In-House Flight Department Using Owned/Leased Aircraft

Most companies engaged in corporate aviation or receiving personal transportation services in larger aircraft use this form of operation. It is self-descriptive, indicating that an owner/ operator uses in-house assets to administer, fund, and operate his or her own aircraft. This form of operation may be most popular because it affords the user the maximum amount of control over the aircraft; i.e., all choices regarding ownership, scheduling, crewing, and training the crew, and maintaining and operating the aircraft remain with the user; outside individuals or companies have no role in operation of the aircraft.

Control over these functions is important to corporate travelers and individuals because they want their aircraft to be available when they want them and crewed and maintained by people they know and trust and they want to have full control over the financial aspects of operation. Seeing familiar faces in the cockpit and around the hangar gives passengers a sense of control and comfort often deemed essential.

These flight departments report to either the principal user or a subordinate regarding aircraft operations. All personnel required to schedule, fly, maintain, and administer the aircraft report to the flight department manager or chief pilot. All functions regarding budgeting, scheduling, and operation of the aircraft are the responsibility of this manager. Ultimate control of the aircraft and its personnel rests with this individual.

The decision of whether to own or lease an aircraft is a matter of company fiscal policy or personal preference, with no common de-nominator driving this decision. Carrying a major capital asset like an aircraft on the company books and enjoying the benefits of depreciation appeal to some companies, whereas others prefer to expense all costs associated with aircraft operation through a lease, thereby avoiding the carrying costs of a major asset. Synthetic leases provide the combined features of both owned and leased aircraft, creating a third option.

Management Company

In this option, a company or individual owns or leases an aircraft but contracts with a management company to provide all functions required to operate it. This includes scheduling, crewing, maintaining, bill paying, and management oversight of the entire operation. This option is often used by those new to corporate aviation or not wishing to be bothered with the details of aircraft operation. Typically, a fixed management fee is paid to the management company, and all operational, administrative, and employee compensation costs are passed through without additional fee to the owner/operator.

Management companies have the advantage of providing their services to a number of companies and benefiting from the resulting economies of scale. For instance, the flight crew assigned to Company X's aircraft also can be used on Company Y's aircraft when the former aircraft is not active. Management companies typically operate an air taxi operation to complement their management operations, offering to use managed aircraft for their charter operation to provide their owners with income that can be used to offset their operating costs. However, in doing so, users relinquish significant amounts of control over their own schedule and how the aircraft is to be used.

Joint Ownership

In this arrangement, two or more companies or individuals jointly own an aircraft. One of the owners employs and furnishes flight crew and arranges for scheduling, maintenance, and administrative services. The other owners pay their share of all operating, employment, and administrative costs.

Joint ownership lowers operating costs for owners who do not fully use an aircraft; however, the method of determining who can use the aircraft at a particular time can prove difficult if clear scheduling guidelines are not established up-front. Yet a significant number of companies and individuals operate well under this method, especially if the nature of their use patterns complements one another. For instance, if a company typically only uses the aircraft during the week and a private individual normally uses it on weekends, the two requirements tend to be mutually exclusive and noninterfering.

Managed companies also provide their services to joint owners, providing them with fully outsourced operations.

Interchange

This option involves an entity leasing an aircraft to another person in exchange for equal time, when needed, on that person's aircraft. No charge, assessment, or fee is made except that a charge may be made equal to the difference in the operating costs of the two aircraft.

Those entering into interchange agreements do so primarily to provide themselves with a locally positioned backup aircraft for use when their aircraft is either scheduled or in maintenance. The ability to occasionally use, at relatively low cost, an aircraft with appreciably different capacity or operating characteristics is another reason to enter into this type of arrangement.

Time Share

This approach is really a method of lending one's aircraft for discrete periods to a second party and recouping a portion of the operating costs. In this arrangement, a company essentially leases an aircraft, with flight crew, to another person for the cost of operating the aircraft for that flight. The real advantage of this method is that it is not considered to be a commercial venture in the United States and other countries and therefore is not subject to air charter regulations.

A formal time-sharing agreement must be established between the parties involved. While this is an excellent method of providing transportation to company employees who desire to use the aircraft regularly for personal purposes and to favored individuals, some tax and liability consequences are incurred by the party sharing the aircraft.

Charter

Aircraft charter is the normal on-demand air transportation entry point for most users. Charter companies offer a wide range of aircraft types with a corresponding range of prices designed to suit a variety of tastes and uses.

Legitimate charter companies must possess an air taxi operating certificate issued by their national regulatory authority to legally offer their services to the public. With the same philosophy that requires the airlines to provide a higher level of safety to the traveling public than afforded by general aviation, charter operators must operate to a higher set of regulatory standards too.

Typically, a person desiring air charter services will find a provider meeting his or her requirements and call them to determine availability and the costs involved. Normally, few other formalities are required except for telling the charter operator when and where the service is needed. Arranging for this service is normally a simple transaction; however, determining the right charter provider to use can be difficult if high-quality, reliable, and cost-effective services are to be found.

A small number of in-house and managed flight operations, perhaps 10 to 15 percent, offer their aircraft for charter use as well. Either they obtain their own authorization (operating certificate) from the regulatory agency to provide charter service, or they enter into an agreement with another company possessing an operating certificate to make their aircraft available. Offering one's aircraft for charter is done primarily to recoup some of the aircraft operating costs from the charter revenue, especially if the aircraft is not being fully used. However, most people who operate their own aircraft do not offer them for charter because of additional regulatory requirements, interference with their schedules, and added wear and tear on their aircraft.

Fractional Ownership

This form of transportation delivery is essentially a large-scale joint-ownership operation. Instead of buying into just one aircraft, the owner buys into a fleet of similar aircraft. While the owner buys a share (as little as one-sixteenth) of one aircraft, the fractional provider also makes a fleet of identical aircraft available to fractional owners through an interchange agreement. This provision allows owners to lease their aircraft to another person participating in the fractional program in exchange for use of the other's aircraft. In this manner, the chances of a like aircraft being available when the owner wants to use it are very good, as opposed to waiting in line for a single aircraft in the case of a single jointly owned aircraft. In fact, availability is guaranteed, utilizing contract charter aircraft when necessary.

Owners purchase a fraction of an aircraft and are entitled to all tax and depreciation benefits of their share of ownership. Additionally, they pay a monthly management fee and an hourly operating charge in exchange for guaranteed response times (4 to 12 hours) and utilization levels (a one-eighth share receives 100 occupied hours annually) in either their aircraft or an identical one. The minimum term of the contract is normally 5 years, at which time owners can renew or sell their share for a predetermined value. All of this occurs as an all-inclusive service with all required support functions provided, virtually invisible to the owner.

This form of transportation is the fastest-growing form of on-demand air transportation, with more than 8000 fractional owners of approximately 1000 aircraft as of mid-2012. It is attractive because of its joint-ownership features, guaranteed availability and response times, uniform service levels, and hassle-free operation.

Fractional ownership is considered private (personal) transportation in the United States but a commercial operation in most of the rest of the world.

Choosing the Best Method

Each of the preceding methods of supplying on-demand air transportation has advantages and disadvantages for prospective users. However, the number of variables involved in determining the ideal form of transportation are many and interact to make the choice difficult.

To evaluate their needs properly, prospective users must conduct an analysis of their transportation requirements. This analysis does not have to be extensive or detailed, but the more detailed it is, the better is the set of requirements that results. Chapter 2 explores this subject in detail.

As a rule of thumb, if a company intends to use an aircraft less than 100 to 150 hours annually, charter is probably the best option from a cost standpoint. For between 100 and 300 hours of use, joint or fractional ownership makes good economic sense. For above 250 hours, efficiencies of scale begin to make full ownership or leasing more attractive.

Again, if seeing familiar faces in the cockpit and around the hangar is important to on-demand users, a wholly owned or leased aircraft is the way to go. While fractional ownership gets a participant closer to the service provider, personnel flying the aircraft may be different for every trip.

Yet the most important factor is the degree of control exercised over the aircraft operation—scheduling responsiveness, flight crew quality, level of training, amount and type of maintenance, administrative and financial controls, and level of customer service (see Chap. 2). Most individuals and companies are very particular about levels of service; in-house operation usually provides the best opportunity to receive the highest levels. Conversely, some travelers care little about the details of their on-demand air transportation services, regarding it in the same way they do their limousine service—a means to an end. Therefore, it is important to place the personal imprint of the principal passengers onto the services rendered.

The airport runway is the most important main street in any town.
—NORM CRABTREE, OHIO AVIATION DIRECTOR

References

1. National Business Aviation Association. *NBAA Business Aviation Fact Book 2012*. Washington: National Business Aviation Association, 2012.
2. Winant, John H. *Keep Business Flying*. Washington: National Business Aircraft Association, 1989.
3. General Aviation Manufacturer's Association. *General Aviation Statistical Databook, 2012*. Washington: General Aviation Manufacturer's Association, 2012.
4. International Business Aviation Council. *IBAC Business Aviation Outlook 2012*. Montreal: Flight International, 2012.
5. Deloitte and Touche. *A View from the Top: Perceptions of Travel Management and Business Travel*. Washington: Deloitte and Touche, 1999.
6. National Business Aviation Association. *The Real World of Business Aviation*. Washington: National Business Aviation Association, 1997.
7. National Business Aviation Association. *Business Aviation Utilization Strategies*. Washington: National Business Aviation Association, 2000.
8. Anderson Consulting. *Business Aviation in Today's Economy: A Shareholder Value Perspective*. Washington: Anderson, 2001.
9. NBAA *National Business Aviation Association Fact Book, 2012*. Washington: National Business Aviation Association, 2012.
10. IBAC Business Aviation Safety Brief, 2012.
11. Business and Commercial Aviation "Operations Planning Guide, 2012."

CHAPTER 2

Determining the Need

The airplane has made us faster and better. We're cutting our time in airplanes and airports down by probably a third.
— CORPORATE CEO

Air Transportation Needs

Individuals seeking on-demand air transportation often select a method as a consequence of either a recommendation from an acquaintance or minimal research. The resulting service may serve many of the individual's needs, yet another type of aircraft or method may be better suited to their needs. Therefore, a comprehensive analysis of one's travel needs and alternative methods of meeting those needs will provide a more satisfying and efficient transportation process.

Similarly, companies spend a large amount of resources on employee travel. As travel costs increase, companies become increasingly aware of the tremendous number of resources expended on getting their employees from point A to point B. At some point, reliance on one's secretary or administrative assistant to make reasonable, economical travel arrangements becomes too much for an individual or group of individuals not well versed in the subject of travel. This is the time that companies set up their own travel department or outsource the function to travel professionals.

While this event marks a milestone in a company's existence, turning over the travel arrangements to professionals does not solve the unique travel requirements of certain employees. Senior executives and teams of marketing, sales, or quality control personnel or troubleshooters requiring rapid, short-notice travel are often overlooked by travel departments. If these individuals are to accomplish their unique missions and goals, they must fend for themselves, often stretching, or even breaking, existing travel policies. In short, these employees' travel needs are not well taken care of.

Normally, these special classes of employees begin to realize that scheduled air transportation does not serve their needs. They may start exploring features of the most readily available and affordable alternative, aircraft charter. Yet, having never before contracted for these services, such employees are at a disadvantage in determining which charter operator will provide the most cost-effective, safe, reliable, and high-quality service. The selection then becomes a hit-or-miss proposition, exposing such employees to a range of service levels that may prove to be a turn-off or excellent, based more on the laws of chance than on intelligent choice.

Alternatively, contact with an aircraft salesperson or broker or the opportunity to take a trip on another company's aircraft may provide sufficient impetus to investigate the possibility of a company aircraft further. Whatever the introduction to business aviation, the desire to investigate further is usually strong. Unfortunately, prospective users of on-demand air transportation seldom perform a thorough analysis of their needs and options. Perhaps this is because the concept is new and the initial impetus is compelling. Whatever the reason, a rush to embrace one method of on-demand air transportation without investigating the full range of options inevitably will leave the user unsatisfied. There are better ways of doing this.

Why Individuals and Companies Use On-Demand Air Transportation

A number of the following issues were covered in the last chapter, but viewing them in a slightly different way will prove useful in opening this chapter.

Create One's Own Schedule.   The airlines have created a transportation network that connects the world's major cities and, through regional airlines, many of the smaller cities. Yet few would say that using the airlines is an easy task. Inconvenient schedules, multiple connections, and departure processing delays waste much of a busy executive's time. When you want to go or arrive, there is seldom the right flight schedule or combination of schedules; even when the schedules accommodate your needs, seats may not be available; you can only get to within a 2-hour drive of your ultimate destination due to lack of airline service—sometimes it seems that you can't get there from here!

While we have all endured some or all the phenomena just listed, the on-demand transportation option is always available, 24/7, to as many as 10 times the number of airports served by the airlines. Creating one's own schedule and destination is truly a liberating experience. Further, it often makes good business sense; having the tools to determine what makes sense will easily lead to the liberating experience.

The Value of Time.   We all have just three resources available in pursuit of our goals: time, people, and money. While the latter two are certainly important, the asset over which we usually exercise the least amount of control is time—time to research, time to plan, time to react, time to produce. Therefore, the conservation and allocation of time are important to everyone, even more so to those who make the decisions and initiate actions.

Conforming to an airline schedule, waiting in airline terminals, and being subject to the whimsy of airline delays do not conform to a philosophy that will permit the efficient use of available time.

The Need to Be Face-to-Face.   The communications revolution of the past few decades has made personal and business contacts easier and faster. Cell phones, fax machines, the Internet, e-mail, teleconferencing, and electronic whiteboards have all enabled individuals to be more productive and aware of their environment. Yet the big decrease in travel predicted as a result of using these modern marvels has not materialized. In fact, these electronic tools actually have increased our need to travel simply because they have increased our productivity and improved our view of the marketplace. Moreover, these electronic marvels have increased the speed with which we receive information about markets to the point where rapid and decisive action is required to take advantage of the information received.

While many issues can be handled remotely, people still subscribe to the following truisms:

- Deals are made in person.
- Individuals like to be courted in person.
- People learn more when they can see, touch, and hear in person.

Increased Productivity.   Everyone strives to do more with less and to be at the right place at the right time—in short, to efficiently take advantage of what the environment and markets have to offer. Increased productivity has become the holy grail of the modern world.

This requires individuals and companies to use their scarce resources effectively and efficiently. In doing so, information databases, communications tools, and intelligent delegating are provided as tools to empower and enable people. Yet these same valuable individuals and employees spend countless hours waiting in airports, languishing in hotel rooms, and driving hours to far destinations, all because the airlines cannot get them rapidly to where they want to go, when they wish, and back again in the shortest possible time. The millions spent on productivity gains are often squandered on lost productivity during the airline travel process.

Safety, Security, and Confidentiality.   We all wish to protect our loved ones, especially when they are away from us. Similarly, key employees are highly valued by their companies because of their ability to make a competitive difference in the environment. Extraordinary methods are taken to protect them from hazards, to put them in the finest hotels, to hire bonded limousine services to carry them around in distant cities, and to contract with security services to protect them from harm. Yet, with all these protection methods, our valued acquaintances and executives are exposed to the vagaries of airline travel.

The ability to take advantage of superior information about the markets presupposes that only your company knows about this valuable information and that the competition will not find out about it until you have taken advantage of it. However, travel schedules of key personnel attempting to take advantage of unique opportunities may be readily found out by a determined competitor. Moreover, the same competitor can overhear sensitive conversations between your employees and observe their work papers and laptop screens during the travel process; airline travel is neither secure nor confidential.

Intangibles.   While the following issues seldom create a compelling reason for the use for on-demand air transportation, experienced travelers can instantly associate with and appreciate their negative impact on both their productivity as employees and the quality of their private lives:

- Airline travel hassles
- Travel-related stress
- Nights on the road
- Fatigue

All these quality-of-life issues have an impact on lifestyle issues and the ability of companies to keep employees relatively happy and productive. Burnout as a result of repeated and excessive doses of the items listed above is a very real concern; anything that can be done to alleviate these issues creates greater peace of mind and relaxed individuals and employees. None of us wishes to subject our loved ones to these unattractive consequences of airline travel.

Defining the Requirement

Individuals and families who travel frequently normally can forecast their transportation requirements at least 6 months and often 1 year in advance. Vacations, trips to residences, meetings with friends, and a certain number of pop-up trips are usually well known sometime in advance of the travel date. Some time spent with a simple monthly planning calendar should yield a reasonably accurate view of their on-demand travel requirements. Going through this exercise is useful in determining the correct type of on-demand method to be used.

Individuals who use on-demand air transportation for their personal needs often have business reasons to travel as well. Board meetings, property inspections, and consulting with advisors all provide business purposes to their travel calendar. Estimating these events can be made easier by using elements of the air transportation analysis that follows.

A company may concentrate on the travel needs of just its top three to five executives, or it may spread its net wider, including a variety of employees, some of whom may be classified as middle managers or below. It is important for a company to define its needs accurately before commencing an air transportation analysis. If this is not done, important requirements may be overlooked that may yield the wrong type of aircraft, or too many aircraft may be purchased; these features tend to create a negative impression on all concerned.

While the most common use of business aviation involves the travel needs of a top few executives within a company, these needs seldom fully utilize the aircraft. Most business jets operate on average about 450 hours per year but have the capability to operate as much as 700 to 800 hours annually. Well-defined repetitive schedules, such as corporate shuttle operations, can provide 1200 to 1300 hours from a single aircraft. Therefore, making the aircraft available to just a few people in the company will unnecessarily restrict its utility and reduce its efficiency.

Corporate Culture. The personality of a company determines how it conducts its business. Its level of aggressiveness in its markets, its risk behavior, its relations with its employees, its communications style, and its degree of formality all determine its culture. All these factors are important determinants of how a company approaches opportunities and problems, reacts to stimuli, and in general, makes its mark on the landscape.

Corporate culture is in large part a result of the personality and style of the chairperson or chief executive officer (CEO). This person's personality is often strong enough to be transmitted into the entire organization through subordinates, including level of participation desired in decision making and degree of aggressiveness in approaching the market. All these features spread throughout the organization via goals and objectives, policies and procedures, and informal communications. These elements combine to define the *nature of the company*; a lasting culture soon follows.

Aggressive companies seize opportunities rapidly and extend themselves throughout a wide range of market possibilities; analytical companies usually examine potential markets carefully and with in-depth analyses; entrepreneurial companies are more extensions of the nature of the founder/president than they are true corporate entities. Each of these company types reacts differently to opportunities and challenges. Knowing the corporate culture is essential in determining the type of aviation operation, aircraft, and aviation personnel best suited for a company.

It is important to note that this culture changes over time. The entrepreneurial, aggressive company of yesterday may have evolved into a more cautious, measured entity in a 5- or 10-year period. Realizing the new culture and its changing needs is necessary to ensure the proper fit between company and flight department. Also, new senior management inevitably brings changes in the culture, although these changes may take years to make.

Situational Needs.   Certainly, priorities must be established that recognize the relative importance of higher-level executives. In doing so, arguments regarding who should use the company aircraft are rarely encountered. The primary point of this discussion is to recognize that not just a few select individuals should be permitted to use a company aircraft. Situational demands placed on a variety of individuals may easily make on-demand aircraft the preferred method of travel.

All processes of a company must be examined to determine eligibility to use on-demand air transportation:

- Marketing and sales often send a number of representatives to one or more locations to work with prospective or existing customers. If the destinations are remote and airline schedules are incompatible with the teams' workload, on-demand transportation may make sense.

- A remote manufacturing facility, either the company's or a customer's, experiences a technical problem. Sending engineering and technical teams in the company aircraft will get the problem solved faster and get the team back to their regular duties more rapidly.

- Customers can be brought to see your display area and manufacturing facilities in a minimum amount of time while transmitting the message to them that they are VIPs.

- A team of company attorneys must take multiple depositions in a remote location. The company airplane may cut 2 days off their travel time.

A list of possible aircraft users and uses should be compiled before starting the air transportation analysis process. Without this list, determining aircraft usage will prove difficult. Chapter 1 discusses a number of potential uses.

Transportation Solutions.   Obviously, an owned or leased aircraft may not be the means of on-demand air transportation chosen for primary use within a company or for private users. Charter, fractional ownership, or joint-use operations may prove the best method for either long- or short-term use. Becoming wedded to the concept that a single aircraft is the only on-demand method that should be considered for travel, regardless of how many users/uses are justified, unnecessarily restricts the range of travel options available. In doing so, decision makers miss the big picture of what they are attempting to accomplish by using on-demand air transportation.

Companies traditionally list profitability as their principal goal. How they achieve profitability within their chosen niche requires a variety of methods and plans. The same is true for transportation.

Just because the principal airplane is booked or that the preferred charter operator does not have an aircraft available when someone wants it should not mean that the individual is doomed to use the airlines for his or her trip. Rather, creative scheduling may permit the use of either aircraft for a multimission role on the day in question, another charter company may be used, or a prearranged interchange agreement with another operator on the field can be used to supply the desired trip.

On-demand air transportation is a method designed to support some business or personal requirement; it is not an end unto itself. All methods of on-demand air transportation must be viewed as means to an end—as *transportation solutions*.

Air Transportation Analysis

Companies must conduct a survey to determine existing and anticipated travel patterns, thereby correctly determining their actual requirements. Determining these needs based on anecdotes and gut instinct often will overlook essential company travel needs.

An air transportation analysis examines three aspects of a company's travel:

- A historical view of airline and on-demand travel
- Planned and expected travel needs for the near future
- Solutions for these needs

The historical view is designed to provide the company with a look backwards at its past 12 months of travel, permitting a view of recent travel. While this will not necessarily predict future usage, understanding existing travel patterns is useful when setting the stage for the future.

Planned and/or expected travel is determined through interviews with key company personnel, reviewing strategic plans, and a knowledge of future products and services to be offered.

Suggested solutions use information from the first two sections of the analysis and develop a range of possible solutions to provide recommendations to decision makers (Fig. 2.1).

The object of this analysis is to create a picture of anticipated travel of key players and to describe the benefits of on-demand air transportation in terms of convenience, time savings, and access to markets. The National Business Aviation Association (NBAA) has some excellent materials available to facilitate the travel analysis and aircraft selection process. While aircraft manufacturers and brokers also will perform this analysis or assist with the process, realize that they are selling a product or service and that this may bias the outcome of the analysis. Independent consultants will perform this analysis for a price; their work may be less biased. Alternatively, company or private analysts should be able to perform the basic analysis with little difficulty.

Travel History

The travel history analysis phase should look at all company travel, not just executive travel. This is done to determine the total travel requirements for the company so that the on-demand solution can be used to fit as many possible transportation needs as efficiently and economically possible. A major selling point for a business aircraft is that it will serve the needs of the entire company, not just the senior executives.

Start the analysis by examining airline travel records from the past year (more that 1 year's worth may be counterproductive because of the changing nature of corporate travel patterns). These records usually are available in a database either from the company travel department or from the company travel agent; such an analysis is usually known as a *city-pair analysis*. (Table 2.1). Similarly, Table 2.2 provides aircraft charter destination information. Use available analytical features associated with the database to determine the following:

- Most popular destinations
- Trip frequencies
- Level of personnel traveling to those destinations
- Cost per trip (ideally, cost per seat-mile)

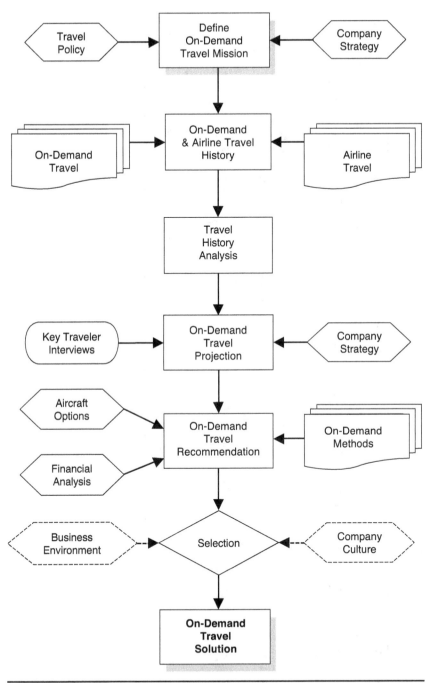

FIGURE 2.1   Air transportation analysis process.

MSP-LGA   **921 mi**   Between Minneapolis and NY LaGuardia

Airline	First		Business		Economy		Other		Totals				
	Fare	Flights	Fare	Flights	Fare	Flights	Fare	Flights	Fare	Flights	Avg	CPM	%
DL	$3000	2			$14,300	22	$4500	9	$21,800	33	$660	$0.72	97
UA					$3,500	5			3,500	5	$700	$0.76	13
	$3000	2			17,800	27	$4500	9	$25,300	38	$666	$0.72	100

MSP-LGW   **4295 mi**   Between Minneapolis and London Gatewick

Airline	First		Business		Economy		Other		Totals				
	Fare	Flight	Fare	Flights	Fare	Flights	Fare	Flights	Fare	Flights	Avg	CPM	%
DL	$7000	1			$14,400	12			$21,400	13	$1646	$0.38	80
AA					$3,500	3			$3,500	3	$1166	$0.27	20
	$7000	1			$17,900	15			$24,900	16	$1556	$0.36	100

TABLE **2.1**   Sample City-Pair Analysis

Irvine	Trips	Greenville	Trips
Houston	16	Owensboro	4
Greenville	8	Houston	3
Washington	6	Baltimore	3
Calgary	5	Hilton Head	3
Chicago	5	Washington	2
San Francisco	4	Peoria	2
Bartlesville	3	Gary	2
Atlanta	2	Savannah	2
Dallas	2	Columbia	2
Teterboro	2	Philadelphia	2

TABLE 2.2   Sample Charter Destinations

From this information, a graphic representation for destinations and frequency of travel versus distances (Fig. 2.2), a map of these destinations, and some qualitative description of relative convenience may be developed. This should be done for both key executives and other selected company employees. Obviously, executives will be the principal beneficiaries of the aircraft, but it also may be desirable for others within the company to use the aircraft. Employees from the CEO down through middle managers should be considered in this research. While all these employees may not be granted regular use of on-demand air

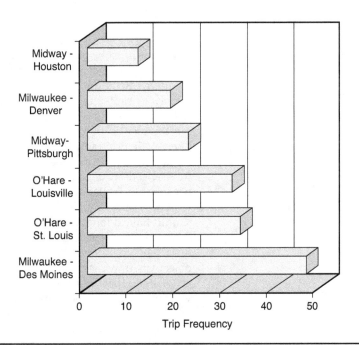

FIGURE 2.2   Sample airline city-pair graph.

transportation, knowing how a range of managerial and technical employees travel will produce a more complete picture for use in the analysis.

This analysis also should be done for on-demand transportation used by company employees during the same period considered for the airline city-pair research. Records of aircraft charter, fractional ownership, and company aircraft travel, if any, will complete the travel picture for employees.

An essential part of the analysis is a section, usually toward the end, that compares on-demand air transportation with the alternative modes of transportation and includes trade-off features arising from not using on-demand methods, including

- Extra nights on the road
- Lengthy travel days
- Poor airline connections causing circuitous routings or delayed travel
- Lost productivity due to additional hours and days on the road
- Poorly located airline airports requiring lengthy rental car trips
- Intangibles—frustration, employee satisfaction, security, airport hassles

These stories are best told anecdotally, bringing the human element to bear—everyone has an airline tale of caution and woe to relate. While the numbers developed in the transportation history section are useful in depicting how employees have traveled in the recent past, how people actually use and react to the travel experience become invaluable when contemplating the future.

Strategic Plans

These are the broad-brush plans for coming events within a company. While lacking in detail, they provide a sense of direction for the company. Items of use for the air transportation analysis include

- New products and services
- New or closing plant/office locations
- New geographic or business-line markets
- Competitive forces
- Strengths and weaknesses
- Resource availability
- Timeliness
- Contingencies

Importantly, these items will serve as entry points for the interviews to be conducted.

The Future

Emphasis must be placed on *anticipated* travel needs for the analysis to be of value to the company. Reliance on past travel needs or merely extrapolating those needs incrementally—"We probably will be on the road in the Northwest 15 percent more next year"— likely will provide an inaccurate or biased requirements picture. Determining the future travel needs of a company is often difficult because future plans may not be well defined.

Each company (and flight department) should have a long- and short-range view of its future, articulated through a series of strategic and operational plans. The goals and objectives flowing from these plans become action items that drive the company toward its view of the future. These plans often will contain the information necessary to gain insight into upcoming travel needs. Unfortunately, these plans often are not well developed or articulated, instead relying on chance and personality to drive the organization forward. And even if they are well formed and documented, they may not tell the full story.

The rate of change of the business environment has given rise to a more ad hoc planning style, with many companies preferring to remain flexible enough to react to a rapidly changing marketplace. Yet, within each chairperson, chief executive officer (CEO), chief operating officer (COO), or other top executive resides a central focus, a vision of where the company will be in 1 year, 3 years, and if they are brave, 5 or more years. It is this vision that will provide the impetus for the corporate engine to receive its power and direction. Therefore, it is this direction that must be discovered to determine the air transportation needs of the company.

The company vision likely will include new products and services, new or different locations to produce or provide those commodities, and certain people to bring shape and form to the dream. It is the people who will provide the details of the vision.

An important part of the air transportation analysis involves interviewing those

Transportation Analysis Interview Questions

- Interviewee's position and duties

- Recent company travel history

- Views concerning travel convenience (ask for specific examples)

- Concept of the company's and interviewee's department future growth (1- and 3-year projections, if available)

- New products and services planned (and travel implications)

- Personal and departmental (especially direct reports) travel plans for the next 18 months

- How can on-demand air transportation help achieve department goals?

who make corporate decisions and those who will actually do the traveling. In addition to putting flesh on the bones of often hastily written plans, interviews serve as a mechanism to focus attention on the on-demand air transportation issue throughout the company. The act of asking about travel needs often will serve to stimulate interest in the subject and drive it forward to some conclusion. Without the opinions of the decision makers and travelers, the data analysis portion of the study may prove a meaningless exercise.

Certainly the CEO, COO, corporate officers, and senior managers should be asked about their views of the future to provide an overview. However, the real focus should be on what may be considered the company's "movers and shakers," those who will bring the corporate vision to reality. These are typically senior managers or officers charged with applying substance to the ideas of top-level executives. It will be their hard work that will bring projects to fruition, and it will be these individuals and their direct reports who may be most in need of on-demand air transportation.

The interviews should be distilled into a report providing principal and anticipated destinations, frequency of visiting them, and broad justifications for the trips (Table 2.3). An essential part of the report must speak to the efficiency and convenience gained by using on-demand air transportation. Information gained during the interviews may be used as the principal source of future travel needs within the company. The picture gained from the interviews may not be precise, yet it constitutes the best possible view of the company's future travel plans provided by the people who know more about the subject than anybody else.

Company Air Transportation Requirements—5/10–11/12				
Department	**Destination**	**Frequency**	**Mission**	**Remarks**
Manufacturing	Ft. Wayne	6	Plant construction	With two direct reports
Manufacturing	Springfield	2	Product conversion	With QC, facilities
Manufacturing	Jefferson City	3	Plant addition	With engineering
Manufacturing	Rockford	2	Merger discussions	With finance, marketing, legal
Manufacturing	NYC	2	Financing presentations	With CEO, finance, legal
Manufacturing	Ft. Smith	2	Plant shutdown	With facilities
Manufacturing	(Midwest)	2	Location search	With CEO, facilities
Legal	Rockford	4	Merger discussions	
Manufacturing	Chicago	4	XYZ lawsuit	
Marketing	(Round-robin)	4	New product intro	Six Midwest cities
Manufacturing	Rockford	1	Product family	Team of six

TABLE 2.3   Sample Interview Results

Solutions

Once all research and interview information has been assembled and analyzed, a total picture of air transportation requirements must be constructed. Much of this travel will not warrant the use of on-demand air transportations; rather, a majority may be better suited to airline travel.

At some point the criticality of the mission, time value of the trip, value of employees, number of employees traveling together, and need for rapid action will tip the scale in favor of use of something more rapid, reliable, and confidential than the airlines. Finding this point will vary widely with the type of company, stage of corporate development, market forces, future plans, and so on. However, such a point must be identified and well defined if the decision to use on-demand air transportation is to be made. Accurately defining this point may be the most important aspect of the analysis because it is this point around which the whole issue of mass versus private transportation revolves. Therefore, significant effort should go into defining this point. (See the discussion of NBAA Travel$ense software in Chap. 1.)

Once this point is defined, a table can be developed that provides the composite requirements for on-demand air transportation for the company (Table 2.4). The table

Utilization Factors

Typically, aircraft are only available for service 90 to 95 percent of the time due to scheduled maintenance requirements. This and other factors make it difficult to claim that a single aircraft is available 24/7/365. Factors that reduce aircraft availability include

- Multiple trips to different destinations (schedule conflict)
- Multiday trips
- Need to reposition the aircraft from a previous one-way trip
- Scheduled maintenance
- Flight-crew duty time limits
- 700 to 800 hours per year are the effective maximum for most applications

Projected On-Demand Air Transportation Requirements for 2012				
Destinations	**Distance (mi)**	**Frequency**	**Passengers**	**Remarks**
Chicago, Des Moines, Rockford	Up to 300	22	2–6	Plant related, personal
Winnipeg, Kansas City, Grand Rapids	300–400	36	2–4	Merger, acquisition, expansion
Ft. Wayne, Springfield	400–600	28	2–8	Construction, new products
Buffalo, Memphis	600–800	6	2–4	Acquisition
Houston, New York	800–1200	12	2–6	Financial
Phoenix, Seattle	> 1200	5	2–4	Partners
(Undefined, round-robin)	—	22	2–6	Marketing, site search

TABLE **2.4** Sample Consolidated On-Demand Requirements

should portray sample destinations, distances to those destinations, anticipated number of passengers for the trips, and any relevant remarks that will assist in understanding and justifying the projected trips. Since interview information serves as the source for this table, justification for projected trips should prove relatively easy.

This table normally assumes that just one aircraft will supply all the on-demand air transportation needs of the company. However, what if high-priority missions destined for widely separated destinations are required on the same day? Or what if the aircraft is unavailable due to an extended maintenance inspection? How are high-priority trips accommodated? Can a single aircraft provide more than 700 to 800 hours per year of service? How many multiple-day trips will be required? The point is that one or two aircraft may not be sufficient to meet all legitimate travel needs. Therefore, either more than one aircraft will be required to meet those needs or a number of transportation delivery methods will be needed. (Alternative delivery methods, including charter, fractional ownership, and interchange, will be discussed later in this chapter.) These facts must be included in the decision analysis to provide a complete picture regarding transportation needs; to do less sends an incomplete and incorrect message to decision makers.

Presentation of these trips may be made more understandable and graphically pleasing if a map covering the destinations can be created that displays the proposed trip frequencies with arrows denoting trip frequency by varying the width of the shaft. This presentation will figure prominently in the decision on the type of aircraft to be used in providing the service (Fig. 2.3).

For a more complete list of items involved in the air transportation analysis, see the "NBAA Air Transportation Needs Analysis Checklist" in Appendix B.

Evaluation of Aircraft Types. The majority of trips shown in Table 2.4 are less than 800 miles, a range easily attainable by most turboprop and turbojet aircraft carrying six passengers, the passenger load factor called for in a majority of the trips. Aircraft range and payload are not independent of one another; rather, they are closely linked. Most airplanes are designed so that if their fuel tanks are full, they cannot also fill all passenger seats due to weight limitations. Therefore, with all seats full, the maximum range of the aircraft is reduced. When flights exceed 800 miles, a number of factors, such as departure airport temperature and altitude, winds aloft, and destination weather, may make an en route fuel stop necessary.

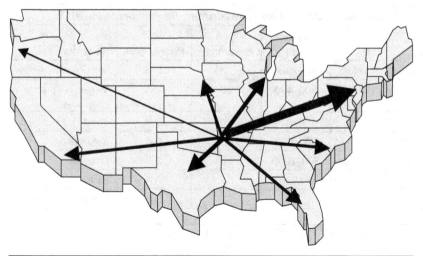

Figure 2.3   Depicting frequently flown routes.

Some passengers prize the ability of an aircraft full of passengers to make a transcontinental flight nonstop, whereas others either do not care or relish the thought of a stop every 3 to 4 hours. The extra range required to make a nonstop trip may double or even triple the purchase price of an aircraft. This capability seldom may be required; therefore, desired aircraft performance characteristics should be defined carefully prior to selection (see the "Aircraft Characteristics" sidebar on the next page).

Much is made of an aircraft's speed, although this performance characteristic must be placed into perspective. While an aircraft may be capable of speeds in excess of 500 knots, it must climb high, normally above 35,000 ft, to achieve this speed. If many of the expected trips are 300 miles or less, the aircraft will either spend much of its time climbing and descending or be unable to reach its maximum speed. Therefore, its effective speed between departure and destination, *block speed,* is often considerably less than its maximum. Air traffic control congestion and delays also will reduce the effective speed of an aircraft, especially on the heavily traveled East and West Coast routes. And the difference between a 500-knot and a 400-knot aircraft over a flight of 350 miles is just 11 minutes; the difference in purchase price between the two types is $13 million.

Cabin size figures prominently into aircraft choices too. The smallest turbojets have a cross-sectional diameter of barely 5 ft and are 11 ft long; a large-cabin aircraft cross section is more than 6 ft and 50 ft in length. The difference can accommodate many additional amenities, including a full-size lavatory, seats that recline into beds, and a full galley, including crystal and silver. The difference in cost of the two aircraft types is $20 million. Moreover, the larger aircraft cannot operate into smaller airports, an important feature to some companies. Conversely, the larger aircraft will carry double the passenger load for several times the range of the smaller aircraft.

Airport Factors.   Not all airport runways are created equal: Some are shorter, narrower, and have less weight-bearing capacity than others. Many on-demand aircraft users take advantage of the many nonairline airports available to them, since many airline airports are not close to their desired destinations. For instance, there are more than 5000 airports usable by business aircraft in the United States but just 500 available to the airlines. More realistically, even small turbine-powered aircraft are limited to not many more than the 4500 public-use airports in the United States, still a major increase over the number available to the airlines.

Larger, faster aircraft normally require longer, wider runways due to their performance characteristics. The required takeoff field length is defined by runway performance tables contained in the airplane flight manual, taking air temperature, airport elevation, and weight of the aircraft into account. These are mandatory minimum values under a country's operating regulations. For instance, a small-cabin turbojet may require just 3000 ft of runway for takeoff; a large-cabin jet may require twice this amount. However, these runway performance requirements may be reduced if the aircraft is not operating at its maximum permissible weight, but they may be increased if there are high temperatures and/or a high-elevation airport is used.

The "Annual Purchase Planning Handbook" (normally the May issue) of *Business and Commercial Aviation* and Conklin & de

Aircraft Characteristics

Selecting an aircraft depends on choosing among a number of factors, some of which are interdependent:

- Seating capacity
- Cabin dimensions
- Cabin amenities—lavatory, galley
- Baggage capacity
- Weight—empty, maximum, useful load, payload
- Speed—maximum, normal, long range
- Range—maximum, seats full
- Ceiling—certified, service, engine out
- Runway performance—landing, takeoff
- Costs—purchase, operating

Decker's *Aircraft Cost Evaluator* are excellent sources of detailed aircraft operational and cost data.

Some of these decision points are performance-related, some are financial, and still others may be considered lifestyle choices. Certainly, a small seven-seat turbojet with a range of 1000 statute miles will meet the needs of many companies' and individual's on-demand air transportation needs, but additional seating, performance characteristics, and interior appointments have created a large market for more than 20 smaller aircraft and 30 turbojet aircraft designed for business purposes. The trade-offs must be considered carefully to ensure that the company or individual user is provided with the right mix of features that will serve his or her needs.

An important feature in making this choice is talking with others who currently use on-demand air transportation for business or private purposes. Although each will value aircraft characteristics differently, their experience will prove invaluable to assisting you and/or your decision makers to find the right aircraft and delivery methods to provide suitable transportation solutions. Finding representatives of these companies can be accomplished easily by asking local fixed-base operators (aircraft service companies), aircraft brokers, and business acquaintances for their opinions and experiences regarding their company aircraft and/or service providers. Affiliates of the International Business Aviation Council (see Appendix A) can supply contact points.

Making the Decision. Deciding which aircraft type or combination of types will adequately serve the needs of the individual or company may be a simple process of presenting a number of recommended solutions to the proper decision makers. Or it can prove complex and labored, depending on whether the proper decision makers have been chosen, internal politics, and the financial state of the company or individual. Ensuring that all these factors have been covered carefully prior to the ultimate presentation will make the decision process smoother and relatively painless.

Gaining a consensus of middle- and upper-level employees and executives often will provide sufficient weight of opinion to sway top-level decision makers regarding aircraft type and delivery method. Yet many chairpersons and CEOs may have strong opinions about air transportation methods because they will be the principal passengers, foreseeing many hours riding in the aircraft. Knowing how the boss perceives air transportation will ease the decision process

There is no substitute for allowing company representatives to try a number of aircraft types before the final choice is made. If a clear choice is not made after presentation of the air transportation analysis, the finalists in the competition may be tested via several methods. The primary method of doing so is by chartering a specific aircraft for an actual trip. This will enable a number of decision makers to experience the actual environment they may choose to travel in for some time to come. If the type chosen does not meet decision-maker preferences, another type can be tried. And charter may prove to be the best method of providing air transportation for the company's on-demand needs. Unfortunately, a full range of aircraft types may not be readily available from local charter companies, and therefore, this method may not be accomplished easily. It is possible to find virtually any aircraft type available for charter, but the distance it must travel to provide the demonstration may be considerable, effectively increasing the cost of the charter, due to the costs of transit time with the aircraft empty.

Another method of experiencing different aircraft types is to arrange for a demonstration flight in a new or used aircraft through the manufacturer's sales force or an aircraft broker. While these flights are often not very long, they will allow passengers to get the feel for the aircraft. Alternatively, demonstration flights often can be arranged to accommodate actual business trips, with the company receiving the transportation paying only the direct operating costs for the aircraft. This type of demonstration is preferable because it shows the aircraft in its operational mode and allows the passengers to actually work and confer on the flight, experiencing its positive and negative attributes.

Healthy Flyers

A health care provider operated eight different hospitals from Minneapolis to Oklahoma City. The company was based in St. Louis, and its executives and managers became TWA frequent flyers, attempting to keep up with the growing organization. Marketing, finance, and the executive officers were the most traveled within the company, so much so that three key managers left the company because of the amount of travel involved. The company began using occasional charter aircraft to ease the travel burden and soon discovered the advantages of on-demand air transportation.

A brief air transportation analysis showed that the company could easily justify an aircraft based on the fact that more hospital acquisitions were anticipated and increasing numbers of people needed to travel. Soon thereafter the company leased a used Beechjet and was soon putting more than 500 hours per year on it. The good news is that two of the managers who had left the company were lured back into the fold because of the better travel arrangements.

Choosing the Method

As previously discussed, on-demand aviation services may be supplied by one of the following methods:

- Employee/owner-flown aircraft
- In-house flight department
- Management company
- Joint ownership
- Interchange
- Time share
- Charter
- Fractional ownership

Each of these will be discussed in this section, but it is useful first to examine what customers really want from flight service organizations.

What Users Want in On-Demand Air Transportation

At first glance, the answer seems simple: providing high-quality on-demand air transportation service to the passengers. This is easy to say, but what are the components of this service? Even better, what do the principal user, CEO, senior executives, and just plain passengers really want from a flight services provider? If they are asked directly, they may not provide either an honest or a meaningful answer. "On-time service" or "Just keep on doing what you're doing" will not help to define the essential components and level of service to be delivered.

After having spoken with many individuals and company executives regarding their expectations for their flight operations, impressions are formed that probably represent a majority opinion. Importantly, many of these passengers did not precisely articulate desires for high levels of safety and value, yet the specific examples they give point directly to those issues.

Safety.   While the principal passenger will almost never grab the flight department manager or other provider by the lapels and say, "You'd better be safe!" this is an inherent expectation. On-demand aircraft are used as an alternative to the airlines; passengers expect their alternative to be at least as safe. While many may say that such an expectation "goes without saying," it may be useful to remind all concerned of the primacy of safety on a regular basis. The key feature to look for is that every decision made within a flight operation begins and ends with a realistic assessment of the risk factors involved. If so, the outcome will probably prove safe.

Reliability.   If a provider says that he or she will pick up passengers at 9 a.m. on Friday, this forms a contract. Not being there and ready to go as promised jeopardizes the reputation of the provider and perhaps future contracts. Granted, every provider experiences mechanical or weather delays, but these should be precious few. Once dispatch reliability gets below 98 or 99 percent, passengers may adopt a more surly attitude toward the service provider. Good contingency plans and backups help prevent these embarrassing moments.

Good News About the Provider.   If other customers say nice things about an on-demand aircraft provider, this tends to build confidence about that provider. Fortunately or unfortunately, bad news seems to travel twice as fast as good news, so getting the word on one provider or another is not difficult. Yet the news may not come automatically; ask for references on a regular basis. Providers who deliver knock-your-socks-off service will cause all concerned to sing their praises. The occasional delay will pass easily if the service provider receives mostly favorable comments.

Good Service.   Passengers paying up to $7500 per hour for upper-end aircraft services deserve and usually demand the highest levels of service. On-time departures, immaculately clean and well-maintained aircraft, courteous schedulers and flight crew, excellent meals or snacks, continuous information about the progress of the flight, and on-time arrivals—as permitted by air traffic control and weather—are all standard expectations in the on-demand aircraft business. Moreover, superb service in all aspects of the operation are a *given*, especially if the service provider desires a long-term relationship with the contracting company. Moreover, the provider should be supplying added value for the user, not just charging for indifferent service.

Value is a combination of level of service and cost. That is, the lowest-cost operation does not necessarily provide the best level of service. Instead, a desired level of service is provided at a reasonable cost. There are no blank checks here; the cost of operation is not the primary consideration in providing the service.

In essence, users look for three elements in an on-demand air transportation service provider:

- Safety
- Service
- Value

The ability to provide excellent on-demand air transportation service while also providing good value can best be accomplished if the user has total *control* over the operation. The best means of ensuring the highest levels of control are to maintain a completely in-house flight department—scheduling, operations, maintenance, and personnel are an integral part of the company. Let's take a look at each type of service provider to examine pros and cons.

It's All About Control.   In the final analysis, companies seek the highest level of control over their on-demand flight operations. Whether it be schedules, personnel qualifications, service quality, or level of maintenance, these and other factors drive companies to choose one form of transportation over another. Table 2.5 lists the various factors of on-demand air transportation and grades them by the level of control each method provides. It is important to note that the grading system does not imply that one form of transportation is better than another, merely that the level of control over the factors listed is either higher or lower.

For instance, just because the in-house operation shows four asterisks in the service consistency category does not necessarily mean that any in-house operation is necessarily better or worse than a charter operation showing two asterisks. Rather, this means that the user has the opportunity and means to control this issue better than he or she would have when contracting with a charter company.

Control over the aircraft schedule is perhaps the most important issue in on-demand air transportation. An in-house operation clearly has the edge on this factor because the aircraft operator/owner has the ultimate level of control over this feature. Although a

Factors	In-House Flight Dept.	Managed Aircraft	Joint Ownership	Fractional Ownership	Charter
Aircraft					
Availability	****	***	***	****	**
Response	****	***	***	****	**
Maintenance	****	***	***	***	**
Security	****	***	***	***	**
Service	****	***	***	***	**
Personnel					
Qualifications	****	***	***	***	**
Training	****	***	***	***	**
Costs					
Capital	****	***	***	***	n/a
Operational	****	***	***	**	*
Liability	****	***	***	**	*

Note: The number of asterisks indicates the ability to control the factor mentioned, not the absolute value of the factor itself.

TABLE 2.5 Flight Operations Control Factors

scheduling conflict or unscheduled maintenance on the aircraft may reduce the availability of the aircraft, scheduling remains firmly in control of the company operating the aircraft. Note, however, that fractional ownership provides similar or perhaps greater availability and response capability due to the additional pool of aircraft. Yet the intelligent shuffling of trip priorities by the user may still provide a higher level of control over this feature.

Service quality and consistency are prized by many on-demand users as the most important features offered by this form of transportation. The in-house flight department clearly has the edge on this method of transportation because the company dictates its standards and performance desires to the flight department directly with no middlemen interfering with the communication. Other passengers (probably a minority) prize on-time operations more than service quality. In reality, most users demand and receive both. Again, corporate culture or the preferences of the principal passengers set the tone for on-demand air transportation activity.

The ability to dictate and control overall aircraft use policies and procedures is very important. Delegating such factors to a management company or fractional provider or sharing them with another in joint-ownership arrangement is often not favored by users. Similarly, knowing who is maintaining and flying the aircraft provides reassurance to many travelers. Yet some users consider the availability of an on-demand aircraft just another support service, such as a limousine or information technology service; ultimately, all services can be outsourced to one degree or another.

Each of these features comes with a cost and gives rise to trade-offs with other features mentioned. Yet cost alone is rarely the principal determinant of satisfaction among on-demand air transportation users. Deciding which features are important and their relative importance before you embark on choosing a transportation method will ensure happier travelers over time.

Methods

Each of the following methods provides a slightly different type and level of service. Some of these differences carry a fee; others do not. No one method is ideal for any class or type of company; individual preferences and unique needs dictate the proper methods. *Note: Each of these methods involves a number of financial, regulatory, and legal issues that cannot be covered in this section because of the complexities involved. Prior to entering into any of these agreements, experienced legal and financial counsel should be sought. Additionally, some of these methods, such as interchange and time share, may not be available in all countries.*

Private-Use/Employee-Flown Aircraft.   This is the most basic form of air transportation used by companies and individuals. It is normally initiated by the company owner or individual, either bringing a love of aviation (and an aircraft) to the job or acquiring a desire to learn to fly for private purposes. In either case, the desires and preferences of one of the users determine an air travel policy, with this person flying his or her own or a company aircraft. The key factor in defining this type of flying is that the individual flies the aircraft as an adjunct to his or her normal duties; he or she is not compensated for transporting either himself or herself or other employees/friends.

The best reason for using individually flown light aircraft is the flexibility they provide. The ability to create one's own travel schedule and then to modify it as conditions and opportunities dictate makes good personal and business sense. Rapid reaction to changing events is accomplished easily with this type of operation. The efficiency generated by this travel mode can be great.

While many aircraft used for this purpose are small single- or twin-engine piston-powered aircraft, a growing number of turboprops and turbojets are being employed. Comfort, convenience, and speed must be balanced against the utility desired and the costs involved.

In many business applications, the CEO or other senior employee uses the company airplane for their near-exclusive benefit, making company employees seek other forms of on-demand transportation. This is not necessarily a bad concept because in small and entrepreneurial companies it is the principal who makes most of the deals and whose time is most critical. It is common that companies begin with an owner-flown aircraft and then progress rapidly to the point where a corporate turbojet flown by professional pilots can be justified.

There is some concern regarding the safety of this type of aircraft. However, business flying is statistically much safer than the more inclusive group, general aviation. This is so because the equipment used is usually superior to most general aviation aircraft and the pilots are better trained and are more mature and cautious. Companies that insure sophisticated aircraft generally require higher pilot qualifications and training standards than they do for general-purpose flying.

Costs vary widely depending on the type of aircraft used. A single-engine piston-powered aircraft capable of carrying four passengers over ranges of up to 800 miles will cost $500,000 to purchase new and roughly $160 per hour to operate. A twin turboprop capable of carrying five passengers up to 1200 miles would cost $6 million new and $1000 per hour to operate. Purchasing either of these aircraft on the used market easily could reduce the acquisition cost by half.

Liability exposure is usually no higher than that encountered with any other venture. Insurance policies should name the company or any associated enterprise as an additional insured if the aircraft is privately owned; sufficient liability limits should be purchased to ensure company viability in the event of a serious accident. Again, pilot qualifications and training and aircraft maintenance practices should be better than required under national aviation regulations to ensure safe operations and create the image of concern for corporate liability.

Employee-flown aircraft occasionally may reduce the productivity of the employee if a proper balance is not struck between work and travel. This situation should be monitored carefully, and a full-time professional pilot should be hired when flying interferes with the work of the employee pilot.

These types of operations usually begin as a consequence of a personal motivation to fly. The love of flying is seen as a means to support travel activities. If done with the business or personal travel needs as the principal motivator, these operations make good sense. For companies, they also may lead eventually to upgraded business aviation operations using larger aircraft and professional flight crews.

In-House Flight Department.   This is the most common form of on-demand air transportation in the world. In this method, an owned or leased aircraft is flown by full-time professional pilots to support the user's on-demand air transportation needs. The aircraft are based at a conveniently located airport and housed in a rented, leased, or wholly owned hangar that contains office space for flight department personnel and, often, maintenance shops for aviation maintenance technicians. However, operators sometimes opt to have all maintenance services outsourced to a local contract maintenance facility or number of facilities.

Passenger scheduling is done by either a part-time company employee or a personal assistant of the principal user; occasionally, a full-time scheduler/administrator, normally located with the other flight department personnel, may be employed. Company human resources, information technology, accounting, security, and legal personnel support flight department operations; individual users must rely on contractors for these services

This method affords users the highest level of control possible of any method of providing on-demand air transportation and, potentially, the highest-quality transportation. Direct policy and procedural control over scheduling, response time, aircraft utilization, flight operations, maintenance, and expenses should ensure that the operator receives exactly what is desired from this form of transportation. No other provider interposes his or her standards and procedures that may tend to reduce service quality. However, depending on one's own limited number and type of aircraft and personnel, this form may reduce options that may be available via a fractional ownership or charter operation. These other types may be used to supplement one's own aviation resources, when needed. Additionally, interchange and time-share agreements can provide additional aviation assets at little additional cost.

Capital and operational costs associated with an in-house operation are borne by the operator. The aircraft may be either wholly owned or leased, depending on a company's willingness to assume debt and/or carry major assets on its balance sheet or the preference of the individual user.

While users appreciate the total control of their aviation assets afforded by this method, administration and management of the aircraft are borne fully by the user. These details are tracked and accomplished by an aviation department manager who is normally also a pilot or aviation maintenance technician. This person usually reports to the principal, CEO, or other senior manager. Ideally, the flight department becomes an integral support function of the user's overall operation, contributing to its success by providing safe, effective, and efficient on-demand air transportation to its passengers.

Management Company.   Under this method of operation, a company or individual contracts with an aircraft management company to provide all necessary elements to operate the aircraft. For a monthly fee (typically $3000 to $7000 per aircraft, depending on size and complexity), the management company provides personnel, training for the personnel, a base of operations, passenger scheduling, and aircraft maintenance services. Salaries, benefits, fuel, hangar rent, aircraft maintenance, and other operational costs are all collected by the manager and passed through to the owner. This is essentially a turnkey operation, which relieves the aircraft owner of the great majority of the responsibilities of aircraft operation.

The management company often manages aircraft for others and may be able to provide some operational efficiencies through the use of pilots and technicians employed for the other aircraft. This procedure provides a ready pool of trained and qualified individuals, but the same flight crew members may not be available to operate the owner's aircraft on a daily basis, a feature prized by some.

Additionally, management companies often provide air charter services using their client's aircraft. In essence, the client's aircraft is leased to the management/charter company for the purpose of providing charter services; in return, the client whose aircraft is used receives a portion of the charter revenue that can be used to offset some of the fixed operating expenses associated with their aircraft's operation.

The use of a managed aircraft for charter services reduces the owner's scheduling flexibility because charter trips scheduled in advance make the aircraft unavailable to the owner for last-minute trips. In addition, additional aircraft utilization created by the charter trips accelerates the aging process of the aircraft and exposes it to users who may not treat it as kindly as the owner does. Attempting to regain major portions of the fixed expenses associated with aircraft operation may prove difficult unless the charter revenue is substantial. In attempting to increase charter revenues, the aircraft becomes less available to the owner. If this method is chosen, a balance must be struck between availability to the owner and desired revenue levels.

Management companies can provide excellent services to owners not wishing to become involved in the details of operating a flight department. In so doing, however, control over policy, procedure, costs, and desired quality levels is shared with the management company; absolute control over the asset is no longer possible. While many of the technicalities and complexities associated with operating a sophisticated aircraft become the responsibility of the management company, full understanding and appreciation of available options may not be possible. However, the use of management companies has increased in popularity within the past few years, principally as a method for aircraft owners to rapidly enter the ranks of business aircraft owners with little or no knowledge of the subject required.

Joint Ownership. Under this arrangement, two or more entities (individuals or companies) become registered joint owners of an aircraft. One of the owners employs and furnishes the flight crew, and each of the owners pays a share of the charge specified in an agreement. Administrative, financial, and operating details usually are assumed by the owner employing the aviation personnel. Otherwise, this type of operation is quite similar to the more traditional in-house flight department. The principal advantage is having others share the fixed costs of the aircraft, which, depending on a number of factors, may comprise one-third of the total aircraft expenses.

This method works well when the owners, normally two, use the aircraft infrequently and at times that complement one another. For instance, a company may use the aircraft mainly during the week for business trips, and the coowner, who uses the aircraft for personal trips, may use the aircraft primarily on the weekends. Scheduling conflicts may be the main problem area for these arrangements, although many partners use this method with little conflict.

A detailed joint-operating agreement is an essential part of a successful joint-ownership operation. Scheduling priorities and conflict-resolution methods should be worked out and tested before the final agreement is signed. Successful joint-ownership operations liken their arrangement to the give and take found in a good marriage—an appreciation of the other's needs is the key.

The advantages and disadvantages of taxes, depreciation, and liability are shared by the partners. Control over all aspects is shared by the owners, somewhat diluting the ability of any partner to dictate all elements of the operation. Yet the shared-ownership aspect usually provides motivation to create and maintain a high-quality flight operation.

Interchange. This in an arrangement, available principally in the United States, whereby an owned aircraft is leased to another entity in exchange for equal time in that entity's aircraft when needed. No charge or fee is made for this service, except that a charge may be made not to exceed the difference between the cost of owning, operating, and maintaining the two aircraft. This method requires that each party own their aircraft and exchange their aircraft on an hour-for-hour basis.

This arrangement is often used to advantage by two companies based at the same airport and using similar aircraft. An interchange agreement will allow one company to use the other's aircraft as a backup when its aircraft is unavailable due to an existing trip or maintenance. The two operators "borrow" one another's aircraft when needed, and if there are cost differences between the aircraft, the companies settle up on a periodic basis. Interchange provides the option of infrequently using another's aircraft without the need for a more formal and binding joint-ownership agreement.

Individuals and companies also use this method to provide a different type of aircraft when needed. For instance, an individual operating a twin-engine turboprop strikes an agreement with a company at the same airport that operates a medium-sized turbojet aircraft. The parties then have the option to use the turboprop for short trips and the turbojet for longer ones interchangeably. At the end of the quarter or year, the individual operating the turboprop compensates the company (if required) for the differential costs associated with operating the more expensive turbojet.

Again, a detailed interchange agreement must be signed by the parties to protect them and to ensure that they have a full understanding of one another's obligations and privileges. Tax and liability issues must be addressed specifically. Perhaps most important is an understanding that specifically provides scheduling, cancellation, and conflict-resolution procedures.

Time Sharing. A U.S. time-sharing agreement means an arrangement whereby a person or company leases its airplane with flight crew to another person, to whom direct operating expenses of the aircraft may be charged. These expenses include

- Fuel, oil, lubricants, and other additives
- Travel expenses of the crew, including food, lodging, and ground transportation
- Hangar and tie-down costs away from the aircraft's base of operation
- Insurance obtained for the specific flight
- Landing fees, airport taxes, and similar assessments
- Customs, foreign permit, and similar fees directly related to the flight
- In-flight food and beverages
- Passenger ground transportation
- Flight planning and weather contract services
- An additional charge equal to 100 percent of the fuel and oil used

Since this arrangement does not compensate the owner for a major portion of the aircraft operating costs, specifically its fixed expenses, such an arrangement is used in only limited cases. Among such cases are agreements to provide company executives with personal use of the company aircraft or the occasional use by friends of the owner of an individually owned aircraft. These uses traditionally constitute a minor portion of the total operation of the aircraft because the owner wishes to use the aircraft for normal travel purposes. Tax and liability considerations must be addressed when entering into a time-share agreement.

Time share, interchange, and joint ownership all tread a fine line with regard to other Federal Aviation Administration (FAA) and Internal Revenue Service (IRS) regulations in the United States regarding private and common carriage; similar restrictions exist in other countries where these options are available. *Common carriage* means that an operator holds itself out to provide transportation of persons or property from place to place for compensation—airlines and charter operators provide common carriage. *Private carriage* differs from common carriage in that the private operator does not *hold itself out* (advertise, market, etc.) as a provider of air transportation to the general public.

Aircraft Charter.   Chartering an aircraft is just about as easy as calling for a limousine—find a charter company in the Yellow Pages, tell them where you want to go and when, complete a few formalities such as a credit check, show up at the appointed time, and you are on your way. This may be the easiest and least expensive means of obtaining on-demand air transportation, especially in the short term. Moreover, using a number of charter providers and a variety of aircraft will provide you with a good sampling of what the market has to offer in terms of aircraft types. This is the easy, low-risk way to enter the air transportation business. Yet the critical nature of providing on-demand air transportation services for compensation should make one more discriminating about the operator selected.

In virtually every country, national regulatory authorities certify and oversee air charter or air taxi operators in an effort to protect the traveling public. The United States will be used as an example to relate what is required of these operators.

Air taxi operators offer on-demand air transportation services to the public and by law must be certified by the FAA by meeting stringent operational, maintenance, and safety rules. In addition, pilots must be specifically qualified in both the type of aircraft to be flown and the type of operation.

The regulations for air taxis provide for a high level of safety and control. They address flight operations, maintenance requirements, and crew member training and testing. They also address crew rest and physical examinations and mandate a drug testing program. The FAA closely monitors air taxi operators to make sure that they conform to the established standards of performance. Your safety depends on flying with a legally certified air taxi operator.

Yet, even with federal oversight of air taxi operators, there are wide variations in the levels of service provided. With more than 3500 air taxi operators available in North America alone, it may be seen that checking their background and reputation makes good sense to find the best available. Also, realize that the FAA regulations covering air taxi operators specify certain minimum operational standards that do not relate to the level of service provided.

Asking a few basic questions of the air taxi operator is all that is needed to be certain that you are dealing with an FAA-certificated air carrier (this is the official name for an air taxi operator) and that the operator is authorized to provide the type and kind of service you require (see "Aircraft Charter" in Chap. 3 for methods).

Available charter aircraft range from small single-engine piston-powered models up through large intercontinental airline-type airplanes. Each charter aircraft has unique speed,

Charter Aircraft Rates (per Hour)

Single-engine piston, 3 to 4 passengers	$500
Twin-engine piston, 5 passengers	800
Twin-engine turboprop, 7 passengers	1200
Twin-engine turbojet, 7 passengers	2200
Twin-engine turbojet, 10 passengers	3000
Intercontinental turbojet, 14 passengers	7000
Single-engine turbine helicopter, 4 passengers	1900
Twin-engine turbine helicopter, 7 passengers	2500

(North America, late 2012.)

range, payload, and comfort features that should be evaluated carefully prior to selecting an aircraft type and charter operator. Each of the features mentioned has an associated cost, some modest and some not so modest; the careful consumer will consider all desired aspects of the prospective air transportation prior to choosing. In addition, it pays to shop around for these services. Considerable savings can be realized by a bit of extra inquiry and planning.

Note that all hourly charter fees are determined by block time, which includes taxi time (taxi time averages 12 minutes per leg flown). If the aircraft must be sent to your location from another airport, you pay for the amount of flight time required to position it both before and after your flight.

Most charters are subject to a 2-hour daily minimum charge; time spent waiting for passengers may be subject to charges of up to $350 per hour. If you wish the aircraft and crew to remain overnight, a fee of approximately $1000 will be charged. Federal excise taxes of 7.5 percent and domestic segment fees of $3 per leg are applicable to most flights. Extraordinary catering, airport use fees (landing/parking), and international customs fees normally are billed at the operator's cost.

Is charter a good choice to provide you with on-demand air transportation services? If your travel needs do not exceed 100 to 150 hours per year of block time and you have found a good, reliable provider, then air charter may be right for you. However, getting the same aircraft or flight crew from trip to trip may be difficult, unless you have a good relationship with the provider. And consistently high-quality service may suffer due to the inability to get the same aircraft and flight crew for every trip. Yet most charter operators strive to provide these service features as a matter of good business practice.

Fractional Ownership.   Fractional ownership is a relatively new concept in the world of on-demand air transportation. It was first introduced in 1986 by Executive Jets Aviation and named *NetJets*. The founder recognized that there was a portion of the population whose aviation needs were not being met through chartering or full aircraft ownership. Charter users wanted their own aircraft that they could count on when they wanted it, yet some could not justify the full purchase price of an aircraft; the compromise was fractional ownership.

Fractional ownership offers the option, instead of buying an entire business aircraft, to purchase shares of one. As little as one-sixteenth of an aircraft can be purchased, which will

Fractional Ownership

The pros:

- Lower purchase price for the aircraft
- Depreciation and tax benefits of ownership (limited to the fraction purchased)
- All costs stated up-front
- No waiting or positioning fees
- Guaranteed rapid response time—normally 4 to 8 hours
- Consistent service standards
- Ability to trade up or down for other aircraft types

The cons:

- High hourly costs when compared with charter
- Rapidly escalating hourly fees after fractional share hours are used
- Built-in positioning fees—normally two-tenths of an hour for each flight hour
- Schedule "blackout dates"
- Rarely ride in the same aircraft or use the same flight crew
- Aircraft maintenance preformed by a variety of contractors
- Penalties apply for early termination of contract
- Accelerated reduction in aircraft value due to heavy use

provide around 50 flight hours of usage per year, or as much as one-half of an aircraft can be purchased, depending on how many hours are needed. The most common amounts purchased usually are one-eighth and one-fourth.

Fractional owners pay a fractional percentage of the purchase price for an aircraft commensurate with the amount of hours they want available per year; e.g., one-eighth share receives 100 occupied hours, one-fourth share receives 200 hours, etc. Most fractional providers offer a variety of aircraft to choose from, ranging from small turboprops to large intercontinental turbojets. Hourly and monthly fees are charged over the period of the contract, normally 5 years.

Fractional owners are guaranteed that their aircraft or another aircraft of the same model will be constantly available to them with as little as 4 hours' advance notice. The fractional provider assumes responsibility for scheduling, staffing, flight planning, weather, communications, maintenance, catering, and insurance. At the end of the initial contract, the owner's fractional share will be purchased by the provider at "fair market value," or the contract may be renewed or upgraded.

The last point in the sidebar requires some explanation: A wholly owned corporate aircraft normally is used at a rate of 450 hours per year. A fractional aircraft accumulates 1200 to 1500 hours per year, thereby reducing the book value of the aircraft at an accelerated rate.

The subject of liability remains a vexing problem for fractional owners. Although an aircraft may be insured for a single-limit liability of $200 million, a one-sixteenth owner may only be insured for $12.5 million. Depending on the development of case law associated with fractional ownership, all owners of a particular aircraft may be held liable for an accident, even though those owners were not on board at the time of the accident.

Each fractional share sold requires signing up to five separate contracts, which contain the purchase and management contracts, interchange and resale agreements, and an acknowledgment of operational control obligations of the owner. This last issue is important because under the law the fractional owners incur certain responsibilities and obligations regarding the operation of an aircraft over which they have control. These include an obligation to comply with applicable aviation regulations and the exposure to significant liability concerns.

Is fractional ownership right for your company? Do you plan to fly between 100 and 300 occupied hours per year? Is the guaranteed availability attractive? Is part ownership a desirable feature in your corporate financial plan? If you can answer "yes" to these questions, take a close look at a program that has captured the interest of more than 8000 companies and individuals worldwide (Table 2.6).

Multiple Methods.   Just as one type of aircraft or method of providing on-demand air transportation service does not fit the needs of all users, one method or type of aircraft may not fit the total needs either. Having a single small turbojet may satisfy the majority of the needs but may not be suited for long-range flights, ones transporting more than seven passengers, or those operating into short runways. These specialized trips may be handled by charter, interchange, or fractional ownership methods.

Provisions should be built into an on-demand air transportation plan to provide for the full range of known needs. Too often users attempt to solve all their on-demand air transportation needs with a single aircraft or aircraft type, which is normally not possible, given the wide range of transportation needs. The flight department and those setting transportation policy must be trained to seek solutions for their transportation requirements from a range of providers, not just their own flight department.

Charter may be used as a backup for multiple trips in a single day, unscheduled maintenance on the owner's aircraft, or supplemental lift for trips in which the number of passengers exceeds available seats. A small fractional ownership share may solve the occasional long-range trip requirement or be used for multiple trips in a single day. An interchange agreement with another flight department based at your airport may solve all the extra requirements already mentioned.

Aircraft	Purchase $ (million)	Annual Fee $ (000s)	Hourly fee $
Medium twin turboprop	.9	168	1450
Small turbojet	1.2	380	2100
Medium turbojet	2.5	535	2800
Large turbojet	6.9	850	4400

TABLE 2.6 Typical Fractional Ownership Costs—Quarter Share—200 Hours per Year—2012 (*Courtesy of Conklin & de Decker*)

International Operations

A growing publishing company gradually began acquiring European periodicals and book publishing houses. To support these operations, company executives and managers used readily available international flights between New York and London, where the majority of the business was located. However, as the European interests grew, companies in Frankfurt, Madrid, and Warsaw were purchased, making additional intra-European flights necessary. These routes often were not served by readily available flights that met the schedules of the headquarters teams. An aircraft capable of flying to Europe and onward to other locations on the continent appeared to be an attractive alternative.

Since the company already used two Citations for domestic operations, management was aware of the advantages accruing to the use of business aircraft. While the purchase of an aircraft capable of regular transoceanic flights was deemed to be too expensive, further investigation revealed that a quarter-share of a Gulfstream IV would provide the limited access the company wanted to European markets at a more reasonable price. The ocean-spanning aircraft permits en route conferences and shaves approximately a day off the traditional European trip.

Running the Numbers

Each aircraft type and transportation delivery method comes with a cost. Each user has a philosophy regarding the desirability of carrying capital assets on the balance sheet, asset control, debt-equity ratios, and operating cost controls. Making a match between these sets of features often yields more than one option; choosing an option may be considered an art form rather than a cold financial decision. Achieving closure on an option requires knowledge of the market and an ability to evaluate the financial consequences of the options. Yet evaluating on-demand air transportation methods based on financial considerations alone is insufficient to make an informed decision.

Even though the air transportation analysis may have determined that a medium-sized turboprop or small turbojet will be the ideal candidate, the principal's desire for a midsized turbojet obviously deserves consideration. After all, it is he or she who will be riding in the aircraft most of the time, plus realizing that the aircraft has many uses and applications. Sometimes the principal's instincts are good and contain a vision of what will come instead of what is happening at the moment. This section will look at the costs associated with the several types of air transportation discussed earlier.

Cost Components

As with any enterprise, the associated costs include both capital and operational expenditures. Capital costs are associated with both the initial purchase of an asset and subsequent upgrades or any improvements that will extend its useful life or increase its value.

Purchase versus Lease.   Capital expenditures are only incurred if an aircraft is actually purchased for an in-house flight department, managed operation, or fractional ownership. However, the aircraft need not be purchased outright; leasing is also a possibility for aircraft acquisition. The decision to lease or purchase traditionally is based on a core financial philosophy and willingness to carry significant amounts of debt.

Length of Ownership (yrs)	Residual Value—% (no inflation)		
	5	10	20
Turbojet	75	57	32
Turboprop	60	42	18
Piston aircraft	50	36	16
Helicopter	64	51	35

TABLE 2.7 Residual Values Table for New Aircraft (*Conklin & de Decker*)

Users not familiar with high-end aircraft often are surprised to learn of the high residual values associated with these assets (Table 2.7). Because of these high residual values, the net cost of either ownership or leasing is actually less than the cost normally associated with large capital assets. The resulting differential between tax and book depreciation can prove useful in budgeting as well. In good economic times, some popular aircraft actually appreciate in value over the first 10 years of ownership.

Leasing the aircraft protects cash reserves and keeps the asset off the balance sheet. Aircraft leases take advantage of a user's ready cash flow without adversely affecting its debt-equity ratio. Lease amounts are based on the difference between the asking price and residual value; high residual values for popular aircraft tend to reduce the overall cost of a lease. Leases may be attractive for companies unable to take advantage of accelerated tax depreciation advantages.

The decision to lease or buy is extended by the concept of the *synthetic lease*. This method of financing permits a user to receive many of the benefits of ownership while keeping any associated financing costs off the balance sheet, thus reducing apparent debt. The synthetic lease allows an operator to treat lease payments as an operating expense and to take the tax advantages associated with depreciation.

As with any major capital expense, the decision to purchase an aircraft must take into account the value of money used in the purchase. If the aircraft is purchased for cash, the net present value involved must be computed; if the aircraft is financed under a loan, the interest charged also must be accounted for. In either case, the cost of ownership must account for these money costs, effectively increasing ownership costs over time.

The purchase of a fractional share of an aircraft is usually arranged through the fractional provider and financing can be arranged, if necessary. Aircraft brokers and manufacturer's sales personnel will present a number of purchase options designed to fit an operator's financial condition and philosophy.

Operating Costs. Operational costs are pure expenses and are divided into fixed or indirect and variable or direct expenses. Fixed costs accrue to the aircraft whether it flies or not; variable costs are incurred only when it flies. The expenses shown in Table 2.8 are typical for a single small turbojet aircraft operation. Note the effect of increased flight hours on cost per hour; this is due to the effect of spreading the fixed costs over a greater operating period.

Aircraft maintenance consists of both scheduled and unscheduled types. Scheduled maintenance is required by regulatory authorities to ensure that the aircraft remains airworthy; it occurs at regular intervals, based on hours flown, engine start cycles, and elapsed time. Unscheduled maintenance occurs when a discrepancy with the aircraft, its engines, or its accessories is discovered during normal operations or an inspection. Both types comprise

		Flight Hours per Year		
	Cost Item	200	400	600
Variable/Direct Costs				
Fuel	$1,200	240,000	480,000	720,000
Maintenance—labor and parts	150	30,000	60,000	90,000
Engine restoration	300	60,000	120,000	180,000
Landing/handling fees	20	4,000	8,000	12,000
Crew expenses	70	14,000	28,000	42,000
Catering/supplies	30	6,000	12,000	18,000
Total Variable Costs	$1,770	354,000	708,000	1,062,000
Fixed/Indirect Costs				
Crew salaries/benefits	$200,000		→	
Hangar	35,500			
Aircraft insurance	30,000			
Training	40,000			
Weather/navigation data	6,000			
Maintenance database	3,500			
Refurbishment	20,000			
Total Fixed Costs	$335,000	335,000	335,000	335,000
Total Costs		$689,000	1,043,000	1,397,000
Total Cost per Hour		$3,445	2,608	2,328
Per Hour Cost Savings		—	−24%	−32%

TABLE 2.8 Sample Annual Aircraft Operating Costs—Small Turbojet (*Conklin & de Decker*)

the total maintenance and parts costs shown in the table. New aircraft typically come with a 5-year warranty for parts and some labor costs; this substantially reduces these costs for a new aircraft.

Engine reserves refer to the advisability to accrue funds, normally on an hourly basis, to perform required engine, auxiliary power unit (APU), and thrust-reverser overhauls. Alternatively, engine operating insurance and overhaul services will provide most required and unscheduled engine maintenance for a set hourly fee.

Landing and handling fees are charged by airports and service facilities to accommodate the aircraft as it operates at those facilities. Crew expenses are primarily for living expenses while on a trip and for overnight stays.

Under fixed expenses, the largest item is that associated with crew compensation; note that benefits are included in the item. This compensation will vary considerably depending on the size of aircraft, cost of living for the operating base, experience of the pilots, and longevity with the company.

Hangar costs are for either maintenance on an owned structure or rent for a leased facility. Utilities and associated costs usually are included in this amount.

Insurance covers personal and property liability associated with the operation of the aircraft and the value of the aircraft itself (hull value). Rates vary depending on use of the aircraft, nature of the operation, and hours flown. Financed and leased aircraft normally incur provisions for minimum coverage amounts.

Training expenses are for just flight crewmembers in this case, who normally undergo recurrent simulator training twice annually. If aviation maintenance technicians and other support personnel are included in the flight department, their training costs will be shown here.

Flight planning, weather services, and required navigational reference publications/databases are both customary and necessary expenses for such an operation.

More detailed charts of accounts are employed by some companies to better understand the expenses of a flight department. A more complete chart of accounts may be found in the *NBAA Management Guide.*

Management Fees.   Aircraft management and fractional ownership companies charge a monthly management fee to cover their costs. This fee should be negotiated with management companies to obtain the best possible price. However, the ability of the management company to provide fuel, maintenance, and other discounts should figure prominently into the overall costs of a managed operation. The management fee charged by fractional providers is not normally negotiable.

Defining the duties of the management company should be a detailed part of the management contract. Virtually every aspect of the aircraft operation and the expenses associated with those functions should be included in the management fee. Oversight and management of the entire operation, scheduling, operations, maintenance, administration, and finances should all be performed for the owner and accounted for in detail (see Chap. 3 for more detail).

Evaluating the Options

Many people make the mistake of looking at just a combination of the purchase price of an aircraft and its first-year operating costs. Unfortunately, this may provide a skewed view of the total cost of owning/leasing an aircraft. Table 2.9 shows the costs of aircraft ownership over a 5-year period. The cost of capital and depreciation expenses figure prominently into the overall cost of ownership, modifying the total-cost picture significantly.

The purchase of a new versus a used aircraft will alter the cost picture as well. New aircraft come with a comprehensive warranty, normally covering most parts and labor costs for the first 5 years of ownership. However, this is offset by the high depreciation cost incurred during the first few years of asset ownership. Conversely, purchasing a used aircraft will increase maintenance costs while avoiding high depreciation rates. *A word of caution:* While corporate aircraft hold their value well and, if well maintained, will remain serviceable for 30 to 35 years, aircraft more than 20 years old may incur higher maintenance costs while sometimes yielding lower reliability rates. A good broker/salesperson should be able to lay out the trade-offs associated with the purchase of a new versus a used aircraft. However, once the trade-offs are understood, a life-cycle cost analysis always should be performed to the anticipated holding period of the aircraft.

Single-Year Cost of Ownership, No Depreciation

	In-House Flight Dept.	Managed Aircraft	Joint Ownership	Fractional Ownership (1)	Charter (2)
Hours flown	400	400	400	400	400
Variable cost/hour	$1,756	$1,756	$1,756	$2,783	
Fixed costs/year	$500,000	$500,000	$250,000		
Management fee/month		$5,750		$61,967	
Charter rate/hour					$3,100
Cost/year (3)	$1,202,400	$1,271,400	$952,400	$1,856,804	$1,333,000
Cost/hour	$3,006	$3,179	$2,381	$4,642	$3,333
Cost/nautical mile (390 knots)	$7.71	$8.15	$6.11	$11.90	$8.54

Five-Year Cost of Ownership, with Depreciation (no sales/use taxes)

	In-House Flight Dept.	Managed Aircraft	Joint Ownership	Fractional Ownership (1)	Charter (2)
Purchase price	$9,000,000	$9,000,000	$4,500,000	$4,400,000	
Operating expenses	$(6,012,000)	$(6,357,000)	$(4,762,000)	$(9,284,020)	$(6,665,000)
Depreciation (4)	$(8,481,600)	$(8,481,600)	$(4,240,800)	$(4,146,560)	
Operating (loss)	$(14,493,600)	$(14,838,600)	$(9,002,800)	$(13,430,580)	$(6,665,000)
Loan interest rate	8%	8%	8%	8%	
Interest expense (5)	$(1,559,402)	$(1,559,402)	$(779,701)	$(756,569)	
Residual resale %	80%	80%	75%	70%	
Sale proceeds	$7,200,000	$7,200,000	$3,375,000	$3,080,000	
Brokering fee	2%	2%	2%	7%	
Brokering cost	$(144,000)	$(144,000)	$(67,500)	$(215,600)	
Taxable gain	$6,681,600	$6,681,600	$3,115,800	$2,826,560	
Before tax (loss)	$(11,315,402)	$(11,660,402)	$(7,859,201)	$(12,896,189)	$(6,665,000)
Income tax bracket	40%	40%	40%	40%	40%
Income tax benefit	$4,526,161	$4,664,161	$3,143,680	$5,158,476	$2,666,000
Net expense	$(6,789,241)	$(6,996,241)	$(4,715,521)	$(7,737,713)	$(3,999,000)

Notes

(1) One-half share
(2) No positioning fees included
(3) Included federal excise tax 7.5% for charter/fractional fuel surcharge amounts
(4) Five-year MACRS
(5) 20% down payment, five-year finance

TABLE 2.9 Sample Costs of Aircraft Ownership (*Conklin & de Decker*)

Note in Fig. 2.4 that as aircraft use increases, the relative advantages of one type of on-demand air transportation give way to another. The owned aircraft costs are initially high due to the fixed costs incurred, but with increased use the per-hour cost decreases significantly. And, while charter costs appear attractive in comparison, the advantages of the fully controlled owned aircraft must be considered (see Table 2.10). Joint ownership and managed operations will shift this equation one way or another based on the lower or higher fixed costs involved. Again, the operating costs form just one dimension of a more complex calculus involved in evaluating the proper type of services to obtain; control over those services and life-cycle costs also figure prominently into the decision process. A supplementary value of this figure is that it shows that the more an owned aircraft is used, the lower will be the unit operating costs.

The acquisition of an aircraft should not be taken lightly because entering most types of on-demand air transportation agreements requires long-term financial and management commitment. Often companies (especially entrepreneurial companies) jump at aircraft ownership more on whim than sound judgment. The elation associated with a few good contracts and two good financial quarters leads many to believe that this performance will continue forever. The urge to commit to the purchase or lease of a whole or fractional aircraft is often irresistible, given the obvious advantages of this form of transportation. The rush of the two good quarters can rapidly give way to two bad quarters, cooling one's enthusiasm for the significant expense and commitment associated with the corporate aircraft.

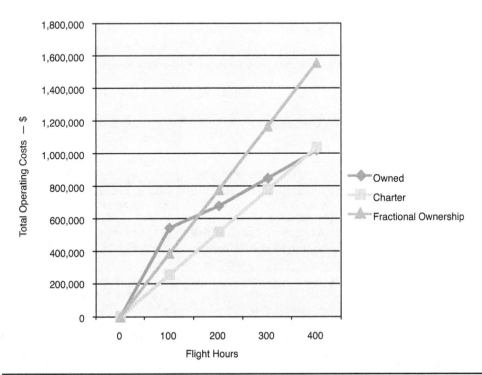

Figure 2.4 Sample operational costs versus hours flown—small turbojet. (*Conklin & de Decker.*)

	In-house Flight Department	**Aircraft Management Company**
Service quality	• Highest level of control/service possible • Immediate availability likely • If aircraft not available, may use charter, airlines, time share, or interchange • Best possible confidentiality/privacy	• Potentially excellent; may be customized • Immediate availability likely • If aircraft not available, may use charter, airlines, time share, or interchange
Aircraft administration	• Owner maintains total control over and manages aircraft operations • Personnel are owner's employees	• Owner maintains total control but delegates management of aircraft operations • Personnel are management company employees
Crew quality	• Consistent; owner-controlled • Owner controls training of crew and maintenance personnel	• Consistent, owner input, crews may be assignable to other activities • Owner delegates control of pilot and mechanic training
Security	• Dictated and controlled by company policy	• Subject to management company interpretation of company policy
Operating costs	• Variable; utilization-dependent; lowest cost of operation at reasonable utilization levels • Subject to positioning expense	• Variable; utilization-dependent • Annual costs may be higher than an in-house flight department (due to management fee) • Subject to positioning expense
Cost offsets	• Charter option may offset some costs	• Charter option may offset some costs • Possible fuel & crew training discounts
Liability	• Completely liable	• Liability shared with management company
Capital commitment	• Full negotiated purchase price	• Full purchase negotiated price
Tax consequences	• No commercial federal excise tax applicable • Noncommercial fuel tax applies • Maximum depreciation benefit	• No commercial federal excise tax • Noncommercial fuel tax applies • Maximum depreciation benefit
Aircraft acquisition and disposition	• Lease, purchase, or sale of any aircraft at negotiated price • Full authority of aircraft make/model, interior and exterior selection	• Lease, purchase, or sale of any aircraft at negotiated price • Full authority of aircraft make/model, interior and exterior selection

Note: Concept Courtesy of NBAA.

TABLE 2.10 On-Demand Air Transportation Comparisons

Joint Ownership	Fractional Ownership	Charter
• Same as in-house department • Availability requires coordination and planning • If aircraft not available, may use charter, timeshare, or interchange	• Service levels may vary • Aircraft availability guaranteed • Advanced notice required (4–8 hours, blackout dates imposed) • Charter aircraft may be substituted	• Service levels may vary • Possible inconsistent service vendor to vendor • Availability depends on market demand; no guarantee
• Owners maintain control and manage operations • Personnel on owner's payroll	• Owners maintain control, but delegate aircraft operations management • Provider supplies people	• None, not applicable • Provider supplies personnel
• Consistent, owner-controlled • Owners control training of crew and maintenance personnel	• Crew changes likely, specific crew possible • Owners delegate control of pilot and mechanic training	• Crew changes likely, rotating from pool • No control of pilot or mechanic training
• Dictated and controlled by company policy	• Security program in place, standardized	• Federally required security program in place
• Variable; utilization-dependent • Subject to positioning expense	• Fixed fee plus hourly charges • Higher hourly charges when exceeding yearly allotment	• Uniform hourly charges • Miscellaneous charges; catering, landing, etc. • Aircraft positioning expenses charged
• Charter option may offset some costs	• None	• None
• Liability shared with co-owner	• Provider/co-owners share liability	• Charter provider completely liable
• Owners share of purchase price	• Fractional share of purchase price	• None
• Shared tax liabilities/benefits • No federal excise tax • Fuel tax applies • Owners share depreciation benefit	• Shared tax liabilities/benefits • Federal excise tax on direct operating costs • Owners share depreciation	• Federal excise taxes imposed on charter rate • No depreciation tax benefit available as no aircraft are owned
• Shared lease, purchase, or sale of any aircraft at negotiated price • Can jointly select aircraft make/model, interior and exterior	• Shared purchase/sale • No aircraft customization • Remarketing fee charged for aircraft disposition • Early withdrawal penalties	• None, no ownership

Decision Factors

Since a variety of factors, both quantifiable and unquantifiable, are used in making the decision regarding the acquisition of an aircraft, it makes sense to set out some decision-making criteria. The following is a list of factors:

- Air transportation needs
- Normal trip load factor
- Average/maximum trip/leg length
- Trip frequency
- Multiple trips in a single day
- Payload versus range trade-off
- Minimum airport size anticipated in normal operations
- Cabin size and amenities
- Debt structure and sensitivity to debt
- Level of control desired over operations
- Anticipated aircraft retention period
- Top-level management preferences/commitment

One should first carefully evaluate transportation needs and base the correct solution on a long-term commitment. Many companies use charters for an extensive period to more fully understand the operational features of on-demand air transportation while exploring ownership options with a variety of sources. These sources must include a representative sample of similar operators who have used their form of transportation for a number of years. It is these testimonials that provide the real insights into the pluses and minuses associated with a particular form of ownership, management, and aircraft type; research with peer users is essential. Table 2.10 summarizes the features of the principal forms of on-demand air transportation.

I was the one who, for many years, said, "No, we're not going to get a jet," but it's really paid off—it's doubled our productivity.   —CORPORATE CFO

Flying has torn apart the relationship of space and time; it uses our old clock, but with new yardsticks.
                                                                              —CHARLES LINDBERGH

CHAPTER 3

Getting Started

First Things

Once a decision has been made to take advantage of on-demand air transportation methods, it is essential to understand one's motives for this in some detail. Doing so will complete the analysis for the chosen method and communicate guidelines for its use to the entire organization.

Failure to set the scene will give rise to confusion among passengers, managers, support personnel, and the flight services provider. However, a surprising number of companies and individuals put their multimillion-dollar asset in place and literally just begin flying trips for all comers or a select, yet undefined portion of the comers. Doing so creates anxiety, ill will, and discontent among all who have to deal with the aircraft. Even if the intent is to use the aircraft for the transportation needs of just principals or top management, defining who those individuals are should be the first step.

The easiest method of setting the scene is to develop an aircraft use policy for the company. (Note that this chapter discusses just the basic elements of getting a flight department up and running; administrative and operational details and more comprehensive issues of flight department operations are covered in subsequent chapters.)

Aircraft Use Policy

The use policy should be a simple, relatively short document that provides the central repository for

- How the aircraft is to be used
- Who can use it
- How the aircraft is scheduled
- Special cases
- Operating restrictions

The document serves as an essential tool to communicate the basic operating rules for the entire company, whether the reader is the owner, chairperson, or the accounts receivable clerk.

We operate our aircraft and use charter air transportation as a management tool in improving efficiency and effectiveness in the marketplace. The rationale for use of on-demand air transportation must be consistent with the overall management policies and practices of the corporation.
—CORPORATE CEO

The Purpose.   Without stating the basic purpose for the company's chosen method of on-demand air transportation, this essential business tool may be easily mistaken for a high-level perquisite or executive toy. While it also may be a perquisite for some company employees, its basic mission is hopefully more utilitarian. Therefore, a succinct and credible purpose must be listed at the beginning of the policy document.

A use statement may not seem important for an individual user, but creating one will serve to continuously remind you of why you purchased the aircraft in the first place. Not doing so may give rise to indiscriminate use or underuse of this valuable asset.

The preceding quote is a good example of such a statement. However, both more lengthy and more terse examples are available. Perhaps one of the most pithy is the following one from a chief executive officer (CEO) who was obviously a true believer: "The company airplane helps us beat the competition—use it wisely."

> *Ace Widget operates a Cessna Citation aircraft to provide air transportation services for its employees and customers. This aircraft is capable of carrying up to seven passengers for distances of up to 1200 statute miles. Certain other restrictions such as minimum runway length and maximum weight apply to the aircraft. Additional on-demand air transportation is available via aircraft charter.*

Some basic operating characteristics or operational limitations of the aircraft should be listed to create the proper expectations among passengers. The preceding example may be a bit elementary. For instance, a typical Citation cannot always carry seven passengers and their baggage for a full 1200 miles if the runway used for takeoff is short, significant headwinds prevail, or the weather at the destination is such that a alternate airport with better weather is necessary. Yet it is impossible to detail all operational limitations in such a brief policy document. Some operators provide examples of typical missions that demonstrate some of the more significant limitations encountered.

Note in the preceding quotation that aircraft charter is listed as an option. Doing so will create the expectation that use of this method of transportation is an option when the company aircraft is not available or incapable of accomplishing the assigned mission. However, it also should be stated that the user must pay the bill for the charter services and that such arrangements must be made through a central point within the company, normally the flight department itself. This latter point is necessary to ensure that only high-quality, safe, and economical charter services are used.

> *Individuals occupying the following positions within Ace Widget may authorize the use of on-demand air transportation:*
>
> | | |
> |---|---|
> | *Chairman* | *Strategic planning* |
> | *Chief executive officer* | *Corporate counsel* |
> | *Chief financial officer* | *Administration* |
> | *Group presidents* | *Treasurer* |
> | *Government affairs* | *Marketing* |
>
> *These individuals may authorize on-demand air transportation for use of their direct reports and other activities directly related to essential business purposes, when required.*

Eligible Users.   While it may be that only the top three or four people in the company or family group are eligible to use the company aircraft, a wider range of personnel traditionally is authorized for use. In the preceding quotation, specific individuals are not named as being eligible; rather, named positions are shown as being authorized to schedule the aircraft. This leaves the decision of who may use the aircraft up to them, based on circumstances, providing greater flexibility and utility.

Some companies choose to list specific employees and/or divisions within the company as being eligible to use the aircraft. This method leaves little doubt as to who may schedule the aircraft; unfortunately, it does not allow for contingencies or unusual situations that may occur. Many different methods of designating who may use an aircraft are employed by companies; the important issue is that all concerned know what the rules are.

Although not specifically mentioned in the preceding example, many companies use their aircraft for transporting customers and potential customers. This can be a powerful marketing and sales tool to bring customers to your location to view new products and processes or just to discuss proposed or ongoing business. Sending the company airplane to pick up customers sends a strong message—that they and their business are important enough to warrant the use of the company's "magic carpet."

> *Those authorized to use on-demand air transportation will supply the following information via e-mail/fax to the Aircraft Coordinator as far as possible in advance of the requested use date:*
>
> - *Dates desired*
> - *Business purpose(s)*
> - *Type of aircraft desired*
> - *Detailed itinerary, including specific times and locations*
> - *Passenger manifest, including applicable cost centers*
> - *Special requirements*
> - *Ground transportation required*
> - *Catering requests*
>
> *Travelers will make their own lodging arrangements.*

Scheduling. Specific details regarding a proposed flight must be obtained from the passenger in order to ensure that the flight is planned correctly. The list shown above is a good start; however, some additional specificity is needed. For instance, although the term specific location is used, passengers or their assistants may not realize that several airports in a locale and locations on those airports may be available. The more information available about the exact business destination desired, available ground transportation, and time of day, the more the trip will be satisfactory for the passenger.

Type and location of ground transportation designated to meet an arriving flight constitute another example of details critical to the success of a trip. While the principal portion of the trip involves only the aircraft, the passenger sees the process as a seamless event beginning at his or her office or home and ending at the desired destination, either a hotel or business address. Therefore, scheduling ground transportation in conjunction with the flight is quite important to the success of the entire trip.

The preceding examples demonstrate the need for an experienced individual to act as the aircraft scheduler/coordinator. The more information this individual has about the flight operation, aircraft operating characteristics, passenger preferences, and desired destinations, the greater are the chances for a successful travel experience. Many single-aircraft in-house flight departments use a senior executive's secretary or administrative assistant to fulfill this task, often to good advantage. Yet this individual must be able to devote sufficient time to the process to ensure the success of a trip. Since trips seldom are flown as initially scheduled, the ability of the scheduler to follow-up on and track scheduled trips is essential. A dedicated scheduler/flight coordinator working for the company flight department is probably the best form of scheduling control available for users, regardless of the size of the flight operation.

> *Charges for the Citation will be computed at the direct operating cost for the aircraft, currently $900 per flight hour. Charter charges will be actual costs billed by the charter company.*

Chargebacks

For the uninitiated, a corporate aircraft *chargeback* is an internal charge levied on the user of the corporate aircraft. In essence, it is an attempt to recover all or a part of the cost of air transportation services from a corporate user. Chargeback income usually is deposited into an overhead account designed to partially fund the aircraft's operation. Beyond this simple definition, there are literally hundreds of ways to apply this charge.

On the surface, the concept appears to be just another internal accounting exercise, shifting money from one pot to another. Yet the manner and rate with which the chargeback is applied may have a direct effect on the health and well-being of the flight department. More important, chargeback rates communicate a sense of *perceived value* of on-demand air transportation to the beneficiaries of the service.

Charge or Not?   If there is no internal charge for use of the aircraft and the aircraft is readily available, there is a risk that users may not appreciate the value of the service they receive— "If it's free, it can't be worth much." Further, allocation of the aircraft must be established by an organizational policy or some other prioritization scheme because "market" price is not used—a process fraught with the possibility of abuse and misallocation. Happily, the risk involved in not charging internally for an aircraft appears slight because fully one-half of National Business Aviation Association (NBAA) members impose no chargeback fee.

Company employees are accustomed to being charged for transportation, computer services, and in the case of centrally allocated companies, every support service used: human resources, cleaning, mechanical repairs, and maintenance of office plants. Why not the aircraft they use?

Those companies choosing not to levy chargebacks have made the conscious decision that the value of offering on-demand air transportation services to selected employees and customers justifies absorbing the cost of the service in a corporate overhead account. This is particularly true for small companies in which close control is exercised over the aircraft. As the size and diversity of the company grow, the aircraft allocation problem usually becomes more difficult. As a result, a chargeback policy may be imposed both to recover at least a portion of operating costs and to better allocate the asset. Only 30 percent of companies with sales of less than $100 million employ chargebacks; slightly more than half of companies with sales in excess of $100 million charge back some portion of aircraft expenses.

How to Charge?   Two-thirds of the companies charging for their aircraft services do so by the hour. The next most popular method, by aircraft distance flown, is employed by just 20 percent of those reporting. Available passenger seat mile and standard industry fare level (SIFL) methods are negligible compared with the two most popular schemes. While it may be easiest for the accounting department to charge one cost center for use of the aircraft, many companies prefer to charge individual passengers or cost center subaccounts. In this manner, a precise account can be kept of who used the aircraft and, possibly, for what purpose.

How Much?   The most popular basis for chargebacks (44 percent) is a percentage of direct operating cost expressed either as an hourly or per-mile charge. Direct operating cost plus a percentage of fixed cost is the least popular method, used by just 10 percent of companies. One-quarter of companies used SIFL rates.

Clearly, the rationale for using a particular chargeback method goes to corporate culture and, perhaps, the value placed on aircraft use by senior management. That is, if the use of a corporate aircraft is an integral part of corporate life at a company, the company will make it easier for key employees to use it. While this does not necessarily mean that the company will employ the percentage of direct operating cost chargeback method, it will allow sufficient travel funds to be placed in budgets to ensure access to the aircraft.

In a number of companies, chargebacks are used to control aircraft use. In other words, the chargeback rate, which normally is based on a percentage of direct operating cost, is adjusted periodically to ensure a target utilization rate for an individual aircraft. For example, if the target utilization for a Citation Ultra is 500 hours per year and utilization in the past 6 months indicates that the annual rate will reach only 400 hours, the chargeback rate is lowered from $1100 to $950 per hour in an attempt to stimulate use. A company also may supplement chargebacks with written use policies to discourage frivolous use.

At the other end of the spectrum, the fully allocated chargeback method may be so onerous as to stifle aircraft use for the majority of employees. This has the effect of reserving aircraft use for those with the largest travel budgets. If fixed costs and book depreciation are added to the Citation Bravo's direct operating cost, the hourly tab easily can exceed $2500, more than three times direct cost. While this may make the bean counters happy with tidy sums, the company's transportation needs may not be well served.

Significantly, about 75 percent of all companies using aircraft absorb all or part of their operating costs in some sort of corporate general and administrative account.

The "Right" Method.   Sorry, no "silver bullet" chargeback method is available that will suit all companies. Corporate culture and senior management policy control the method used. Unfortunately, these characteristics are subject to internal politics that may tend to defeat the purpose of having corporate aircraft in the first place.

When a company decides to acquire an aircraft, it hopefully does so with the expectation that the asset will add value to the company and provide a useful service. The decision to acquire an aircraft represents a significant commitment of capital or operating funds or both. Whether an aircraft is purchased or leased, a significant amount of money will be tied up in the transaction, money that could be used for other business purposes. Capital and/or fixed costs become sunk costs and therefore unavailable for other uses. Given the expense, the aircraft should be used to some optimal level to realize a maximum return on the investment.

Optimizing Aircraft Use.   Rather than imposing some arbitrary fee on the use of a valuable business tool, why not establish chargeback rates designed to make the best use of the aircraft? This is the conclusion that a majority of companies seemed to have reached, either by default or by design. Determining optimal use for an aircraft is based on examining the intended purpose or mission of the asset. If the aircraft is to be used as a business tool to increase the company's ability to compete in the marketplace, then the essential elements comprising the competitive factors must be studied. Strategic use, customer service, marketing, operational control, and emergency use are just a few reasons to use the aircraft—how does your company value these factors? Enlightened management will facilitate use of the aircraft to optimize these factors by making the aircraft available to the *right* people and purposes. Chargebacks are a means of achieving this end. Alternatively, the realization that on-demand air transportation is an essential corporate tool may be sufficient motivation to eliminate the use of chargebacks.

The knowledge and expertise of senior corporate executives is such that the loss of more than one of them may jeopardize the continuity and smooth operation of the company. Therefore, in an effort to manage the risk associated with such losses, the following travel restrictions apply:

1. *No two persons in the Office of the Chairman may travel together.*
2. *Not more than five corporate officers may travel together.*
3. *Not more than one direct report may travel with a Group President.*

Restrictions. Some companies restrict the travel of key executives on the same aircraft in an attempt to protect their business from interruption or loss of revenue as a consequence of an accident in which one or more essential employees are disabled or killed. This policy is usually applied to all forms of transportation, regardless of relative risk.

The rationale for such policies stems from the concept that a business may be harmed or actually fail if certain key employees are not able to continue their work. Key employee insurance is a prime manifestation of this concept, with companies insuring employees, usually essential officers, against loss of life and/or productive time. However, this concept normally applies to individual employees and not groups. Therefore, companies may restrict simultaneous travel of more than one of their essential employees.

Companies not having travel restrictions fall into two categories: those feeling that they do not need one and those feeling that they do. The former feel that the depth of their executive talent is such that barring a near-total loss of the company hierarchy, they could continue to function quite well after the loss of any group of executives. The other category of companies believes that they have raised the issue of concurrent key employee travel restrictions with senior management and have been told either that it was not important or that it would be "taken up at a higher level."

There have been only a few corporate aircraft and charter aircraft accidents within the past 15 years in which the number of key personnel killed or severely injured materially affected a company's performance. However, anecdotal evidence indicates that airline and automobile accidents may have created similar problems for a greater number of companies. Therefore, having a policy restricting group travel for all forms of transportation may be prudent.

The simpler the policy, the better chance it has of application and survival. More important, a group at the highest level within a company or perhaps the board of directors must decide what types and combinations of key employees warrant travel restrictions. Key factors in this discussion should be

- Which individuals are responsible for the highest levels of strategy formulation and execution?

- Which employees have key technical or process information that, if removed, might cause long-term damage to the company?

- What combinations of employees, if removed from the workplace, would cause significant temporary or permanent damage to future company business and/or revenues?

- How practical is the imposition of proposed restrictions on concurrent travel for these groups?

- Finally, will the imposition of these restrictions create additional hazards for the employees; that is, if one employee has to drive a long distance in lieu of aircraft travel, will this expose the employee to a higher risk of injury?

Whatever the policy desired, making it applicable to all forms of transportation and enforcing the policy are important features to be included.

In general, carriage of candidates for elected office or their agents must be paid for by the candidate prior to the flight. Carriage of office holders is generally prohibited except in cases where fact-finding trips or activities in which substantial participation by the official are involved. Obtain clearance from Corporate Counsel prior to carrying any candidate or office holder.

Special Uses. There are several special uses of company aircraft that require special treatment and are subject to special aviation and tax rules and regulations. While the rules associated with these uses may seem complicated at first, once a system has been established to accommodate them, they are relatively simple. Importantly, these special uses may extend the utility of the aircraft appreciably and prove to be important features for the company.

Carriage of Elected Officials and Candidates for Federal Office. Following passage of the Honest Leadership and Open Government Act in 2007, the Federal Election Commission (FEC) promulgated new rules governing the private carriage of candidates for federal office. These rules came into effect on January 6, 2010, introducing further restrictions, and in some cases prohibitions, on the carriage of federal candidates. The rules also provide a new method for calculating the reimbursement rate for noncommercial air travel. Reimbursement by candidates is required in order to avoid an in-kind contribution that would exceed permissible limits set by the FEC.

Members currently serving in either the U.S. House or Senate must follow strict rules when traveling in noncommercial aircraft. For House members participating in privately sponsored, official travel, a gift rule prohibits travel on a noncommercial private or chartered flight unless exceptional circumstances are demonstrated. Additionally, the House Code of Official Conduct prohibits members from using personal, official, or campaign funds to pay for or reimburse the expenses of a flight on any aircraft, unless one of the exceptions in the rule is met. The major exceptions are for travel on commercially scheduled flights and flights provided by a charter service. However, the use of personal, official, or campaign funds to pay for a flight on a noncommercial aircraft is normally prohibited.

Federal Aviation Administration Rules. The FAA allows noncommercial operators to carry federal candidates under FAR 91.321 if they are private operations. Any reimbursement paid by the candidate should not exceed the rates specified by FEC regulations. Unlike federal candidates, state and local candidates are governed by state and local laws, not FEC regulations. For this reason, it is recommended to consult appropriate counsel before providing or accepting noncommercial air travel in connection with state and local elections.

Anyone considering carrying an elected official or candidate on his or her own or his or her company's aircraft should exercise caution. Even though elected officials and candidates should be knowledgeable about the rules and regulations appropriate to their office and circumstances, any failure on their part to adhere to these various rules and regulations could jeopardize the business aircraft operator. In addition, because the different government entities involved do not always coordinate or clarify their rulings, the business aircraft operator might find other facets of his or her operation jeopardized, even if the elected official or candidate does adhere to the rules.

Many countries have similar rules and regulations governing carriage of officials. Therefore, all business aircraft operators are urged to closely examine all the ramifications of carriage of elected officials, including taxes, insurance, and the possibility of compromising the integrity of an elected official or candidate when considering carriage of these individuals, their families, and their staff. (See nbaa.org/admin/taxes.)

Personal Use. Similarly, many countries provide rules for the circumstances in which an employee is provided a flight on an employer-provided aircraft, and sometimes the flight is potentially taxable to the employee. This also may be true when a nonemployee guest or family member (spouse or dependent children) of an employee is aboard. The value of employee, guest, or family member's transportation may be considered additional income to the employee, not the guest or family member.

For example, in the United States, the valuation rule applies to personal flights on employer-provided aircraft. These flights can be either primarily for business or primarily for personal reasons. If an employee combines personal and business flights and the employee's trip is primarily for the employer's business, the employee must include in income the excess of the value of all the flights that would have been taken had there been no personal flights. However, if an employee combines personal and business flights and the employee's trip is primarily personal, the amount imputed to the employee's income is the value of the personal flights that would have been taken had there been no business flights.

Under the valuation rule, the value of a flight is determined using the aircraft valuation formula (also known as the SIFL formula). This is done by multiplying the SIFL rates by the mileage attributable to the flight, multiplying that product by the appropriate aircraft multiple, and then adding in the terminal charge. The terminal charge is not a landing fee or user charge levied by an airport operator but a base amount to cover certain fixed expenses incurred when providing a seat to a passenger, which is added to the other values. The SIFL rates, expressed as the cost per passenger seat mile, are calculated by the Department of Transportation and are revised semiannually.

If 50 percent or more of the regular passenger seating capacity of an aircraft is occupied by individuals whose flights are primarily for the business of the employer, the value of the flight on that aircraft by any employee (spouse and/or dependent child) who is flying for personal use is deemed to be zero.

Personal use of company-controlled aircraft is a complex issue, with the Federal Aviation Administration (FAA) and the Internal Revenue Service (IRS) often disagreeing on essential points. A full understanding of the issue is essential to protect the interests of both the company and the employee/guest. Yet personal use of a company aircraft is a regular feature of flight department operations and should be covered in the policy document, whether permitted or not. See NBAA website for current information on special uses flights.

Aircraft Operating Standards. There is a lot to know about what creates a safe operating environment and, conversely, what reduces safety. Much of what is done in the company flight department, charter operator, or fractional provider is designed to increase the margin of safety. Standards, procedures, training, and equipment are designed to reduce risk in the environment in which we operate.

It is important to appreciate that business aviation providers live not only by the national aviation authorities' rules but also by a series of standards, limitations, and procedures that they adhere to voluntarily. The national regulations form the foundation for safe operation, but they constitute only *minimum standards*. Following only the basic regulations would be like taking a small boat beyond sight of land with no training, no survival equipment, and no navigation aids—legal, but unwise. Rather, to ensure that flight operations are as safe as possible, additional restrictions are imposed on those operations and placed in a document known as the *Flight Operations Manual*. The national regulations and the manual provide the basis for the total operation.

It is these standards, limitations, and procedures, adopted through years of experience and research, that provide for an in-depth safety shield to protect all operations. Significantly, everyone in on-demand aviation, from scheduler through pilots and technicians, follows these rules day after day. An attempt is made to do things the same way every time. It is this discipline that provides the predictable consequence of safe operations.

Initial and recurrent training prepares aviation personnel for the worst that can happen. Simulator training and ground school received at least annually keep pilots and technicians sharp. The nuclear power industry trains the same way, using simulators to create worst-case scenarios that enable employees to face just about every possible combination of plant and environmental failure know to humanity. Additionally, specialty training in weather radar, international operations, aircraft troubleshooting, and medical emergencies further hone skills so that department personnel may benefit from the best of what the aircraft has to offer. Conferences and conventions focused on on-demand air transportation keep participants abreast of the latest equipment and techniques in the field of high-speed, high-altitude aviation.

On-demand aviation operations, especially those belonging to individuals and companies, meet and exceed the maintenance and inspection requirements imposed by both aircraft manufacturers and the national regulatory authority. These operations rarely scrimp on parts, never leave a worn part to chance, and fix discrepancies before the aircraft can fly again. Technicians receive recurrent training to keep them up-to-date on the latest changes in the aircraft and on repair techniques.

Pulling It All Together.   All the issues just mentioned are set forth in the flight department Flight Operations Manual. Everything the aviation operation does—every policy, standard, and procedure are contained in this manual. Deviation from these items is rare.

It is important for everyone in the company to realize this, for many of the standards and limitations contained in the manual materially affect how operations are conducted. For instance, aircraft normally cannot take off from an airport when the visibility is less than one-quarter mile. This is so because with less than this value, there is insufficient forward visibility to discontinue a takeoff in the event of a mechanical malfunction or to see potential obstacles on the runway. Similarly, most aircraft cannot land at an airport when the ceiling and visibility are less than minimum values prescribed for that approach. And loading the aircraft beyond specified limits is not permitted under national laws.

Finally, the person in charge of the flight, the pilot in command, must be accorded full and absolute authority regarding the safety of a flight. Moreover, he or she is charged with this responsibility under the International Civil Aviation Organization (ICAO) rules and those of the country in which the aircraft is registered.

Because these items contained in the manual are so important, it is recommended that the principal or CEO sign an appropriately worded preface page to lend sufficient weight to its contents and to communicate to the company that all employees must abide by its provisions. For a complete sample aircraft use policy, see Appendix B.

IS-BAO

From 2000 through 2002, the International Business Aviation Council (IBAC) worked with five of its worldwide affiliates to devise a universal set of operating and safety standards for business aviation, regardless of world location or type of aircraft flown. The result was introduced in 2002 as the International Standard for Business Aircraft Operations (IS-BAO). It is a code of practice devised by business operators for use by all business operators around the world, incorporating applicable International Civil Aviation Organization (ICAO) standards and recommended practices (SARPS) primarily for large (greater than 12,500 pounds) and turbojet-powered aircraft engaged in business or charter operations.

The core of IS-BAO is a safety management system (SMS) originally drafted by ICAO, which is now being adopted by countries around the world and made mandatory for business operators. The SMS enables operators to measure their degree of safety through the use of risk assessment techniques. All parts of the standard feed into the SMS, providing a coordinated and well-integrated overview of the entire flight operation, be it big or small.

The IS-BAO standards are primarily performance and goal oriented, setting targets for the operator to achieve and letting the organization devise its own method of achieving them. In this manner, the system is entirely scalable and applicable to both small and large operations. A variety of informational materials is provided with the standards, including sample operations and safety manuals.

I highly recommend the use of IS-BAO to meet your operating standards and best practices. More may be found regarding IS-BAO and SMS in Chaps. 6 and 8.

Oversight

When initiating on-demand air transportation services for an individual or company, those services must be overseen and managed by some person. This entity will be responsible for the following functions:

- Devising and maintaining the aircraft use policy
- Managing the interface with the transportation provider
- Either scheduling the aircraft services for passengers or coordinating/overseeing those services
- Tracking air transportation utilization
- Oversight of the financial issues associated with the air services

These oversight functions are quite important to the overall success of the aviation services effort for the company. If strong, well-developed policies and procedures and positive management and leadership are not imposed on the operation, important service details may be overlooked and financial details not tracked, leaving both passengers and the owner displeased with the overall level of service.

Managed in-house and charter operations often promise seamless turnkey air transportation experiences for clients, but without a company central office or person in charge of the operation, important details will be missed. For instance, when two people in the company wish to use an aircraft simultaneously, to whom will the contractor provide the aircraft? The type and amount of charges imposed for a particular trip or type of trip may not be well advertised or understood by the controlling passenger, leading to misunderstanding and conflict regarding the bill. And complaints regarding the service may not be channeled or articulated properly, given an uncertain relationship with the contractor.

With an in-house flight department, aviation personnel generally are located at the base airport, creating an artificial barrier to good communications and regular interaction not only with passengers but also with the company's or the principal's support personnel. It is this remoteness and autonomy that may create misunderstanding and conflict with the flight department. The same issues that may arise with a managed or charter operation also apply to the in-house flight department unless a strong connection is established between the company and the flight department. Table 3.1 presents a flight department startup checklist.

Getting Organized. While the CEO's or principal's secretary may be able to handle the day-to-day scheduling and billing activities associated with on-demand aviation services, they

---

Assumptions

- Aircraft selected (see "Considerations When Acquiring an Aircraft" Checklist on page 90)
- Ownership method decided (sole/joint ownership, leasing, etc.)
- Chief pilot selected/hired
- Pilots/technicians trained
- Safety program/procedures discussed with senior management

Management

- Reporting senior lines of communications established
- Policy and control procedures established—operations manual drafted

Administration

- Insurance purchased (see Table 3.2)
- Operating base/hangar selected
- Other personnel selected/hired (with corporate human resources)
- Aircraft use policy drafted
- Passenger scheduling system devised

Maintenance

- Line maintenance services selected
- Maintenance record tracking system chosen
- Inspection/maintenance provider selected
- Maintenance logbook records custody decided
- Regulatory compliance issues checked

Operations

- Trip sheet format established for passengers and flight crews
- Flight log sheets devised
- Recordkeeping requirements decided
- Regulatory compliance issues tracked/checked
- Operations manual and safety management system drafted

Accounting

- Develop detailed chart of accounts with accounting
- Draft budget—variance reports system setup
- Billing/invoice handling system established
- Aircraft business/personal use codes and procedures devised

---

TABLE 3.1 Flight Department Startup Checklist

Typical flight department reporting seniors within a company:

- Executive vice president
- VP facilities
- VP administration
- Director of human resources
- Chief financial officer
- Director of real estate

may not be the right person when it comes to authorizing flights, resolving scheduling conflicts, and settling billing disputes and other inevitable consequences of an outsourced operation. Rather, some midlevel or higher manager should be assigned the responsibility for on-demand air transportation oversight. This person will be able to interact easily with the management company or charter provider, director of operations, or chief financial officer (CFO). Daily routine operations may be assigned to a lower-level employee, but at least a "dotted-line connection" should exist between this person and the manager given oversight responsibility to ensure continuity. Typically, a financial or facilities manager is assigned the task of coordinating the contract flight operation.

When an in-house flight department exists, a stronger, more formal tie should exist between the flight department and the company. It is important to have the aviation department manager or chief pilot report to the proper person and level within the parent organization. If the single aircraft is used mainly by the principal or CEO, the department probably should report directly to that individual. In reality, the day-to-day communication and coordination probably will take place between the chief pilot and CEO's administrative assistant or executive secretary.

When a single aircraft is used by more than a few top executives or the department has more than one aircraft, the person to whom the top aviator reports becomes more problematic. If the owner or CEO is not the principal user of the aircraft, the department probably should report to some other high-level executive or assistant for both administrative and operational control. Whatever the job function, the executive should have sufficient power within the company to defend and support the flight department. Most important, the individual should be motivated to accept this responsibility; without this, the department may become an orphan, subject to the whims of other top executives.

The following are desirable features to be embodied in the relationship between the flight department manager and the reporting senior:

- Quick decisions
- Direct communications
- Appreciation for the role of the flight department
- Resolves conflicts and aids scheduling
- Budget and expenses resolution
- Keeps politics away from the department

While all these traits may not exist between the flight department and the company reporting senior, a more useful and stable relationship will exist if they are present.

An Integral Part of the Company. On-demand air transportation probably is quite different from any other function within the company. Because of this, and because of its remote location, the flight department often suffers from the out-of-sight, out-of-mind syndrome. If the aircraft shows up on time and the passengers are treated reasonably well, the flight department is permitted to continue in its autonomous ways. This is a mistake.

The charter operator, aircraft management company, or flight department provides a valuable service to the company, and it must be managed as such. On-demand aviation should be considered to be a vital support service for the company in the same manner as information technology, human resources, finance, and legal functions. If this is true, then it should be held to the same standards and performance measures required of other support functions within the company.

Unfortunately, because of the remoteness of the aviation function and its high level of complexity, it may be left to shift for itself. Few individuals within the company wish to take on something seemingly as complex and different as aviation, and besides, isn't that why we hired the management company/charter operator/flight department manager? Yes, those companies/people are being paid to perform the aviation service function, but they also should be held accountable for their actions and required to provide reports of their performance.

There must be a strong link between the aviation services function and the company if service levels are to meet company expectations and if the flight department is to know and understand the air transportation needs of the company. The aviation services function should provide *transportation solutions* to the company, not just supply an airborne limousine on call. The aviation manager should be an integral part of the company management team and provide ideas and suggestions as to how the company can be made more competitive and productive. In order to do this, the link between the company and the aviation manager must be strong, permanent, and part of the corporate culture.

Staying Informed

Business aviation is a rapidly changing field. New aircraft, regulations, restrictions, products, and opportunities are constantly arising; missing some of these new features can prove costly and inopportune. The principal method of staying informed is through member associations that both represent the interests of members within the industry and with the regulators that keep the membership informed about not only what the association is doing but also what else is happening in the field that is important or of interest.

Joining and participating in a business aviation organization is essential to ensure that the environment in which the aircraft operates remains favorable to business aviation and to stay abreast of events currently affecting this field. The International Business Aviation Council (IBAC) represents a number of affiliated nationally based organizations (15 at the time this book was published) that look after business aviation interests. In the United States, the National Business Aviation Association (NBAA) has represented its members' interests for more than 60 years, carving out a credible and proud place for itself in the aviation industry, U.S. Congress, and FAA. The membership fee is relatively small compared with the level of representation and information received as a consequence of membership. Similarly, other IBAC affiliates perform the same services in the countries they serve.

The IBAC family of organizations represents companies that operate business aircraft, not pilots. The Aircraft Owners and Pilots Associations (AOPA) has been representing the interests of pilots and aircraft owners in more than 70 countries for as long as 60 years. Its reputation is excellent in securing the best possible regulations and operating environment for its members.

Other specialized organizations represent charter operators, fixed-base operators, aircraft brokers and manufacturers, aeronautical repair stations, and aviation maintenance technicians. All have something to offer their niche members. See the Appendix A for a listing of these organizations.

Finally, a number of periodicals are published to meet the information needs of business aviation users. Subscribing to one or more of these publications will keep you current in the ever-changing aviation environment. Most of these are distributed free of charge to qualified operators. See the Appendix A for a listing.

Owner/Employee-Flown Operations

This is the entry level of on-demand air transportation, actually do-it-yourself air transportation. The essential prerequisite for this form of business aviation is a pilot capable of both flying an aircraft and performing some other company function. It is this combination of skills that makes this form work. Significantly, it may be difficult for this multitalented employee to find a balance between flying and his or her principal company function. Such an individual may lose sight of the fact that his or her principal value to the company lies in his or her nonflying job, not the other way around. (*Note:* Since owner/employee-flown business aircraft is a relatively unique operation, the subject will be covered in its entirety in this section.)

Purpose of the Business Aircraft

The on-demand aircraft is a valuable asset. The most valuable benefits provided are time savings and operating flexibility for the owner/operator. Moreover, multiple-stop trips are particularly well suited to small aircraft use. While the owner or senior company officials are the customary users of this type of aircraft, virtually any company employee or acquaintance who has travel needs can make good use of this means of transportation. Expanded access to the aircraft increases its utility and cost-effectiveness.

However, for businesses, the aircraft must be used like any other business tool, with profitability and competitive advantage in mind. Trips that can be completed more efficiently or economically on the airlines should be made using that mode. However, the intangible benefits of business aircraft also must be factored into the transportation decision. The opportunity to set one's own schedule, return home rather than waiting for tomorrow's airline flight, the cachet of flying oneself to visit a customer, and the option of changing schedules at will to take advantage of emerging opportunities are compelling reasons to use an aircraft.

Learning to Fly

Not every person who becomes an employee-pilot comes to the job with a pilot's license or sufficient experience to carry out piloting duties successfully. Sometimes a need is identified and combined with the desire on the part of an employee to learn how to fly; the outcome is often beneficial to both the individual and the company.

Human beings have been intrigued with flight since the beginnings of time. The ability to evade the normal ties to earth and to enter a new realm of transportation has fascinated millions of people to the point of learning to fly—more than a million people worldwide are qualified to fly aircraft. Combining the ability to fly with business and personal pursuits has fascinated people since the early part of the twentieth century, enabling many to combine the joy of flight with practical pursuits.

A Dream Come True

The CEO had always been fascinated with flying, but the press of business and family activities had kept him from learning to fly. A hunting trip in a friend's aircraft convinced him that the time had come to realize his dream of flight.

A local flight training school agreed to fit its schedule to the CEO's, providing an aircraft and instructor, often on short notice, to accommodate his busy schedule. After his solo flight test, flight training was continued, frequently combining instruction with business trips. Within months, the CEO became a licensed private pilot and the proud owner of a new six-place Beechcraft Bonanza.

He now flies the aircraft more than 200 hours per year for business and 100 hours for pleasure, the latter mostly on family vacations.

Learning to fly is not necessarily difficult, but it does require discipline and attention to detail to ensure the safety of flight. However, most people motivated enough to learn to fly have little problem with these details. Every pilot of a conventional aircraft must be licensed by a national aviation authority to operate a specific category (airplane, helicopter, etc.) and class (single-engine, multiengine, etc.) of aircraft. Additionally, each pilot must possess a current medical certificate issued by a physician designated by the national authority. The medical certificate is not difficult to qualify for in the case of a private pilot; most adults who can qualify for a driving license also can qualify for a pilot medical certificate.

After approximately 50 to 70 hours of flight instruction and some academic study, a student pilot takes written, oral, and flight proficiency examinations in order to qualify for a pilot certificate. Once certified, the pilot is able to carry passengers in a small single-engine aircraft during good weather. Flight in clouds requires additional training leading to an instrument rating. Increasing levels of proficiency and confidence for the new pilot are gained through experience, an essential feature when contemplating the use of an aircraft as a reliable transportation tool.

Role of the Owner/Employee-Pilot

The owner/employee-pilot is also the flight department manager. Responsibilities of this position include

- Aircraft acquisition
- Aircraft financial and accounting requirements
- Setting standards for aircraft use and operation
- Scheduling
- Maintenance management
- Regulatory compliance
- Record keeping

While all of these items may sound rather daunting in prospect, most will take care of themselves once a *tracking system* has been devised. The concept of a systematic approach to aircraft management cannot be overemphasized; the personal/business pilot has more than enough nonaviation items to occupy his or her time without considering the recurring details of aircraft operation. The core of such a comprehensive system is flight and financial records (described below). If these are maintained on an ongoing basis and in sufficient detail, the system will work well.

A set of policies and procedures should be developed to accommodate the routine requirements of aircraft operation. Once developed, these procedures should be delegated as much as practicable to the operator's assistants and clerical personnel. However, even though these bookkeeping tasks are delegated, the overall responsibility for management of the aircraft should be retained by the principal operator. Through a series of normal and exception reports, the operator will be able to ensure that all essential required items associated with aircraft operation are accomplished, costs are controlled, and safety is maintained.

Many pilots are not in a position to delegate required functions of aircraft management. For this group, a systematic approach to record keeping and operational oversight is essential. Managing a business/personal aircraft requires adherence to standard procedures and accurate documentation, especially when dealing with the tax consequences of aircraft use for business.

Acquiring the Aircraft

A central theme or idea often drives a company or individual toward the acquisition of an aircraft. Exposure of the owner or senior management to another company's or friend's aircraft can serve as the catalyst to begin the investigation process. More typically, an aviation-oriented person realizes the potential value of an aircraft and begins using his or her own or a rented aircraft to meet personal and/or business travel needs. However initiated, the combination of time savings, flexibility, and efficiency involved is a compelling reason for using an aircraft.

Once the decision to acquire is made, a number of additional questions arise:

- What type of aircraft is suitable/desirable?
- What does it cost to acquire and operate?

Considerations When Acquiring an Aircraft

- Age of the aircraft
- Aircraft limitations [*read* the limitations sections of the Airplane Flight Manual (AFM)/Pilot's Operating Handbook (POH)]
- Paint and interior condition
- Inspection methods used and current status
- Damage history
- Status of airworthiness directives and service bulletins
- Aging aircraft/corrosion requirements
- Noise classification and potential airport restrictions
- Engine status—overhauls, interim inspections, life-limited items
- Anticipated operating costs—fuel, oil, routine maintenance
- Anticipated extraordinary costs—inspections, airworthiness directives, aging aircraft, life-limited items
- Availability of parts and knowledgeable repair facilities (orphan aircraft)
- Desired upgrades

- How should the aircraft be used/scheduled?
- When should the airlines (or charter) be used?
- Should the aircraft be leased or purchased?

These challenging questions require informed evaluation based on facts. A transportation needs analysis derived from past and anticipated travel patterns is the usual method of determining executive air transportation requirements (see Chap. 2).

The travel analysis will provide significant information for aircraft selection. Such factors as passenger capacity, speed, range, and runway requirements will become obvious. Choosing among available aircraft types and models, however, is difficult because no aircraft can do all things equally well, and cost is usually a major consideration. Even the best selection will be a compromise. Be willing to seek advice from knowledgeable sources.

New versus used, turbine- versus piston-powered, cabin size, interior amenities, ability to operate from small airports, and potential resale value are but a few of the variables that must be addressed during the selection process. A variety of opinions from operators of the candidate aircraft will prove very useful in the process.

Reference materials to assist with the aircraft selection process are available from *Business and Commercial Aviation* (Annual Buying Guide), Conklin & de Decker Associates (aircraft performance comparison and cost evaluations), and individual aircraft manufacturers.

The capabilities and experience of the pilots who will fly the aircraft also should be considered in the selection process. A low-time pilot with little multiengine experience may not be a good candidate for a pressurized twin. A natural check on the capabilities of a pilot rests with the broker or underwriter who insures the aircraft, however. Such companies will not insure an inexperienced pilot to fly a sophisticated aircraft or will add a hefty surcharge to the normal premium if the candidate is marginally qualified.

Once the analysis has been completed and provisional selection made, some judgment must be exercised in selecting an aircraft. There is no "right" aircraft for an individual company. Several types will be adequate for a company's needs; final selection is not too different from selecting an automobile or a house. The big difference between these more common consumer items and an aircraft is complexity and price. Thus expert assistance is needed. More than one voice should be heard in the process.

Acquisition Assistance

Once a decision has been made regarding an aircraft type, a broker may be useful in the search for the desired aircraft. Remember, brokers and dealers bring some bias to the acquisition process; therefore, it may be desirable to approach several sellers/brokers and solicit advice or actual bids. Lists of aircraft manufacturers and used aircraft vendors are available from the General Aviation Manufacturers Association (GAMA), the National Air Transportation Association (NATA), the National Aircraft Resale Association (NARA), and NBAA. The recommended method of selecting an aircraft purchase advisor is a word-of-mouth recommendation from an associate you trust or a company operating the type of aircraft you want.

Unless you have considerable experience with the type of aircraft to be purchased, a prepurchase inspection should be done. There are too many hidden limitations, restrictions, costs, quirks, skeletons, and anomalies associated with a business-type aircraft for the uninitiated to consider. Just because an aircraft is presented with a "fresh" set of inspections does not necessarily mean that it is in good condition; experienced eyes must make that determination. A prepurchase inspection should be performed by an *expert* specializing in the *particular type* of aircraft to be acquired. The inspection report should address most of the preceding considerations and come with a guarantee or warranty against unknown or unanticipated surprises about the aircraft. To qualify the potential inspector, ask essentially the same types of questions you asked the dealer/broker. Finally, the prepurchase inspector should have no conflict of interest with regard to owners, operators, manufacturers, brokers, dealers, or the like.

A number of preowned aircraft are usually on the market. Popular aircraft are scarce and command a premium price; unpopular aircraft are plentiful and may be susceptible to heavy price negotiation. The point is that the aircraft market is constantly changing, driven by financial, regulatory, technological, and preference factors. Therefore, market timing is important for both new and used aircraft. Markets that are awash in good, low-time aircraft will cause all prices to soften and be subject to negotiation. Conversely, if the aircraft to be purchased is in short supply, an acceptable one may not be available, regardless of price.

Advisors selected for the purchase process will be reacting to these market features; their advice must be evaluated carefully to get the best possible deal. If the market for the type of aircraft desired is bad at the time the decision to acquire is made, a short-term lease of the desired aircraft type may be desirable. Charter, rental, or a lease also may be used to wait for better market conditions.

Insurance

Insurance should be obtained before commencing any business aircraft operation. Before any aircraft lease or purchase agreement is signed, an insurance binder must be in hand to ensure that coverage is in place. This includes insurance for both aircraft operations and the facility at which it will be based. If the aircraft is leased, or if a lien is attached to it, the lessor or financial institution will stipulate certain insurance limits to be carried. The base facility (hangar) probably will not be included in this requirement.

As with so many other professional issues, several quotes should be obtained for insurance. This process also will provide you with an education about the components and provisions of aircraft insurance. Before the search for insurance begins, however, a risk advisor, facilities director, or legal counsel should be consulted regarding a risk philosophy and insurance. The Appendix of the *NBAA Management Guide* contains an excellent primer on the subject of aircraft insurance (see also Table 3.2).

Coverage
• Hull—all risk, motion and nonmotion • Liability—bodily injury, passengers, and property damage • Medical payments • Lienholders interest endorsement • Guest voluntary settlement • War risk • Foreign operations • Broad form expansion • Hangar keepers/airport liability
Points to know about
• Deductibles for each coverage category • Actual liability coverage and limits • Substitute/loaner engines/accessories coverage • Acceptable pilots/required training • Depreciation effects on hull damage coverage • Loss of use/substitute aircraft provisions • Additional insured possibilities/provisions • Salvage rights • Provisions for lienholders interest endorsement (breach of warranty) • Effect/compatibility of corporate umbrella policy cover • Coverage outside country of registry • Interaction between hangar keepers and aircraft coverage • Legal defense coverage • Exclusions due to regulatory violations • Owner/operator obligations • Areas not covered by the policy
General
• Read and understand the policy *thoroughly*— details are very important. • Have legal counsel review the policy. • Ensure that all members of the flight department know essential provisions of the policy. • Review policy annually with corporate risk manager and broker. • Report minor incidents promptly to your insurance company.

TABLE 3.2   Aircraft Insurance Checklist

Employees using their own, rented, or borrowed aircraft must ensure that the company is adequately protected against liability exposure. This normally means that the company must be a named insured on the aircraft insurance policy and that the coverage is adequate. Additionally, employees require special coverage under worker's compensation insurance; pilots' premiums are usually higher than those of most employees.

Flight Records

A permanent record of each flight should exist, detailing

- Flight times/cycles/landings
- Costs incurred
- Destinations
- Passengers carried
- Business purpose, if any

Maintaining such a record will satisfy the FAA and IRS, your accountant, and your own need for information. A separate record should exist for each day on which the aircraft is flown. The form should be completed whether the aircraft is used for business or not; the continuity of information is important, regardless of the use of the aircraft. Once completed, these records of individual flights should be placed in a file away from the aircraft and kept for a minimum of 3 years after your income tax return is filed. If a record of past flights is needed in the aircraft, carbonless duplicating forms may be devised to provide this function. A typical flight record form is shown in Fig. 3.1.

Date_____   Aircraft _____

Pilot _____

	Leg 1	Leg 2	Leg 3	Leg 4
From				
To				
Hobbs In *				
Hobbs Out				
Hobbs Time				
Takeoff Time				
Landing Time				
Miles				
Passengers				
Flight Purpose				
Expenses				
Remarks				

* Assumes Hobbs meter records engine time vs. time off/on.

FIGURE **3.1**   Sample employee-flown business aircraft flight record.

While some people may wish to include maintenance discrepancy and other airworthiness information on the flight record form, the flight itinerary and passengers carried may be considered confidential and not appropriate for release to maintenance personnel or a contractor, therefore dictating the use of a separate form for these purposes (see Fig. 3.2).

For complete record control, the daily flight records may be entered into a computer database for future reference. This process will allow the data to be manipulated so that information concerning your operation may be obtained readily, and this is especially useful when questions arise regarding the use and utility of the aircraft. While some proprietary software is available for this purpose, a simple spreadsheet program can accomplish almost as much as the commercially available programs. Clerical personnel can be trained easily to do this job (see Fig. 3.3).

Care should be taken to accurately reflect the purpose of each flight. This IRS requirement, which attempts to connect a specific flight with a business purpose, does not impose an unreasonable burden on the individual. Extensive explanations are not required, but a clear purpose should be recorded for each leg of a trip if it is different from the previous leg. If the purpose of the flight is personal or pleasure, it should be so noted on the flight record.

Carriage of passengers also should be connected with the flight purpose. If the passenger is being carried for a different reason than the primary flight purpose, it should be noted

Aircraft _____

Discrepancy		Corrective Action	
Pilot	Date	Technician	Date

Discrepancy		Corrective Action	
Pilot	Date	Technician	Date

FIGURE **3.2** Sample employee-flown aircraft discrepancy record.

	May 2013	May 2014	12 month average
Trips	6	6	5
Legs	17	16	13
Miles	4140	3885	2880
Hours	23	21	16
Passengers	9	7	5
Average			
Hours/trip	3.8	3.5	3.2
Miles/trip	690	647	576
Passengers/leg	0.5	0.4	0.4
Costs -$			
Direct	4225	3570	2640
Indirect	3220	2990	2515
Total	7445	6560	5155
Direct cost - $:			
Hour	185	170	165
Mile	1.02	.92	.92

FIGURE 3.3 Sample owner/employee-flown aircraft report.

on the flight record. Passengers being carried for personal reasons may be subject to additional tax treatment; the *NBAA Management Guide* and other NBAA tax publications contain additional information on this subject.

Standards

An operations manual that provides operational and administrative standards for the operation of an aircraft whether it is used for business or not is essential. The manual will define procedures that enable the operation to comply with federal aviation regulations and establish standards and limits that exceed federal standards when deemed necessary. The operations manual is designed to create a sense of professionalism that will enhance safety, efficiency, and effectiveness.

The process of developing the operations manual is probably more important than the manual itself. This is so because the introspection and research required to create the manual will lead to a realization of the essential factors that make up a well-run, safe flight operation. Once this realization has taken place, setting those factors into policies and procedures will promote a more structured and efficient operation. A suggested table of contents for this manual is shown in Table 3.3. Detailed procedures and limitations will be found in Appendix C and a comprehensive flight operations manual template is shown in Chap. 6. A flight operations manual designed specifically for light business aircraft (LBA) is available at www.nbaa.org/admin/policies.

The resulting manual may be referred to on an ongoing basis to accommodate both normal and unusual situations. If the manual is to be truly useful, it must be reviewed and updated to reflect the changing needs of the individual and the company using it. A brief review at least annually should prove to be sufficient for most operations.

Administration
- Purpose of the business aircraft
- Safety
- Operational control
  - Scheduling
  - Flight following procedures
- Use of the aircraft
- Recordkeeping procedures

Flight operations
- Health and fitness
- Duty time
- Standard operating procedures
- Operating limitations
  - Pilot
  - Aircraft
- Training

Maintenance
- Maintenance program
- Discrepancy control
  - Recording discrepancies
  - Deferred maintenance
- Maintenance provider selection
- Maintenance away from home base

Note: This is an abbreviated outline. For a more complete outline, see the Appendix.

TABLE 3.3 Owner/Employee Operations Manual

The Business of Safety

Every pilot makes a commitment to safe flight operations at one time or another. Experience has shown, however, that the "one time or another" is not sufficient. Safety demands a full-time, conscious commitment to continuous, safe operations. Conforming to good operating practices and exercising good judgment must be accomplished every time you fly, not just on bad-weather days or on more challenging flights. By analogy, a person uses accepted standards and good decision-making techniques daily to survive in the social or competitive business world. If these techniques are good for these applications, they should be good for flying as well.

Unfortunately, the person who typically uses an aircraft in the pursuit of business or targeted personal reasons may tend to put most of his or her effort and concentration into those matters to the exclusion of other activities. Family and personal relationships may suffer as a result of overconcentration on business matters, as well as the individual's physical and mental health. So it may be with flying.

If the pilot views flying in the same manner as he or she does running a core business or family affairs, in terms of risk and profit and loss, a healthy frame of reference will develop.

Flying home in instrument meteorological conditions (IMC) late at night after a full day on the job should be considered risky business in the same manner as expanding into a new market without market research, capitalization, and personnel resources.

The concept of risk assessment and mitigation serves as the core of safety programs (see "Safety Management Systems" in Chap. 8). Identification of hazards, assigning a relative value to the severity of each hazard, and estimating the probability of its occurrence will yield a level of risk. Once proficient at risk assessment, the pilot can quantify it and make more informed decisions designed to reduce risk to the lowest practical level commensurate with the type of mission performed.

Training

A structured initial and recurring ground and flight training program will sharpen your compliance to standards and prepare you for the unexpected. The thought process that drives one to train regularly serves to reinforce other safe attitudes. And training itself will create a mind set that fosters adherence to standards and professionalism. There is no regulatory requirement that pilots in general aviation operations undergo any recurrent training for non-type-rated aircraft. Once pilots receive their initial training for a pilot certificate or rating and perhaps some small amount of transition training for a specific aircraft, little or no training is accomplished by the average business or pleasure pilot. However, airline and charter pilots are required to undergo at least annual flight and ground training, and it a rare corporate flight department that does not send its pilots to recurrent training at least as frequently. Since the mission of the owner/employee pilot is often as challenging as that of the corporate pilot, it makes sense to receive some form of initial and regular recurrent flight and ground training.

The best safety device in any aircraft is a well-trained pilot. —FLIGHT SAFETY INCORPORATED

Initial Training. Before flying any sophisticated aircraft that will be used for business purposes, some form of transition training should be undertaken. Vendors offer type-specific transition courses that provide both ground and flight training. These courses often include a day or more of ground training to introduce aircraft systems and limitations and a number of flights in an aircraft or simulator to practice normal and emergency procedures. Such instruction is often given by instructors who have thousands of hours of experience in the type of aircraft and can impart more than just procedures and limitations specified in the pilot's operating handbook.

Prior to authorizing an individual to act as pilot in command (PIC) in an unfamiliar aircraft type, the aircraft insurer may specify a certain amount of training or flight time supervised by a qualified instructor. The insurer's requirements should be considered a minimum requirement and not a final qualifying authorization, however.

Recurrent Training. Although a pilot may fly a specific aircraft hundreds of hours each year, experience in airline, military, and corporate flying has shown the need for periodic retraining. Recurrent training allows pilots to practice unusual and emergency procedures not encountered in the normal course of productive flying. Moreover, it detects undesirable habit patterns that may have crept into an individual's piloting technique over time. Finally, it provides a review of aircraft systems and limitations that often are forgotten during normal operations.

Because business and personal pilot operations are at least as challenging as those for corporate pilots, it makes sense to undergo recurrent training at least annually. The most desirable form of training is formal classroom and simulator training provided by a recognized training vendor. Since this type of training may not be available for all types of aircraft or at convenient locations, use of local training facilities and experienced individuals may be necessary and certainly is a very helpful experience. At a minimum, the training should be accomplished in accordance with a formal syllabus and should include ground training as well.

Other Training.   Many pilots stop their training activities after recurrent training in the aircraft they normally fly. However, there are other related issues and skills that should be either learned or reviewed on a regular basis. Some of theses subjects include

- Avionics/systems
- First aid and CPR
- Survival
- High altitude (altitude chamber)
- Crew resource management/pilot judgment (see below)
- Advanced weather analysis, including weather radar
- International operations
- Air traffic control/instrument procedures

Constant Learning.   The safe pilot is naturally curious about the environment and should always be probing the edges of his or her existing store of knowledge. One's ability to gain a sufficient variety and amount of aeronautical experience before it may be needed is quite limited; knowledge fills the experience gaps. Use your fellow pilots and especially experienced flight instructors to assist in this process. Staying up-to-date with selected aviation periodicals is another means of constant learning.

Compared with discrete training sessions, the learning process should be continuous and never-ending. Detailed aeronautical knowledge may provide the best insurance policy, and the premiums are surprisingly low.

Duty Time

The term *duty time* comes from national regulations that specify how much time a professional pilot can be in a condition of readiness to fly, regardless of whether or not he or she actually flies. This concept recognizes the fact that being on duty eventually will create an unacceptable level of fatigue.

Airline pilots have strict and complicated duty time limits that are guaranteed by regulations and their union contracts. While the idea is under consideration, most national regulations currently do not limit duty time for corporate flight crews. Most flight departments, however, include duty time limitations within the corporate flight operations manuals. Since corporate flying is closer to business flying, the norms used in such operations will be used for examples and comparison.

Symptoms of Fatigue

- Accepting lower standards of performance
- Decreased reaction time
- Reduced attention span
- Sloppy or rough flying
- Distorted or impaired judgment
- Irritability

Duty Time Factors

- Length and complexity of the business day
- Time zone changes
- Flight time
- Number of legs
- Type of flying, e.g., good/bad weather, day/night, etc.
- Cumulative working/flying days

Fatigue creates a degradation of mental and physical skills that can prove harmful or even dangerous if taken to extremes. If allowed to progress too far, fatigue may conspire to create a hazardous situation. Knowing the warning signs will help avoid the onset of fatigue, but the ideal way to avoid fatigue is to set and abide by personal flight and duty time guidelines.

Because business and personal transportation embraces a combination of flying and other activities, it is difficult to provide strict guidelines for duty time. Each situation and pilot characteristics are quite variable, necessitating an assessment of each set of circumstances.

Perhaps the most significant variable regarding fatigue is the type and amount of business and personal activities conducted either before or after a flight. Intense, contentious dealings tend to drain most people and bring on fatigue earlier than business involving more routine events or participation in a large meeting. Only the individual can evaluate the effects of a full day of activities on the prospect of flying. Therefore, individuals must be honest in making their preflight stress and fatigue assessments.

Guidelines for a maximum-length business/flight duty day:

- *Total day:* 16 hours (waking hours)
- *Business and flying:* 12 hours (productive business + flight-related activities)
- *Maximum flying-only day:* 7 hours (all flight-related activities)
- *Reductions due to type of flying* (reduce flying or business and flying day by 1 hour for each):
  ○ Two legs over an initial two
  ○ Each leg flown during the hours of darkness
  ○ Each IMC leg more than one

Duty and flight time are only half the total picture. Adequate periods of rest must be interspersed between duty periods. Rest, too, is a subjective measure that is difficult to quantify. Time spent sleeping is considered good-quality rest time. However, if "rest" is spent divided between doing business "homework" and fitful sleep in a high-noise and light environment, it may not adequately fulfill the intent of rest time.

Guidelines for minimum rest time:

- 10 hours between duty periods
- If a period of only 8 hours of rest is available, reduce the following duty day by 2 hours
- One period of at least 24 consecutive nonduty hours (including 10 hours of rest) for each 7 consecutive day period

Note that these guidelines are just that—guidelines. All guidelines are subject to modification based on situation and circumstance. These have been formulated from a mix of corporate flight department duty times and the National Aeronautics and Space Administration (NASA) fatigue study guidelines. The most important aspect of these limitations is not the absolute numbers stated but the fact that the individual has taken the time to think through the fatigue problem and arrive at personal restrictions.

Limitations

Just as each aircraft has limitations imposed on it, flight departments and individual pilots often have restrictions imposed on them. Airlines have limitations imposed on them not required of private operators in recognition of the higher level of safety afforded those who pay for their transportation. And the majority of corporate flight departments operate under self-imposed restrictions oriented toward individuals and aircraft alike.

Therefore, it makes sense that business and personal operators, who may be flying less capable aircraft with less experienced pilots, also should impose limitations on their operation. These restrictions do not have to be extensive or draconian, but they should take into account the capabilities of the individuals involved in company aviation operations. Limitations should be imposed before the fact and not based solely on inputs derived on the spur of the moment. See Appendix C for suggested employee-pilot limitations

Maintenance

Responsibility for the airworthiness of an aircraft is shared jointly by the owner/operator and the PIC for each individual flight. Responsibility for required maintenance and inspections, however, belongs solely to the aircraft's owner/operator. In most business flying operations, the PIC and owner/operator are usually the same individual, making this person totally responsibility for the continuing airworthiness of the company aircraft.

Many pilots attempt to transfer responsibility for ensuring that the aircraft is maintained in an airworthy condition to their maintenance contractor or vendor. However, it is *not the responsibility* of the maintenance provider to ensure the airworthiness of the aircraft, only to comply with the national maintenance, repair, and recordkeeping requirements. It is the *owner/operator* who must ensure that the maintenance and inspection are accomplished.

Selecting a Maintenance Vendor.   Maintenance and inspection may be performed on an aircraft by either of two entities: a licensed aircraft maintenance engineer or an approved maintenance/repair organization. [In the United States, these are known as an FAA airframe and powerplant (A&P) mechanic or an FAA-certificated repair station.] Either is capable of performing normal maintenance and inspection tasks on most aircraft and approving it for return to service. While repair stations are held to the same performance standards as the A&P mechanic, or aircraft maintenance technician as they are now being called, they are required to have inspection systems in place and supervision and training for personnel that exceed what the A&P mechanic is required to have. Further, repair stations are specifically rated to accomplish certain tasks and to work on specific equipment, whereas the individual A&P mechanic does not have such requirements.

Whatever method chosen, the principles of good business-consumer relationships apply. A recommended method for determining whether a vendor is suitable is to ask someone who has used the vendor's service in the past. When doing so, it is important to ensure that the person providing the reference has used the vendor recently for a similar service. Obviously, the person providing a reference may not be as discriminating as you regarding quality of work, time to complete work, and cost. Therefore, try to ensure that the person used has similar standards when making his or her evaluation.

Ask national aviation regulatory authority personnel (the local FAA Flight Standards District Office in the United States) if they have any negative information regarding the prospective vendor.

Vendors should have some experience with your type of aircraft. A Beech Baron probably would receive more expert care from a Beechcraft service center than from one specializing in Cessnas. However, the vendor's affiliation with a manufacturer does not necessarily mean that it is an excellent shop.

While cost is a major consideration in the selection process, quality and reliability of the work performed should rank at least as high, if not higher, in the selection process. Ask the following questions of the prospective vendor:

- How long have you been in business?
- How much experience do you have with my type of aircraft?
- What type and amount of initial and recurrent training do your technicians receive relative to my class and type of aircraft?
- Describe the process you use to obtain spare parts for aircraft. (Quality and price are important.)
- Do you use written work orders and provide a detailed breakdown of work completed on your invoice?
- Describe the warranty you provide for both labor and parts.
- Provide the names of three recent customers flying my type of aircraft.

Maintenance Control.   The owner/operator is solely responsible for anticipating airworthiness requirements and dealing with discrepancies that render an aircraft unairworthy. In reality, however, the owner/operator relies heavily on the maintenance vendor to assist with these tasks. A well-functioning relationship between pilot and maintenance provider ensures that required maintenance and inspections are accomplished in a timely manner.

A schedule of maintenance actions should be used to trigger inspections and time-controlled maintenance, whereas aircraft discrepancy reports are used to initiate repairs. Work orders are commonly used both to initiate actions and to determine their progress toward completion. Only after the work has been inspected and signed off on should the aircraft be released to other work or for service. Ideally, when required maintenance has been completed, a maintenance release will be signed by the vendor, certifying that the aircraft is airworthy and can return to service.

Airworthiness compliance should be viewed as a partnership between owner/operator and vendor. Both should consult prior to any maintenance performed on the aircraft, whether scheduled or nonscheduled. It is important to specify in writing exactly the work to be accomplished. This serves as a formal authorization for the vendor to commence work and also should outline the extent of the work to be accomplished. Orders for major work such as upgrades, refurbishment, overhauls, and modifications should be quite detailed and specify completion times and standards. It is this chain of paperwork that provides a firm method of communication between owner and vendor and serves as a series of contractual arrangements between the two parties. This system will create a written record of expectations and promises that will prove valuable many times to both parties.

Aircraft Charter

Charter services are a useful means of providing on-demand air transportation for companies and individuals who need just 100 to 150 hours service per year—this equates to approximately 30 to 50 domestic round trips per year, one every week or so. However, charter also should be considered to supplement other forms of business air transportation, such as an in-house operation or fractional ownership.

There are a significant number of charter providers to choose from in most locations around the world. While each of these operators should be officially sanctioned by operating under an air taxi or charter operating certificate issued by their national regulatory authority, levels of service among charter operators may vary widely. Therefore, it is wise to perform a market survey of charter operators in an attempt to obtain the safest and best possible service. Choosing the first charter operator listed in the Yellow Pages may provide disappointing results—take time to investigate the options.

Using a number of charter operators and different aircraft is a good means of exploring what on-demand aviation services have to offer. Service levels, aircraft types and configurations, and associated aviation systems should rapidly become evident to the traveler, providing a useful background for subsequent choices in the field of on-demand air transportation.

As with any other service, reputation and word-of-mouth recommendations may provide the best indicators of service levels. Asking your peers or peer companies should give you a general idea of what is available in the local area and which may suit your needs.

Yet one charter user's needs and standards may not match your requirements. For instance, a charter user may be quite pleased with a small, older turboprop aircraft obtained from a local provider. If your needs tend toward a larger cabin, faster turbojet aircraft, no matter how good the turboprop provider's service is, it will not satisfy your requirements. Therefore, when asking for recommendations, ask for the following service discriminators:

- What type of aircraft do they have/did you use?
- How long was the trip?
- How many passengers were carried on the trip?
- Did they depart and arrive on time?
- Was the aircraft clean and apparently in good condition?
- How were the passengers treated?
- Was the billing as quoted?
- Do you feel that you received good value from the experience?
- Would you use this company again?

Once you have obtained a good idea of what is available through others' recommendations regarding local services, look in the Yellow Pages for other companies that may provide national charter service and make their services available through a toll-free telephone number. The Internet is also a good source of national and international charter service providers. Additionally, *The Air Charter Guide* is a subscription-based periodical (also available online) that provides detailed information about worldwide charter services. Charter brokers are available who, through their knowledge and expertise, will attempt to obtain the best possible charter deal for you, for a price. Two excellent publications, the *NBAA Air-*

craft Charter Consumer Guide and the *NATA Chartering an Aircraft, a Consumer Guide* will provide useful background information.

Checking the Record

Armed with the recommendations of those who have used the charter service and other information that you may have gained in your research, you need to know some basic facts about the company. The amount of money involved and criticality of the services provided warrant checking into the operator's record with the national regulatory authority.

Since each air charter company has been issued an operating certificate by their national regulatory authority, that agency maintains records on them, gained from surveillance of their operations. First, call the provider to determine its regulatory status and then, using contacts the supplier has provided, contact the regulatory authority to find out how the supplier has performed under the regulations. Here are a few basic items you should know about an air taxi company:

- Air taxi (charter) operating certificate number
- Name of the company as it appears on the certificate
- Location and telephone number of the local aviation regulatory authority having jurisdiction over the company [In the United States, this is the FAA Flight Standards District Office (FSDO).]
- Name of the person assigned oversight responsibility for their operation [This is the FAA Principal Operations Inspector (POI) in the United States.]
- Is the company or has it ever been under investigation by the regulatory authority (FAA) for a violation of applicable national regulations (Federal Aviation Regulations)?
- Has the company ever experienced an incident or accident reportable under national regulations (National Transportation Safety Board)?
- What is the experience level of the flight crew that will fly the aircraft? (The PIC of a charter aircraft must have a minimum total flight time of 1200 hours in many countries; however, most reputable companies provide PICs with 3000 hours or more of flight experience. Helicopter pilots operating under visual conditions may have as few as 500 pilot hours, although 1500 hours is customary.)
- Liability insurance limits carried on your aircraft
- Names and telephone numbers of people who have used the charter services within the past month (to be used as references)

Then contact the named regulatory authority and ask about the supplier's violation history or any other safety-related issues. If the air taxi operator is unwilling or reluctant to provide the answers to the preceding questions or does not want you to contact the regulatory agency for verification, you would be wise to consider another operator to fill your travel requirements.

Charter evaluation services are available worldwide to perform all of these tasks for a fee. Most provide their information as a subscription service, sending an initial comprehensive evaluation followed by periodic updates. These services can be quite useful because they are performed by experts and cover virtually every aspect of the charter provider's operations. Moreover, evaluation services are available for major charter operators worldwide. They may be found by contacting the national regulatory authority, the charter operator itself, or a national organization representing business aviation and/or charter operators (in the United States, see nbaa.org/prodsvcs).

Charges

Many charter companies are not based at a local airport; rather, the charter provider may be located hundreds of miles away and will ferry (position or deadhead) the aircraft to the client's location. The charter client is charged for this positioning, normally at the same rate as for occupied service. These charges may add substantial amounts to the total trip charges, leading the wise consumer to consider carefully the use of a local charter operator who may not have exactly the aircraft desired or the best service terms available.

Additionally, landing fees, catering charges, flight crew expenses, wait fees, and overnight charges also may be added to the final bill. Transportation or use taxes often are added to the fee by local or national authorities, further driving up the total charge. Therefore, a detailed quotation stating all anticipated charges should be obtained prior to finalizing arrangements for the service. Figure 3.4 shows a sample quotation.

Evaluation

While you may be an infrequent user of charter services, having a single person in charge of these services is highly desirable. Tracking the various providers and their service levels and charges will provide service continuity and overall satisfaction for both travelers and the company. This person will be able to observe overall utilization and perhaps even perform a central scheduling function as well.

Determining whether each charter flight met the expectations of the passengers will prove useful in evaluating the various providers' services. The charter coordinator/scheduler should keep a record of each flight, noting both statistical details and quality-of-service comments, both observed and asked of the passengers. Exhaustive questionnaires or in-depth interviews are not necessary for this function; casual comments received from the

September 30, 2012

Quotation for: Ace Widget Company

Date	ETD	Depart	Arrive	ETE	ETA	Pass
10/15/03	8 AM	Wilmington-ILM	Birmingham-BHM	1:30	8:30 AM	6
10/15/03	5 PM	Birmingham-BHM	Wilmington-ILM	1:30	7:30 PM	6

Cessna Citation Bravo Aircraft @ $1,900 per hour x 3 hrs *	$5700
Federal excise tax @ 7.5%	428
Segment fee @ $30	60
Crew wait fee @ $100 per hour x 5.5 hrs	550
Landing/ramp fees	350
Catering – breakfast and evening snack x 6	250
Total	**$7338**

* Total time may vary according to weather and air traffic conditions. Crew and fuel expenses included.

FIGURE 3.4   Sample aircraft charter quotation.

passengers normally are sufficient for this function. Some service quality issues that can be recorded include

- Did the flight depart and arrive on time?
- Was the aircraft clean and apparently in good condition?
- How were the passengers treated?
- Do you feel that you received good value from the experience?
- Would you use this company again?

There are many charter providers available to clients in most areas of the world, and the charter operators are aware of this essential fact of business competition. While you should remain aware of this basic tenet, the most successful services are derived from a mature and open relationship developed over time between the provider and the recipient of the charter service.

Fractional Ownership

Fractional ownership offers travelers the ability to purchase shares of a business aircraft. As little as one-sixteenth of an aircraft can be purchased, which offers 50 flight hours of use per year, or as much as one-half of an aircraft can be purchased, depending on how many hours are needed. The most common amounts purchased usually range from one-eighth to one-quarter shares.

Fractional Comparisons

Ask the fractional provider to compare its services with those of other on-demand air transportation methods.

- Fixed costs
- Variable costs
- Scheduled response time
- Aircraft range versus payload trade-off
- Resale value/options
- Customer service
- Level of control over operational features
  - Scheduling response
  - Service levels
  - Pilot qualifications and training
  - Maintenance
- Contract options/flexibility

Fractional owners pay a percentage of the purchase price for an aircraft, commensurate with the amount of hours they want available per year; e.g., one-eighth share receives 100 flight hours, one-quarter share receives 200 flight hours, etc. Most fractional providers offer a variety of aircraft to choose from, ranging from small turboprops to large intercontinental turbojets. Hourly and monthly fees are charged over the period of the contract, normally 5 years.

Fractional owners are guaranteed that their aircraft or another aircraft of the same model will be constantly available to them with as little as 4 to 8 hours of advance notice. The fractional provider assumes responsibility for scheduling, staffing, flight planning, weather, communications, maintenance, catering, and insurance. At the end of the initial contract, the owner's fractional share will be purchased by the provider at "fair market value," or the contract may be renewed or upgraded.

Fractional owners use the service because they appreciate the following aspects:

- Guaranteed availability
- Consistent scheduling practices

- Predictable service levels
- Usage ranging from 100 to 200 hours per year
- Relatively small investment that fits the corporate financial plan

Investigating the Service

Many potential fractional owners are veterans of aircraft charter services who are seeking a more predictable or higher level of on-demand air transportation service. While charter services can be quite good, especially if one establishes a relationship with a single provider, the desirability of owning at least a portion of one's own aircraft can be a powerful draw for customers. The "always available" nature of the service is also a positive aspect. And the turnkey, no-hassle nature of the service has convinced thousands of travelers worldwide that fractional ownership is the way to go. However, there are details of the service that must be known prior to committing to the program.

Researching the potential providers of fractional ownership service is not difficult. Most advertise in business periodicals and newspapers and often are featured prominently in the charter section of national Yellow Pages. While most popular and available in North America, fractional services are available in many other parts of the world. Note that fractional providers are considered to be private operators in the United States but are considered commercial operators in many other countries.

It is important to have the fractional representative provide comparisons of his or her company's services with other available options, including charter and whole ownership. These comparisons must include both financial and service related features. Particularly important are comparisons that show cost per hour over a range of utilization amounts, ideally between 50 and 400 hours.

Details. The major expenditure associated with fractional ownership is the value of the fractional share of the aircraft itself. Ranging from $500,000 to $20 million, depending on the aircraft type and fractional share, this expense entitles the fractional owner to the rights, benefits, and obligations of any owner of a major asset, albeit for just a fraction of the full aircraft. The fractional owner may depreciate his or her share of the aircraft, as would any other full aircraft owner, to the full limit of the value of his

Fractional Flight Hours per Share

Share	Flight Hours
$\frac{1}{16}$	50
$\frac{1}{8}$	100
$\frac{1}{4}$	200
$\frac{1}{2}$	400
Full	800

or her share. Along with this privilege comes the obligation of owning the aircraft, that of keeping the aircraft airworthy and sharing the asset with the co-owners. Accomplishing these obligations is taken care of under the several agreements signed between the fractional owner and provider, the principal ones being the management, joint-ownership, and interchange agreements entered into at the time of the purchase.

Fractional owners are subject to additional charges above the stated monthly management fee and hourly flight charges, including the following:

- Extraordinary maintenance, handling, and administrative fees may be applicable depending on the nature of a specific flight or the condition of the aircraft.
- Providers are permitted to raise their standard monthly fees based on external indices, such as the consumer price index.

- While normally adequate hull and liability insurance is provided under the provider's contract, war risk liability insurance may be required to be purchased at the owner's expense.

- Early contract termination (normally less than the specified 5-year minimum) will allow the owner to sell the aircraft to the provider at a predetermined value, less a marketing fee.

While all of these items are specified in the several contracts the owner is required to sign, some require careful examination to determine their applicability.

Note that the owner's specific aircraft may never be made available to him or her due to the complex scheduling required to run a large fractional operation. However, the same type of aircraft will be supplied from the provider's stable of aircraft, if available. If the appropriate aircraft is not available, a similar type of aircraft will be supplied from an aircraft charter operator under contract to the fractional provider. These charter operators may not operate to the same standards of service specified by the fractional provider, however.

As with charter operations, the fractional provider supplies a transportation service subject to many variables. Therefore, the fractional owner should know what is being provided in terms of technical, administrative, and service quality. A number of consultants are available to evaluate fractional providers to ensure their compliance with terms of the contract, national aviation regulations, and service quality; they may be found through national and international aviation associations dealing with business and charter aviation. The "Sherpa Report, a Guide to Private Aviation" provides a detailed perspective of fractional ownership operations and may be consulted for more information.

Managing It

One of the greatest difficulties fractional service users report is managing the available number of flight hours specified in the service agreement. Each participant is provided a certain number of hours under the contract, according to the amount of fractional share purchased. A one-eighth share will receive 100 hours under the contract at a specified hourly cost, ranging between $1000 and $7000, depending on the type of aircraft involved. If this number is exceeded, most providers supply additional hours "borrowed" from subsequent years of the contract, up the limit of the contract, normally 5 years. When the total number of contract hours is reached, supplemental hours may be purchased, but at up to three times the normal contract hourly rate.

Similarly, it is possible to trade up or down into larger or smaller aircraft. If the owner's contract is for a midsized-cabin aircraft and a large-cabin aircraft is desired for a specific trip, this is possible by trading a multiple of 2.5 (or more hours) in the midsized aircraft for 1 hour in the larger aircraft; if a smaller aircraft is acceptable, tradedown provisions are also permissible. Upwardly mobile trade-offs can rapidly consume the annual allotment of program hours. Therefore, it is prudent to program and track available hours to ensure that they are not exceeded early in the contract.

This may be best accomplished by channeling all requests for use through a single person or office to track hours used. If a variety of people/offices are permitted to use the service, an allocation of time for each group may be necessary to control the expectations of those using the aircraft. This office also should coordinate closely with the financial function of the company responsible for paying the fractional provider's invoices to ensure that the conditions of the contract are complied with.

In-House Aircraft

This is the most popular of all forms of on-demand air transportation. Perhaps the reason for its great popularity is the level of control provided to the company or individual using the service. These forms of control include

- *Schedule.* The personnel setting policy for use of the aircraft exercise total and ultimate control over who uses the aircraft, where it goes, and when it returns. Priorities are set, trips are evaluated on their merits, and the "winner" is assigned the valuable transportation asset.

- *Aircraft type.* The aircraft or combination of aircraft that best suits the operation has been selected carefully by the user. And if user needs change over time, changing aircraft is relatively easy.

- *Personnel.* Finding and hiring the best manager, pilots, aviation maintenance technicians, and schedulers may be difficult, but with proper assistance, they should prove best suited for user needs.

- *Policies and procedures.* Every flight operation needs well-founded, mature policies and procedures to guide department administration, operations, scheduling, and maintenance. Within the regulations governing flight operations, whatever a company wants in terms of policy and procedure can be theirs for the asking—no compromises required.

- *Cost.* Flight operations normally have wide latitude in controlling their costs but must have direction from the company to ensure that concomitant safety, quality, and reliability goals are met. Companies operating their own aircraft have the ultimate level of control over expenses.

- *Quality.* Although sometimes a secondary consideration, service quality is quite important to many companies. The ability to travel in a style of their choosing, whether strictly utilitarian or opulent, rests solely with the dictates of the users.

So it's all about control—control over the various aspects of the operation, be they technical, financial, or service-related. Companies having their own in-house flight departments, fully managed and controlled by company employees, have the best chance at getting the flight operation they desire. If they do not get it, they have no one but themselves to blame. The factors just mentioned should serve as guiding principles for companies establishing a flight department. While it is easy to get caught up in the technical aspects of aviation, it is the basic elements of any process or operation that are important.

Beginnings are important. We will start the process by assuming that the processes mentioned in Chap. 2 have been accomplished:

- Air transportation analysis completed
- Basic type of aircraft selected
- In-house flight department chosen

As mentioned earlier, a comprehensive checklist for a flight department startup is shown in Table 3.1.

Finding the Right People

Since aviation, especially aviation involving turbine-powered aircraft, is a complex and specialized proposition, the first step in establishing a flight department is to hire or appoint a flight department manager. Hiring a knowledgeable and experienced individual to start the operation makes good sense. If a person with little aviation experience is given the responsibility to purchase an aircraft, establish an operating base, and hire aviation personnel, he or she may easily miss some important aspects of the operation. Without extensive experience in the field, the person making such choices will be relying on a variety of individuals to advise him or her on the various aspects of the operation. While a variety of opinions may be valuable in determining the various aspects of the operation, a cohesive view of the end result should be formed and built by the person who ultimately will be responsible for the flight department; that person will be designated the flight department manager. A good compromise is to hire a manager and contract with a flight department management consultant to provide perspective and a view of industry best practices in the process of establishing the department.

The Leader/Manager.   The aviation or flight department manager is the key person operators should rely on to establish and maintain their flight operation. This is the person who must live with the operation over time, direct its growth, and make policy corrections as the operation progresses. Therefore, this person should be given full responsibility and authority for the operation; without this level of control, the degree of success in establishing and maintaining the operation will be less certain. Choosing the manager/leader of the flight operation is quite important.

Finding the right person can prove difficult if the right objectives are not provided for this individual. Unfortunately, most operators rely on their instincts in finding the right individual, depending on a person's aviation credentials to determine his or her fitness for the job. Running a flight department is no different from running any other venture; however, having a modest amount of aviation experience can help the process. Yet companies frequently are blinded by extensive aviation qualifications and experience possessed by individuals who do not have credible leadership or management skills.

Experience has shown that the best *managers* make the best flight department managers. While this may sound obvious, companies seldom heed its simple truth. Too often aviation personnel with little management training or demonstrated management experience are thrust into positions of control and power; while some survive, and a few even thrive, many are doomed to failure because they are not predisposed to being managers and leaders.

While tradition has placed pilots at the head of flight departments, an increasing number of aviation maintenance technicians have been filling the top slots. Again, it is the experience, training, and preference of the individual that make him or her a candidate for the top job, not necessarily his or her specific type or depth of experience. And once the flight department has been established and has grown in complexity and function, managers with little or no aviation experience have been placed successfully as aviation department managers.

The lead pilot chosen for the operation, whether manager or not, should be experienced in the type of aircraft to be acquired by the flight department. Additionally, 3000 total pilot hours is the normal minimum requirement for the principal pilot in a startup flight department; 1500 hours of PIC hours should be required as well (PIC time means time spent actually flying an aircraft). The pilot also should have least 500 hours in the class of aircraft to be used, e.g., turboprop or turbojet, and preferably be type-rated (a specific pilot rating required for certain sophisticated aircraft) in the actual type of aircraft to be operated, with 100 to 200 hours of pilot time in type.

While titles may not seem important, they make a difference in pay and prestige for individuals. Typically, a startup flight department operating a single aircraft with just two or three total employees designates a chief pilot as the head of the operation. This individual takes care of all aviation functions from hiring people, to selecting contractors, to flying the aircraft—he or she instantly becomes a jack-of-all-trades. Granted, the range of management duties is somewhat limited, but leadership and management skills are important to the success of the operation. Larger flight departments use an aviation or flight department manager in recognition of the additional number and range of duties required as a department grows in number of aircraft, employees, and functions. Flight department managers traditionally are paid 10 to 20 percent more than chief pilots for a given type of operation, based on title alone.

In addition to a solid background of training and experience in management and leadership, operators should look for the following traits in the person they choose to run their flight department:

Curiosity.   The manager must be interested enough in new concepts and ideas to be able to go beyond his or her aeronautical realm and find out what makes the company function. It is this thirst for new knowledge, to understand what goes on at headquarters and at the company facilities, that facilitates the learning and absorption process. Without it, the transition from aviation person to executive is hard, sometimes boring work and probably will never be completely effective. Once the manager has made the transition, holding onto this characteristic will serve him or her well for years to come.

Vision.   The ability to see beyond the airport boundaries, beyond the company's current product line, and to see new applications for corporate transportation will accelerate the transformation to executive status. The willingness to explore new ways of doing business and to be able to connect disparate elements of the entire corporate function will make the new executive more valuable for both the flight department and the company as a whole. And unlike our eyesight after age 40, our *vision* actually should improve with age.

Commitment.   Working long and hard for more than just the aviation function but also to make that function the best possible asset the company has will cement the new manager's position in the upper reaches of the company. Working with the travel, disaster preparedness, and marketing departments, for example, to help perfect those functions demonstrates

Flight Department Manager Duties

- Provide on-demand air transportation services for the aircraft operator.

- Develop and institute department policies and procedures that will accomplish the department's mission safely, effectively, and efficiently.

- Prepare and administer an annual operating budget.

- Establish procedures for aircraft use and scheduling.

- Maintain contacts with user personnel to ensure proper service.

- Regularly advise users on the status and requirements of the aviation department.

- Devise and use metrics with which to measure department performance.

- Prepare short- and long-range plans that include user and departmental objectives.

- Develop and train subordinates.

an understanding of corporate priorities and a willingness to support them. If properly acquired, this trait will grow with time as well.

Innovation.   Working with the company travel agency to coordinate (or even consolidate) its functions with the flight department schedule, prepositioning flight crews at other locations to accommodate long duty times, and instituting zero-based budgeting for the flight department show an innovative and agile mind. And if these efforts are coordinated properly within the company, the resulting changes undoubtedly will attract favorable attention. Innovation thrives on new experiences, fresh approaches, and a wide range of interests.

Marketing.   The manager should realize that he or she always must be selling the company product and the benefits of his or her department and making the right people aware of the value of both in appropriate measure. Market analyses and implementing strategies are essential to this effort, too. Indiscriminate selling to either the inside or outside world may be a waste of effort; more important, the right audience is not getting the benefit of the sales pitch. The tendency to explore markets can be developed.

People power.   With a bit of luck, executives realize the value of people long before achieving their grand status; just in case they made it all on their own, they have an important lesson to learn. Peers, subordinates, and seniors—they all have the power to make, break, or outlast the person with authority. If the manager always invests time, faith, and effort into the people reporting to him or her, life certainly will be easier, if not more successful.

It's really a management job—flying the company airplane is just an occasional perquisite!
—EXPERIENCED FLIGHT DEPARTMENT MANAGER

Where does one find Mr. or Ms Right? Surprisingly, in the same places that all good employees are found, through diligent search and effort. Here are a few places to start:

- *Word of mouth.*   Other companies having flight departments in the area should be able to supply a number of recommended individuals through their networks.

- *Online.*   A number of aviation personnel and specifically flight department websites are available that list a variety of occupational types by experience, location, and specific qualifications.

- *Aviation placement firms.*   Several aviation placement firms exist worldwide that specialize in finding all levels of pilots, technicians, and administrative personnel for flight departments.

- *Advertise.*   Advertisements placed in aviation periodicals such as *Flight International, Business & Commercial Aviation, Aviation Week,* and *Space Technology* often yield substantial numbers of qualified individuals.

Other Personnel.   Many flight departments are able to commence operations with just the aviation manager or chief pilot. If an aircraft requiring just one pilot is used, no other personnel are required immediately. However, support personnel within the company are always required to perform financial and administrative functions. Table 3.4 provides a hiring checklist.

More Pilots.   If the aircraft is to be operated by a single pilot, the flight department personnel complement is complete once the manager is in place. However, many aircraft require more than one pilot, based on the regulatory requirements for specific aircraft types. And

What do you want?
- Develop a *detailed* job description
- When do you need the employee?
- What compensation package can you offer?
- Look beyond technical qualifications:
  - Interpersonal skills
  - Supervisory potential
  - Administrative capabilities
  - Ability to fit flight department/corporate culture
  - Personal presentation (appearance, demeanor)
- What corporate human resources rules apply?

Finding them
- Networks—company, business, aviation
- Ask employees to help
- Corporate human resources
- Headhunters/professional search firms
- Advertise in business and trade publications

Selection
- Screen résumés carefully—ask for references
- Work with human resources on the process:
  - Requirements
  - Procedures
  - Forms required
  - Compensation levels
  - Applicable legal issues
  - When can a job offer be made?
- Develop interview questions beyond the technical aspects:
  - Customer service orientation
  - Time demands of the job
  - Ability to integrate with flight department team
  - Administrative skills (speaking, writing, organization)
  - Job/promotion expectations of the candidate
  - Corporate/flight department culture and values
- Schedule interviews with people inside/outside the department
- Develop a practical knowledge/skills test for pilots/technicians
- Ensure understanding of job demands, compensation, benefits
- Request FAA violation history check for pilots/technicians
- Draft detailed job offer specifying compensation and job description, for clearance with human resources

TABLE 3.4   Flight Department Hiring Checklist

even though only one pilot may be required by regulation, many operators choose to operate with two pilots as an increased safety factor, especially for aircraft operations in high-density flying areas surrounding major metropolitan areas. More than the minimum number of pilots to crew the aircraft may be required depending on how the aircraft will be used and the tempo of operations (see Chap. 4).

An experienced aviation person should need little additional experience in finding an additional pilot through existing or local aviation networks. If specific types of individuals or qualifications are needed, aviation placement firms or aviation periodical advertisements probably will suffice.

Pilots and technicians should be selected primarily for their potential to bring value to the company, not just because they possess good technical skills.   —EXPERIENCED AVIATION MANAGER

Aviation Maintenance Technicians (AMT).   Many small flight departments with just one aircraft depend on outside sources for required maintenance and inspection of the aircraft. Normally, the local fixed-base operator (FBO) will supply traditional nontechnical support services such as aircraft positioning (towing), fueling, and cleaning. Additionally, minor aircraft discrepancies relating to tire pressures, light bulb replacements, and minor adjustments may be available through the FBO AMTs. With this arrangement, major inspections and maintenance traditionally will be accomplished by an aircraft repair station or service center for the specific type of aircraft operated. These facilities may not be located where the aircraft is based, however.

There are advantages of having an AMT on staff. Not only is this person able to comply with required inspection and maintenance procedures for the aircraft, but he or she is also able to ensure that the aircraft is kept in top condition and that maintenance services are available when needed (see Chap. 7). Guidelines for finding a good AMT include

- Aviation maintenance technician/engineer licensed by the national regulatory authority
- At least 1 year (ideally 2) working on the same class of aircraft
- Specialized training for the type of aircraft operated (may be provided after hire)

The AMT's duties include

- Performing maintenance and inspections in accordance with manufacturers approved methods and techniques
- Keeping abreast of and maintaining a current working knowledge of the procedures, parts, and tools needed to perform maintenance and repair work on the company aircraft
- Complying with the manufacturer's technical data, national regulations, and applicable safety rules
- Being responsible for strict adherence to the inspection system employed for the assigned aircraft
- Shipping, receiving, and inspecting aircraft parts

The first AMT employed is also normally designated the director of maintenance in recognition of his or her managerial role in ensuring that the aircraft is maintained properly. The director of maintenance reports to the aviation department manager and is responsible for maintaining department aircraft in accordance with the manufacturer's standards and regulatory requirements. Obviously, the director's managerial responsibilities will be minimal with just one technician on board, but such abilities should be considered when employing the first person in this category.

Others. While a variety of personnel, including line service, aircraft cleaners, receptionists, and schedulers, may be required for a flight department, only the scheduler is required for most flight departments. The others come as the department grows in size and function, but the scheduler is the required regular interface between the flight department and those who desire transportation.

The scheduler receives requests for transportation in the aircraft, recording details of desired trips for the passengers. This person may not be devoted full time to the scheduling task but rather may be the full-time secretary or administrative assistant for a highly placed executive within the company, traditionally the principal or CEO. The scheduler's duties are essentially a collateral duty for this type of person but a good fit given the position and function of his or her immediate superior.

This person does not necessarily have to be particularly aviation literate because the little knowledge he or she requires is learned easily for most routine scheduling tasks. It is important, however, for this person to be familiar with company and personal operations and personnel at middle management and above. The scheduler deals with the principal's family members, top-level employees, and customers on a regular basis and must be customer-oriented and aware of appropriate service levels for the passengers using the aircraft.

Eventually, the scheduler may devote his or her full time to the scheduling task and coordinating the internal functions of the flight department with those of the operator. This is a normal evolutionary process emerging from increased numbers of aircraft and operations.

A full-time scheduler/flight coordinator can prove quite beneficial to any flight department because of the organization's dynamics. Typically, the pilots, technicians, and flight department manager are deeply involved in a wide range of detailed operational tasks. However, these tasks are often fragmented and highly compartmentalized, taking these individuals away from the hangar for significant periods of time or denying them the big picture of the total operation. The scheduler is normally the only person who stays at the hangar and provides a central point of information and reference. Since these tasks require

Director of Maintenance Duties

- Ensure that all maintenance operations are conducted safely
- Provide guidance and direction for those assigned to the maintenance section
- Maintain the company aircraft in accordance with the manufacturer's continuing airworthiness instructions, federal aviation regulations, and the highest standards of airworthiness and appearance
- Assist with flight department planning issues, including budgets
- Ensure that maintenance personnel are properly qualified and trained for their assigned duties
- Maintain adequate spare parts and consumable inventory
- Contract with vendors to provide maintenance services as needed
- Ensure that the hangar and associated facilities are properly maintained

Scheduler Duties

- Scheduling travel for company passengers on the company aircraft

- Scheduling ground transportation and accommodations for both flight crew and passengers

- Coordinating aircraft operational information with pilots and AMTs

- Logging flight times and keeping department flight records

- Keeping passenger manifests and logs for all flights

- Providing flight crews with a flight plan, and manifest

- Arranging for catering services and FBOs

good people and organizational skills, the transition to a full-time scheduler should take these factors into consideration.

Collateral Duties.   There once was a time when all a corporate pilot did was show up 1 hour before takeoff time, fly the trip, and then retire gracefully to his or her abode to await the next scheduled trip or the beep of the dreaded pager. However, an increasing number of flight departments are tapping the many talents of their pilots, schedulers, and AMTs for additional duties that are not related directly to their primary job descriptions.

The advent of the collateral duty probably came to us from military aviation, where no squadron member escaped with just the single task of flying or fixing aircraft. These additional duties were given in recognition that a subsidiary task needed doing and a warm body was needed to accomplish it. To a greater degree, however, the military realized that its drivers and maintainers would not receive the training and experience necessary to accept increased responsibility and become more knowledgeable of the big picture of the many aspects of running the store. The same forces are at work in corporate aviation today, particularly with the demands to be leaner and more efficient.

An informal survey of department managers shows that most of them were not elevated to their position by chance, but by demonstrating their administrative and managerial abilities through collateral assignments. At the other end of the spectrum, the chances of being retained during a downsizing are enhanced if the individual has shown that he or she can do more than fly and fix aircraft.

Instituting collateral duties may meet with some resistance in established departments, but chances are that if the right people are chosen for the other duties, they will welcome the chance to show their stuff. Most departments have individuals who are interested

Collateral Duties

Training	Flight information planning
Safety	Budget/financial
Automation support	Catering
Aircraft type specialist	Standardization
Reports	Flight/maintenance records
Purchasing	Publications

in computers, flight planning, the intricacies of specific aircraft, and administrative matters. Ask them what their hobbies are—hobbies are a good indicator of what people will be good at.

The big payoff in having collateral duties is that the added involvement of these people will make the department work better and it should develop a more satisfied employee. And since the department managers only have so much time to give to the job, more work will get done. Finally, the jobs of pilots and mechanics will be easier to justify to corporate management when their collateral duties are added to their job descriptions. See the Appendix E for collateral duty job descriptions.

Basing the Operation

Where to base an aircraft is a critical decision if full advantage of the aircraft is to be taken. The reason the operator will acquire an aircraft is to have ready access to on-demand transportation. Convenience for passengers should be a major factor in determining where to locate; however, the airport closest to corporate headquarters or the principal's home may not be the right location. A careful balance must be struck between competing factors when selecting an airport (see the following sidebar).

Note that selecting an airport has been placed ahead of the section on aircraft acquisition. This is done because the size and facilities available at the selected airport may figure prominently in the type of aircraft to be used. Actually, selection of the airport and the aircraft should be done almost simultaneously to accommodate their mutual features and restrictions.

Once an airport has been decided on, an operating location at that airport must be found. Typically, a startup flight department hangars its aircraft and bases its pilots at a suitable FBO at the airport. Considerations for selection of an FBO include

- Cost
- Maintenance availability
- Access during nonworking hours
- Passenger facilities
- Security
- Fuel availability/cost
- Aircraft storage/maintenance facilities
- Crew accommodations

Using an FBO's hangar may require the company aircraft to be moved unnecessarily during repositioning of other aircraft, exposing the company aircraft to the possibility of damage. Security of the aircraft while it is in an FBO's hangar is a potential problem as well. Despite the FBO's best efforts, it is difficult to ensure total access control over either the hangar or the individual aircraft. While this is not a common problem, the potential

Airport Selection Considerations

- Passenger accommodations
- Aircraft facilities
  - Hangar space
  - Maintenance
  - Fueling
- Flight crew office space
- Amount of aircraft traffic
- Noise restrictions
- Runway length
- Instrument approaches available
- Operating fees
- Security
- Type and quality of airport management

for vandalism, theft, and sabotage is possible in an unsecure hangar. Therefore, one's own hangar or one shared with a limited number of tenants is preferable for security reasons.

Buying or building a dedicated hangar may be a good investment once it has been determined that an airport is a good fit with one's operation. This capital expenditure may prove quite affordable once life-cycle costs are compared against the renting alternative. Additionally, the resale value of hangars is usually quite good. Airport land use restrictions and covenants should be investigated carefully prior to building because land/structures reversion provisions in airport leases are common.

Ideally, flight and maintenance crews should be provided with office, working, and storage space within the same hangar where their aircraft is located. Using space leased from an FBO makes this difficult because such space is normally limited. Crews often share office space with other corporate operators or work out a combination of their homes and the FBO crew lounge. Since theirs is a nomadic existence under normal circumstances, providing them with dedicated office space should be a priority consideration when basing an aircraft.

If aircraft maintenance is to be performed by the FBO, it is preferable for it to possess a repair station certificate/authorization from the national regulatory authority, appropriately rated for the type of work needed for your aircraft. If company personnel will be used to perform maintenance on the aircraft, arrangements must be made with either the FBO or other leasing authority to perform the maintenance. Provisions also must be made for tools, parts, and ground support equipment storage.

Whatever aircraft storage and servicing options are selected, there are liability issues to be considered. Whether there is a potential for a claim between the FBO, airport, lessor, or colessees and the company, this liability should be examined carefully by corporate risk managers and/or corporate counsel. Appropriate types of insurance should be secured to protect the company from potential liability concerns.

Acquiring the Aircraft

Once the decision has been made to acquire an aircraft, several issues must be resolved:

- Aircraft type
- Purchase or lease
- In-house personnel or outside management

All these issues are too difficult for the layperson to address and may be too difficult for an experienced individual who will head the flight department. These issues can be addressed by aircraft manufacturers, brokers, or acquisition consultants. Each brings some bias to the decision process; therefore, it may be desirable to approach several and solicit advice and actual bids. Lists of such resources are available from the *World Aviation Directory* and business aviation association in the country of residence. In the United States, the General Aviation Manufacturers Association (GAMA), the National Air Transportation Association (NATA), the National Aircraft Resale Association (NARA), and the NBAA are good sources. The best method of selecting an aircraft purchase advisor is a word-of-mouth recommendation from a trusted associate or other company operating the type of aircraft desired. Yet this valuable advice only provides a single data point in what should be a comprehensive search. See the sidebar "Aircraft Acquisition Advisors" in the section entitled, "Owner/Employee-Flown Operations," earlier in this chapter.

Aircraft Selection.   The air transportation needs analysis should have narrowed the choice of an aircraft to just a few. However, the final decision may be difficult given the conflicting advice and sales information provided by manufacturers, brokers, and consultants. Often company executives or family members who have some familiarity with aircraft they have flown in will further narrow the choice. The final decision must be contingent on the decision maker taking a demonstration flight in the actual aircraft to be purchased. And don't forget to take spouses along for the ride, too, if they are to be regular users of the aircraft.

New versus used, turboprop versus turbojet, cabin size, interior amenities, ability to operate from short fields, and resale value are but a few of the variables that will have to be addressed during the selection decision process. A number of opinions from operators of the candidate aircraft are very useful in the process.

Reference materials to assist with the aircraft selection process are available from *Business and Commercial Aviation* (Annual Purchase Planning Handbook, usually the May issue), Conklin & de Decker Associates (aircraft performance comparison and cost evaluations), and individual aircraft manufacturers.

Once the analysis has been completed and provisional selection made, some judgment must be exercised in selecting an aircraft. If the analysis indicates the ideal aircraft to be a small turboprop, but the principal or CEO clearly likes jets, a modest entry-level jet may be a wiser choice. Certainly both types of aircraft should be suggested as solutions to the identified air transportation needs. And if a prominent passenger is mildly claustrophobic, a medium to large jet may be a better choice than a more economical small one. Finally, do not forget, again, the preferences of spouses of the primary passengers; more than a few aircraft acquisition decisions have been made based on interior amenities and the size of the lavatory.

There is no "right" aircraft for any individual or company. More than several types will be adequate to meet their needs; final selection is not too different from selecting an automobile or a house. The big difference between these more common items and an aircraft is its complexity and price; expert assistance is needed. And more than one voice should be heard in the process. See the sidebar, "Considerations When Acquiring an Aircraft," earlier in this chapter.

Unless you have vast experience with the type of aircraft to be purchased, a prepurchase inspection should be done. There are too many hidden limitations, restrictions, costs, quirks, skeletons, and anomalies associated with a corporate-type aircraft for the uninitiated to be familiar with. Just because an aircraft is presented with a "fresh" set of inspections does not necessarily mean that the aircraft is in good shape; experienced eyes must make that determination. A prepurchase inspection should be performed by an *expert* specializing in the *particular type* of aircraft to be acquired. The inspector's report should address most of the preceding considerations and come with a guarantee or warranty against unknown or unanticipated surprises regarding the aircraft. To qualify the potential inspector, ask essentially the same types of questions you asked the dealer/broker. Finally, the prepurchase inspector should have no conflict of interest with regard to owners, operators, manufacturers, brokers, dealers, or the like.

Purchase or Lease.   This decision goes more to the realm of the accountant than to the aviation expert. The capital debt structure and philosophy of the company or principal will be the main determinant for this decision, although brokers, manufacturers, and financial institutions may provide alternatives and insights that could affect the decision.

Leasing may be the most attractive option to an operator new to on-demand aviation because it may afford the opportunity to evaluate an aircraft over a relatively short term without making a long-term commitment. However, financial and tax considerations should be used as a primary factor in making the lease-purchase decision. In addition, a wide variety of creative leasing schemes is available that may combine the best features of both purchase and lease methods.

And whether or not you are well-heeled enough to write a check for the purchase price of the aircraft, it is a good idea to use or obtain the advice of a finance company or bank familiar with aircraft purchase/lease. Their knowledge of markets, aircraft values, resale values, title issues, insurance, and other paperwork items make their counsel invaluable. While their caution exercised in the purchase process may seem unwarranted, the time taken up-front may save grief later.

Aircraft Markets.   At any one time, a number of preowned aircraft are on the market. Popular aircraft are scarce and command a premium price; unpopular aircraft are plentiful and are subject to heavy price negotiation. The point is that the aircraft market is constantly changing, driven by financial, regulatory, technological, and preference factors. Therefore, market timing is important for both new and used aircraft. Markets awash in good, low-time aircraft will cause all prices to soften and be subject to negotiation, even those of new aircraft. Conversely, if the aircraft to be purchased is in tight supply, an acceptable one may not be available, regardless of price.

The advisors selected in the purchase process will be reacting to these market features; their advice must be evaluated carefully to get the best possible deal. And if the market for the type of aircraft desired is bad at the time the decision to acquire is made, a short-term lease of that type or charter may be used to wait for better times.

Insurance.   Before commencing any type of flight department operation, insurance should be obtained. That is, before any aircraft lease or purchase agreement is signed, an insurance binder must be in hand. This includes insurance for both aircraft operations and the facility at which it will be based. If the aircraft is leased, or if a lien is attached, the lessor or financial institution will stipulate certain insurance limits to be carried for the aircraft; the base facility (hangar) probably will not be included in this requirement.

As with so many other issues holding significant financial consequences, several quotes should be obtained for the insurance. This process also will provide an education about the components and provisions of aircraft insurance. However, before the search for insurance

Father/Son Handoff Expands the Fleet

An Oklahoma entrepreneur built a drilling equipment business in the 1970s and 1980s by specializing in niche well-support equipment. The successful business soon discovered the virtues of on-demand air transportation and acquired a Westwind jet to support its business. The Westwind was replaced by a Learjet Model 35, and it, in turn, was replaced by new Lear 60. The entrepreneur's son joined the business and learned to fly. As the son took over more of the business, the father found other pursuits that occasionally began to take the Learjet away from the business. The son bought a CitationJet to fill the gap, which he regularly pilots along with the company pilots. The son finally encouraged dad to release the Lear 60 to the company and to get his own aircraft.

begins, your risk manager, facilities director, or legal counsel should be consulted regarding risk and insurance philosophy. Table 3.2 provides an aircraft insurance checklist. An Appendix to the *NBAA Management Guide Appendix* also contains an excellent primer on the subject of aircraft insurance.

At a minimum, two quotes should be obtained for initial insurance coverage and every 3 years thereafter to ensure that the company is getting adequate coverage at the best possible price. The flight department manager should become sufficiently familiar with aviation insurance terms, coverage, and fees to act as a consultant to a risk manager or legal counsel on aviation issues. This often can be accomplished by tapping the manager's network of fellow managers or by going to a number of broker/underwriters to become familiar with the desired information.

Management Company

Under this arrangement, an aircraft management company provides all personnel and services required to operate an owner-supplied aircraft. This method of aircraft operation is often used by companies new to aviation or those not wishing to become involved in the aviation department management process itself. In essence, an aircraft is "given" to the management company to operate for the owner as a turnkey operation.

The management company charges a monthly fee and provides personnel, training, a base of operations, passenger scheduling, and aircraft maintenance services. Salaries, benefits, fuel, hangar rent, aircraft maintenance, and other operational costs are all collected by the manager and passed through to the owner, usually with no markup. These comprehensive services relieve the aircraft owner of most of the responsibilities of aircraft operation.

Hundreds of aircraft management companies are available worldwide, ranging from the giants that have hundreds of aircraft in their stables to the very small operators who may be, in reality, small charter companies engaged in management activities as a sideline or complement to their main activities as charter operators. This is so because many charter operators "borrow" aircraft from corporate flight departments when the company is not using the asset for use as a charter aircraft. The management-charter symbiosis works to the advantage of both the charter company and the corporate operator if the asset is managed properly.

The use of a managed aircraft for charter services reduces the owner's scheduling flexibility because charter trips scheduled in advance make the aircraft unavailable to the owner for last-minute trips. The additional aircraft utilization created by the charter trips accelerates the aging process on the aircraft and exposes it to users who may not treat it as kindly as will the owner. Attempting to regain major portions of the fixed expenses associated with aircraft operation may prove difficult unless the charter revenue is substantial. In attempting to increase charter revenues, the aircraft becomes less available to the owner; if this method is chosen, a balance must be struck between availability to the owner and desired revenue levels.

The services provided by management companies are valuable, yet the act of outsourcing these services reduces the company's control over policy, procedure, costs, and desired quality levels; the management company introduces an additional level of control and a filter to use of the asset. While many of the technicalities and complexities associated with operating an aircraft become the responsibility of the management company, full understanding and appreciation of available options may not be possible—the management company may recommend expediency over cost control or service levels. However, the use of management companies has increased in popularity within the past few years, principally as a method for aircraft owners to rapidly enter on-demand air transportation with little or no knowledge of the subject.

These companies may be found at most airports featuring significant numbers of either charter providers or corporate operators. The Yellow Pages in large metropolitan areas are a good starting point. The *World Aviation Directory* contains a comprehensive listing of charter operators and management companies.

The Contract[1]

The instrument binding the aircraft owner to the management company should be a comprehensive and detailed contract that sets out the duties and responsibilities of both parties. Efforts to incorporate precision and completeness in the contract should leave little room for interpretation or complaint between the parties. In addition to normal concerns of services rendered in return for a fee, legal and liability considerations also must be included to protect both parties.

For instance, it must be made clear that the management company is simply providing management services and not transportation. There can be substantial adverse national regulations and tax consequences if the management company is determined to be providing transportation services to the owner. In the United States, the FAA focuses on which party has "operational control," and the IRS looks to which party has "possession, command, and control" of the aircraft. Factors such as who employs and/or controls the pilots, who controls the scheduling and availability of the aircraft, and who procures the aircraft insurance are all relevant.

Liability considerations are important to both parties because each has obligations and responsibilities based largely on what is contained in the contract. Issues to be covered should include

- Insurance requirements
  - Applicability—operations and geographic coverage
  - Aircraft hull and liability
  - Personnel injury (worker's compensation)
- Rights of subrogation and breach of warranty
- Management company's standard of care
- Liability for losses caused by negligence, misconduct, etc.
- Hold harmless or indemnification provisions.

If the aircraft and associated personnel are to be used for charter flights, additional provisions must be included:

- Management company responsibilities normally include
  - Obtaining necessary certifications and approvals to operate
  - Obtaining the owner's prior authorization for all charter flights
  - Ensuring that the crew meets regulatory qualification and training requirements

Management Company Contract Issues

An enumeration of services to be provided must be included:

- Identification of the aircraft and its equipment
- Operational, administrative, and financial policies and procedures to be used
- Scheduling practices, especially when the aircraft is used for charter as well as owner personnel
- Aircraft maintenance and inspection program to be used and who will perform the work
- Employment and supervision of the flight and maintenance personnel
- Rights of both the management company and owner for hiring, reviewing, and firing personnel
- Minimum experience levels for each employment category
- Personnel training requirement
- Personnel exclusive use (optional)
- Alcohol and drug testing requirements

- o Drug and alcohol testing program
- o Invoicing of charter customers for transportation and applicable taxes
- Owner responsibilities normally include
  - o Making the aircraft available for certification and training flights
  - o Paying the cost of any aircraft modifications required for charter qualification
  - o Paying certain expenses related to initiating and maintaining the aircraft on the management company's charter operating certificate
- Additional considerations include
  - o The formula used for allocating charter revenue and expenses between the parties
  - o Specifying which party is responsible for bad debts from charter customers
  - o Ensuring management company "possession, command, and control" when it operates the aircraft for charter
  - o Allocating applicable taxes
  - o Building in performance measures/reports

Because of the complexity and liability considerations of the management contract, an attorney experienced in these matters should be consulted prior to finalizing any agreement with a management company.

Performance Measures/Reports

An aircraft management contract is typically a lengthy and complex document that concentrates on who is responsible for what and the associated legalities. Yet the company initiating the contract should have a method of ensuring that it is getting what it pays for and that components of the services are received at the best possible price. Table 3.5 presents a list of fee/expense elements associated with aircraft management operations. In subsequent chapters these same performance measures and report formats will be recommended for in-house and other types of on-demand operations, but they are just as important to the recipient of managed services.

The management company is at a distinct advantage in procuring and passing through the many expenses required to operate the client's aircraft. The manager is the expert; the client knows little about aeronautical issues, especially the financial aspects. Yet a very important subset of the management services supplied should be in obtaining the best possible deal for the client. Yet how are clients to know, given their relative naiveté on the issues?

Therefore, a series of reports and insights should be supplied to the client outlining the efforts the management company has made to obtain the best possible price on parts, fuel, training, cleaning services, etc. Anytime a new or different article, commodity, or service is purchased by the manager for the client, an explanation should be supplied. Table 3.6 provides a list of suggested reports.

Fixed management fee for each aircraft.

All other fees associated with this contract are normally passed through to the client with no markup. These may include but not be limited to

- Direct (variable) operating costs associated with aircraft operation to include
  - Fuel—contractor-provided/controlled fuel to be supplied at best available prices
  - Maintenance (nonscheduled and scheduled)—provided at preferred shop rates
  - Aircraft parts—best negotiated price
  - Engine overhaul and parts reserves/insurance programs
  - Landing/parking/hangar fees
  - Catering
  - Flight crew expenses
- Indirect (fixed) costs associated with aircraft operations to include
  - Salaries for a negotiated number of pilots, flight attendants, schedulers, aviation maintenance technicians, and aircraft cleaners, as applicable; other personnel costs must be approved by client
  - Aircraft Insurance
  - Required or approved training for core client personnel
  - Flight information and aircraft maintenance publications
  - Aircraft flight planning and aviation weather information systems (pro rata share if used by other management company clients)
  - Hangar rental/aircraft movement and cleaning services
- All other legitimate fees/charges associated with client aircraft operations (must be submitted to client for approval)

TABLE 3.5   Typical Aircraft Management Company Fee/Expense Elements

The management company should maintain databases and devise reporting systems to prove compliance with the terms of this contract and to inform the client about general aspects of flight operations and financial activity. These reports should be provided to a designated employee in the company within 10 days after the end of a reporting period and should include:

- Monthly flight activity—company and charter
  - Date
  - Aircraft
  - Origin
  - Destination
  - Duration
  - Passengers carried
  - Crew members
  - Significant remarks
- Monthly flight crew utilization—normal and contract pilots/flight attendants
  - Date
  - Origin
  - Destination
  - Flight and duty times—company and other use for normal crews
  - Training activity
- Monthly maintenance activity
  - Routine inspections
  - Significant maintenance actions
  - Unusual unscheduled maintenance
  - Upcoming scheduled maintenance activity
- Monthly expenditures—list all expenses associated with the operation and maintenance of the aircraft and flight crews
- Irregularity reports—any irregular event required under national air charter operating rules (Federal Aviation Regulations Part 135 in the United States) and any event in which a passenger is inconvenienced or makes a negative comment about a flight to a flight crew member or other contractor employee will be made verbally to the client company aviation representative within 24 hours of occurrence
- Ad hoc reports—operational and financial databases should be sufficiently comprehensive to provide on-demand ad hoc reports requested by the company

TABLE 3.6   Suggested Aircraft Management Company Reports

Joint Ownership

Under this arrangement, two or more persons and/or companies become registered joint owners of an aircraft. One of the owners employs and furnishes the flight crew, and each of the owners pays a share of the charge specified in an agreement. Administrative, financial, and operational details of the aircraft operation usually are assumed by the owner employing the aviation personnel. Otherwise, this type of operation is quite similar to the more traditional in-house flight department. The principal advantage is in having another company share the fixed costs of the aircraft, which, depending on utilization, may comprise one-half the total aircraft expenses.

This type of operation is normally considered legal in most countries; however, tax considerations in some jurisdictions may make the arrangement less attractive than other types of ownership and control.

This method works well when the owners, normally two, use the aircraft infrequently and at times that complement one another. For instance, a company may use the aircraft mainly during the week for business trips, and the co-owner may use the aircraft primarily for personal purposes on the weekends. Scheduling conflicts may be the main problem area for these arrangements. A preexisting relationship between parties or an intimate knowledge of the other's business operations is almost mandatory if a peaceful and long-term joint ownership is to be enacted.

Selection of a common aircraft type is an important first step for the partners in establishing their operation. While casual investigation of several types of aircraft by principals may prove an adequate method of choosing the type, an independent air transportation analysis of each party's needs should be conducted. A good fit for the aircraft to the parties' requirements is important to getting the agreement off to a good start. Further, basic policy issues of how long one party may keep the aircraft on the road, will company personnel be permitted to use the aircraft for personal use, and scheduling priorities must be discussed prior to deciding what aircraft will be purchased and where it will be based.

A detailed joint operating agreement is an essential part of a successful joint-ownership operation. Scheduling, priorities, and conflict-resolution methods should be worked out and tested before the final agreement is signed.

Issues to consider:

- Duties of the managing partner
- Compensation for managing partner
- Payment procedures/considerations
- Recordkeeping responsibilities
- Scheduling procedure/conflict resolution
- Administrative and operational controls/standards
- Tax considerations
- Command, control, possession issues—legal and operational
- Risk management/insurance

Successful aircraft joint-ownership operations are like a good marriage—give and take makes it work.
—EXPERIENCED JOINT-OWNERSHIP PARTY

The advantages and disadvantages of taxes, depreciation, and liability are shared by the partners. Control over all aspects is shared by the owners, somewhat diluting the ability of any partner to dictate all elements of the operation. Yet the shared-ownership aspect usually provides motivation to create and maintain a high-quality flight operation. Regular communication between those controlling aircraft operations and the partners is a critical element in a successful joint-ownership operation.

Other Methods

Interchange and time-share methods were mentioned in Chap. 2 as methods of supplementing existing in-house, managed, or joint-ownership provisions. These two methods permit a flight department to temporarily augment its fleet by reliance on other flight departments' aircraft. Yet specific provisions must be instituted to achieve a legal and equitable agreement. Note that these methods may not be practical in all countries.

Interchange is an arrangement whereby an owned aircraft is leased to another entity on a permissive basis in exchange for equal time in that entity's aircraft when needed. No charge or fee is made for this service except that a charge may be made not to exceed the difference between the cost of owning, operating, and maintaining the two aircraft. This method requires that each party own his or her aircraft and exchange his or her aircraft on an hour-for-hour basis.

A time-share agreement means an arrangement in which a company leases its airplane with flight crew to another person, and up to twice the direct operating expenses of the aircraft may be charged.

Detailed agreements must be signed by the parties to protect them and to ensure that one another's obligations and privileges are fully understood (see NBAA website for suggested agreement formats). Tax and liability issues must be specifically addressed. Perhaps most important is an understanding that specifically provides scheduling, cancellation, and conflict-resolution procedures.

References

1. NBAA Operating and Leasing Package, www.nbaa.org/member/admin/options/Operating-Leasing.pdf.

CHAPTER 4

Running the Business

It Really Is a Business

Too many flight departments view their activities as being principally aviation-related and not business- or service-related. It is all too easy to get caught up in the technical and, yes, the romantic aspects of flying to the exclusion of the more mundane yet practical aspects of running a business. Those depending on the flight department for services are more pragmatic about them and find it hard to understand the technical appeal of aviation as a subject unto itself.

Many flight departments are run like an executive taxi service or as the principal's personal limousine. Unfortunately, they see their mission as a very narrowly defined activity that requires few of the traits of an actual business. These departments are lulled into feeling that they provide an on-demand service to whomever may call for it, with little regard for the concept of other essential service functions. Marketing, sales, profit and loss, customer service, product development, and research and development are the essentials of normal businesses and many business units within large corporations. These "you call, we haul" departments are missing a major and essential portion of their activity, that of considering themselves as a complete self-contained business unit.

Businesses are marked by certain characteristics. They

- Have a product to produce and sell
- Market the product
- Have customers who use their product
- Receive compensation for their product
- Exercise control over all aspects of the enterprise
- Anticipate and respond to market forces and trends

Certainly, flight departments have a product—on-demand air transportation—and customers—passengers—but do they work under and accomplish the remainder of the business objectives? Unfortunately, for many flight departments, the answer is no.

Marketing, control, and response to markets are not usually pursued by flight departments because, for the most part, the department enjoys a monopoly status in the executive transportation field and is protected from competition from outside agencies. So why bother with these unnecessary features not germane to flying airplanes? Because the privileged, protected status now enjoyed may evaporate with a changing passenger base and less profitable times.

The flight department must be run as a business, a venture that will stand on its own because dependence on artificial factors and protections constitutes false hope for better times. In essence, the flight department is a small business or business unit that must stand alone.

Whether it makes a profit is immaterial; it must make *sense* to the powers requiring the service. Ideally, the flight department manager must act as an entrepreneur or a small business owner to ensure the life and future of the venture. There has to be a sense of ownership not only for the manager but also for the members of the department. The first step in ensuring the success of the venture is to be sensitive to the needs and concerns of the customer.

Successful flight departments are run as a business, with a service to be provided as the product of their labors. They look toward being both effective at what they do and efficient as well. To do this, they employ accepted business and management practices, especially those used by the company or person for whom they work. It is this mind-set that allows them to become integrated into their principal's or corporate structure and to measure their performance as a service. This allows them to become a part of the larger organization and still let them do what they came to do in the first place—operate aircraft.

Flight departments are often isolated from their parent organization because of their need to be located at an airport. They are further isolated from the company by the autonomous and technical nature of their operations. Some flight departments are considered to be the private domain of a privileged few, often as providers of transportation perquisites. These factors tend to promote further isolation and concentration on technical aspects of aviation, often to the detriment of business-support concepts that should be driving them.

Human resources, information technology, risk management, and accounting all fulfill essential support roles for their companies—the company could not exist without their services, either in-house or outsourced. Why shouldn't company flight departments be considered as important or essential? In many companies they are, but achieving this institutional status has taken many years and much effort to achieve. The key ingredient of these institutional flight departments is that they have become an integral part of the company, actively following company strategies and providing *transportation solutions* to help the company achieve its goals and objectives. In short, it is a true business orientation that allows flight departments to become fully integrated, whether they work for an individual or a company. Successful flight departments provide high-priority transportation services to their clients; they just happen to use aircraft instead of some of some other form of transportation.

Flight departments normally operate as follows:

- A demand or request is placed on the department for a trip.
- Feasibility for the trip is determined (operational issues).
- Availability of a flight crew and suitable aircraft is determined.
- The trip is confirmed and scheduled.
- Detailed trip information is gathered and disseminated.
- The trip is flown.
- Flight information is recorded and stored.

In order for this procedure to be accomplished, a structure must be in place that will make the process as routine as possible. Policies, procedures, and recordkeeping systems must be available to accommodate a variety of actions performed by the department. Personnel must be in place to perform the required functions within the structure. All these items (and more) must be devised and implemented to make the department run with some semblance of order and discipline.

Planning

We all make plans, consciously and subconsciously—retire by age 55, buy a house in 5 years, pick up a bottle of milk on the way home. Plans define our lives, provide direction, and communicate intentions. Unfortunately, most people do not communicate their plans well to others who may be directly involved or be able to help. Flight departments are also victims of this deficiency, often to their detriment.

Plan the flight, fly the plan.   —ANONYMOUS

Mission Control

Mission statements have less to do with how well you see who you are as they do with how clearly you see where you are going. Every organization has a vision, a concept of where it fits into the grand scheme of things and, most importantly, a sense of direction. The statement may not be memorialized on parchment and framed in prominent locations around the building, but every employee should know what that vision is and *live* it every day.

The statement of vision may exist only in the head of the top person in the organization and may not be a particularly well-formed statement either. However, it is he or she who must be the "keeper of the flame" that constitutes the reason for the organization's existence. This is true for each unit within the organization, too.

The flight department not only must have a view of how it fits into the overall scheme of things, but also it must have its own sense of direction and worth. The aviation department manager or chief pilot must conceptualize, develop, nurture, and *communicate* the vision to every person within the flight department. The vision statement becomes the daily reason for wanting to come to work and the rallying cry for which superior performance and customer service are delivered.

Unfortunately, many vision statements are grandiose, unrealistic, and unattainable. Because of this and the fact that whatever vision statement that comes from the top is frequently poorly supported at all levels, these statements are held up to ridicule and largely ignored. The key to making the statement work is to make sure that it reflects reality and is supported by all management levels *on a daily basis*. Without this level of support and dedication, department members will view the statement as just another dream delivered from on high. Management not only must talk the talk but also must walk the talk.

The vision statement must be well-focused, simple, and realistic. It should not be a wish list; it must be rooted in terms of attainable and well-defined issues. It should answer the following questions about the department:

Missions

- "At Avis Rent-A-Car, our business is renting cars; our mission is total customer satisfaction."

- "[Boeing strives] To be the number one aerospace company in the world and among the premier industrial concerns in terms of quality, profitability, and growth."

- "The mission of Southwest airlines is dedication to the highest quality of customer service delivered with a sense of warmth, friendliness, individual pride, and company spirit."

- "The mission of the flight department is to provide safe, reliable, and efficient on-demand air transportation to our employees and customers."

- Where does it fit within the company?
- How does it want to be seen by the company and by other departments?
- What dreams does it hope to realize?
  - Where are we going?
  - What are our goals?
- What reputation will it establish?
- Where do you intend to be positioned in your industry?
- What desired standards of excellence will be used?

No department is too small not to have a vision statement. Without it, the employees will use their own values and standards, but most important, they will not all be working toward the same goal. Without it, wherever the department goes is just fine, for it has no sense of direction. Importantly, a view of the organization's destiny is needed before launching any plan.

A goal without a plan is just a wish   —ANTOINE DE SAINT-EXUPERY

The Plans

Large companies used to live and die by their plans—strategic, operational, contingency, backup, 5-year, 10-year, long-range, short-range, near-term—all had their place in corporate life. No move to the future was attempted without ensuring that it was done in accordance with the applicable plan or plans; plans ruled the corporate course. Changing a plan was difficult and regularly encountered resistance from the corporate hierarchy or bureaucracy. Carefully measured change within the planning process was the key to stability and predictability for the company. Alas, those days are a part of corporate America that have gone the way of jobs for life, gray flannel suits, and unlimited expense accounts.

The flight department manager of a large multinational company recently told me that the company's long-range planning horizon was *1 year* and that operational plans extended only to 6 months. He laughed at the idea of aligning the flight department's plans with those of the company. The flight department purchased, outfitted, trained for, and began operating a large-cabin aircraft in less than 3 months as a consequence of an emerging corporate interest in Asia. His comment: "How can you plan when the program changes on a daily basis? I haven't heard of a strategic plan around this place in years."

Welcome to the world of *Future Shock* and *Megatrends*. Corporate America has learned to use rapid communications and comprehensive information systems to gain a competitive advantage. In doing so, corporations have had to become

Plans

Strategic:

- Long-range
- Examines contingencies
- Open-ended
- No timeline

Operational:

- Specific
- Detailed
- Timeline
- Resource-specific

agile and flexible not only within their traditional markets but also within unrelated markets, as opportunity presents itself. Smart companies do this within a broad planning framework that may read more like an extended mission statement than a plan. The less insightful companies become reactionaries to the winds of change, altering course at the whim of a few and making it difficult to keep following their lead. Regardless of the disdain for strategic plans in some quarters, such plans are still a good idea because they bring future *possibilities* into clearer focus.

A basic strategic plan that looks out to a 3- to 5-year horizon containing broad objectives and alternative courses based on possible economic, market, or environmental moves seems to be the plan of choice. The big difference between this type of plan and those of previous times is their inherent flexibility; major changes to the plan are possible in as little as *1 day* due to the very responsive processes built into the plan.

Plans for the Flight Department

Strategic:

- Air transportation analysis
- Aircraft replacement and refurbishment
- Long-range budget projection
- Facilities

Operational:

- Annual training plan
- Scheduled maintenance
- Flight schedule
- Projects
  - Vendor selection
  - Fuel purchase plan
  - Equipment purchase

Operational plans are inherently shorter term and are more specific, containing milestones and resource requirements (see sidebar). Strategic plans should be used to create operational plans—this creates planning continuity within the department and the company.

Regardless of your company's plans and planning processes, the flight department should have a basic but flexible plan ready to accommodate what the company may need, short of going to the moon. It need not be elaborate, but it should be heavy into "what ifs," looking at possible alternatives to changing demand, destination, and capacity needs. Some thoughts:

- Confer with the flight department's reporting senior on a regular basis, attempting to gain a sense of what the corporation is planning to do.

- Sensitize flight department personnel and friends of the department to listen for trends and possibilities that may affect its future. But do not take every scrap of information as gospel; rumors abound in times of change.

- Devise a 3- to 5-year plan incorporating aircraft capability and service life, potential upgrades, and personnel considerations. Share it with the boss and revise it at least annually (see Fig. 4.1).

- Ensure that major changes within the department are made in accordance with "the plan." Making changes outside the planning framework will invalidate it; change the plan, and then act.

- Draw as many members of the department as possible into the planning process; they can implement change better if they understand it. Be ready for what the company may want.

Action	2013	2014	2015	2016	2017
Personnel					
Hire	AMT	Scheduler		(Pilots)	(AMT)
Upgrade	Pilot		AMT		
Aircraft					
FMS	Upgrade				
EGPWS	Upgrade				
Engine overhaul			Both		
72-month inspection		XX			
Paint		XX			
Interior		XX			
Replace				Analysis	Purchase
Additional aircraft			Analysis	Purchase	
Facility					
Offices		Add			
New hangar				Analysis	Build

Figure 4.1 Sample flight department mid-range plan.

Organization

Flight departments are generally semiautonomous organizations owing to their location and technical complexity. However, since the goal of the flight department is to be fully integrated with the parent organization, a strong link to it should be cultivated. Typical corporate flight department reporting seniors and the rationale for these choices are listed under "Getting Organized" in Chap. 3.

The department organization should be both functional and flat. That is, the three principal functions within the department—administration, operations, and maintenance—should be represented by key individuals; their relationship should be quite close, with little intervening bureaucracy. Figures 4.2 and 4.3 show typical flight department organizations, keyed to department size.

Note that in smaller departments, an aviation manager is generally not needed; instead, a chief pilot is used to manage the operation. This is typical of single-aircraft operations. Also, reporting to the chief executive officer (CEO) or principal allows the chief pilot to have direct access to the individual who can both appreciate the flight department's value and provide necessary support. In this size department, there may or may not be an aviation maintenance technician. If assigned, this person takes care of all aircraft maintenance duties, except when large inspections are undertaken or specialized maintenance is required; in such cases, either contract employees are hired to assist or the aircraft is sent to a service center.

In larger flight departments, a manager is designated to serve as the focal point of the aviation operation and to supply most of the administrative and financial expertise for the group. The chief pilot is used to provide operational expertise and represent the interests of

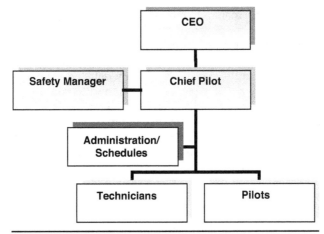

FIGURE 4.2   Typical small flight department organization.

pilots and flight attendants. Similarly, the maintenance manager is usually an experienced technician who has demonstrated administrative, leadership, and managerial skills to enable his or her segment of the organization to operate effectively and efficiently. Typically, the chief pilot acts as a pilot in command (PIC) in one or more types of assigned aircraft, flying somewhat less than a normal line pilot due to administrative and management duties associated

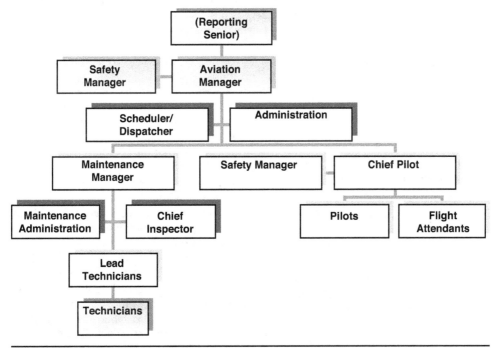

FIGURE 4.3   Typical large flight department organization.

with the job. While maintenance managers remain qualified as technicians, they are not often used for actual maintenance and inspection duties in large flight departments.

Other administrative, financial, operational, facilities, and ground support personnel may be assigned to provide specialized services to larger flight departments. All personnel, regardless of classification, should be provided with a detailed written job description to ensure that they understand what is expected of them and to coordinate their tasks with others in the department. A few job descriptions are listed under "Finding the Right People" in Chap. 3; a more comprehensive set may be found in the *NBAA Management Guide*.

The Safety Manager is important enough to have a direct line of communications with the head of the department, even though the person filling that position may also be a pilot, technician, or administrative person.

Scheduling

The scheduling function within the department may be thought of as the point of sale or ordering point. The availability and capabilities of the department (the marketing function) are known among the principal passengers—through the joint efforts of both the parent organization and flight department agents. Now a customer must place a demand on the system for it to perform its primary function, that of providing on-demand air transportation service to the company. Therefore, the scheduling function is a critical link between passenger and flight department and should be treated as such.

Personnel

Schedulers in single-aircraft departments are typically administrative assistants assigned the collateral duty of scheduling the aircraft and its passengers. This person works closely with the chief pilot to ensure that as many passengers as possible can be accommodated, given the availability of the aircraft and limitations of the flight crews and maintenance personnel. This arrangement works well with some organizations because it places the administrative and decision burden on people outside the flight department. Conversely, this feature may mean a loss of control for the flight department and, more importantly, a loss of contact with people within the company or principal's organization who may be politically useful to the department. At best, keeping the scheduling function within the administrative structure of the parent tends to isolate the flight department from the company, denying it useful information and contacts.

In larger departments, a dedicated individual is normally located with the flight department to both schedule and coordinate all departmental operations. Some large flight departments use dispatchers to assist flight crews with the operational planning process and to schedule flights and passengers as well. Weight and balance, route and weather plan-

Scheduling Elements

- Name of person making the request and authorizing the trip
- Trip purpose
- Dates of trip by segment (leg)
- Destinations—by address, nearest airport, if possible
- Passengers (by leg)
- Chargeback authority
- Ground transportation
- Catering desired
- Special information (unusual baggage, medical concerns, special ground handling requirements, etc.)

ning, crew duty time limitations, and air traffic control (ATC) and airport restrictions must be known and accommodated by this person. While all operational factors ultimately are the responsibility of the PIC, the scheduler/dispatcher can make his or her task easier and provide an independent source of critical information. The shared responsibility of pilot and dispatcher in airline operations has become an accepted practice that yields economic, operational, and safety benefits; the same partnership is possible in corporate flight operations. Some of these dispatchers hold dispatcher licenses issued by the national regulatory authority.

Policy

Passenger scheduling policies must be devised and published to provide prospective customers with information regarding use of the air-

Trip Sheet Elements

- Trip dates
- Crew names
- Aircraft assigned
- Departure times, by leg
- Destination airports
- Passenger manifest, including lead passenger
- Catering
- FBO/fueler
- Ground transportation—passengers and crew
- Accommodations—passengers and crew

craft. These measures should be published in a company memorandum or on-demand air transportation policy and procedures document (see Appendix C) and authorized by the principal or high-level executive within a company. In some organizations, these issues are contained within a subset of the department's operations manual specifically prepared for use outside the department itself. These company aircraft procedures manuals provide potential passengers with the guidelines for aircraft usage and procedures for scheduling them. See "Aircraft Use Policy" in Chap. 3 for a more complete discussion of the aircraft use policy document.

Employees above certain levels or specifically named positions within the company may be authorized to use the aircraft. These individuals either may have scheduling authority or may have to seek that authority from an even fewer number of executives within the organization. Sometimes a single individual such as the principal's or CEO's assistant may act as the de facto scheduling authority, only deferring to the boss in cases of conflicts. A few principals and CEOs prefer to authorize all aircraft use as a matter of control and information about business operations. Priority to use any aircraft or a specific aircraft usually follows hierarchical lines, with juniors deferring to seniors.

Use by nonemployees or those outside the principal's family, whether they are customers or potential customers, family members of employees, political candidates, or friends, should be addressed in detail within the policy to preclude misunderstandings and to protect against potential liability concerns. Customers or potential customers normally are not a concern; however, many companies stipulate that a company employee be present in the aircraft anytime a nonemployee is on board. Family members and friends should be specifically authorized to travel on board a company aircraft by a corporate sponsor and may be subject to tax considerations for the individual or sponsor. [The Internal Revenue Service (IRS) rules concerning this usage are complex; see the *NBAA Management Guide* and other National Business Aviation Association (NBAA) tax publications.] The decision to use the aircraft for nonbusiness purposes should be well documented in use policy documents considering the tax consequences of this usage.

Procedure

A formal procedure should be developed to accommodate the passenger scheduling needs of the company. This procedure is normally tied closely to the aircraft usage policy and promulgated throughout the company. The procedure may include provision for informally checking on the availability of an aircraft but always should require the person requesting the trip to provide certain information in writing (facsimile, e-mail, or linked computerized scheduling system are the normal methods) to confirm the request.

After scheduling information is received, the scheduler will research aircraft and crew availability, suitable airports and flight-based operations (FBOs), and special considerations (ATC and airport slots and restrictions, airport hours of operation, noise considerations, etc.). Once it has been determined that the flight is possible, it should be placed on the flight schedule and confirmed with the requestor in writing, preferably by a detailed trip confirmation sheet. If the trip is routine, the scheduler may be allowed to schedule the trip without further consultations within the department. In some instances, the department manager or chief pilot is the only individual who can schedule a trip and assign a specific aircraft or flight crew. These individuals retain this right to ensure that they maintain operational command and control of each flight.

After the trip is authorized, a trip sheet is prepared providing both requestor and flight crew with information required for the flight.

Designating a lead or principal passenger on the trip sheet is useful for the flight crew. This is the person with whom the crew communicates concerning the conduct of the flight, while en route, regarding schedule and passenger changes. This person should be provided with information on how to contact the flight crew at any time during a trip, regardless of the length of layover or location.

Scheduling Software

Many small flight departments create, modify, and maintain their flight schedules, using a simple spiral-bound monthly planning book or dry marker white board. This system appears to work well for these departments, yet the many inevitable changes made to scheduled flights prior to departure test the erasing and communications skills of the scheduler. Thus only the scheduler has a completely up-to-date copy of the schedule; others, including the chief pilot, rely solely on the scheduler for changes to flights made prior to departure. Further, dissemination of accurate schedules to flight crews, maintenance personnel, and other company personnel is made more difficult under this method.

Flight Scheduling Software

- Provides accurate schedule data
- Allows a variety of individuals to share rapidly changing data
- Promotes a professional image with passengers
- Generates important management reports
- Reduces administrative load
- Provides an operational information database

Flight scheduling software is available that not only keeps the schedule up to date but also captures virtually every detail of the flight, before and after, in a database. This type of software provides flight operations management features for the entire department by scheduling, tracking, and reporting on all operational activities of the organization. While many people consider this software a luxury, especially for smaller operations, a long-range view of what it can do for the department should be investigated.

As a trip is "built" by the scheduler, a number of details must be captured and tracked through a number of inevitable changes occurring prior to flight. Having passenger, flight crew, aircraft, airport, and FBO data preloaded into the software greatly speeds the process. The schedule can be made available online to anyone possessing proper access to the system, making multiple and invariably inaccurate versions of the schedule held by several people unnecessary. Passenger confirmation and trip sheets and flight logs are prepared easily from the database electronically or on paper. After the flight, flight times, corrected manifests, trip fees, chargeback information, and pilot flight log data are captured in the database, completing the planning/recording process.

Perhaps the best feature of the software is its array of standard management reports available at the touch of a few computer keys.

Flight Operations Software Reports

- Aircraft utilization
- Flight statistics
- Destination/city-pair summaries
- Flight crew utilization/flight logs
- Chargebacks
- Passenger utilization/load factors
- Fuel usage/costs
- Department expenses
- Operational performance measures
- Total department activity

These reports (see the accompanying sidebar) can be used to provide essential management information for both flight department and company personnel to determine the best methods for supplying on-demand air transportation services. Without these reports and their underlying database, most flight departments do not measure important elements of their operations, thereby depriving themselves of essential management tools. At some point in each flight department's existence, a call will come for an analysis of some feature of its operations; deriving any of the reports listed in the sidebar manually, using paper flight schedules and logs, involves much tedious effort and may yield results of questionable accuracy. While some operators may be reluctant to purchase this software because of the costs involved, it inevitably proves to be a good investment over time for all flight departments.

A number of software programs are available that should suit just about any flight department's needs. Since these are detailed and complex systems, a significant amount of research should be invested in determining the right type for your operation. *Business & Commercial Aviation* magazine occasionally reviews this type of software, and the NBAA website, Products and Services section (nbaa.org/prodsvcs), lists manufacturers.

International

International flights require additional planning and consideration because of the different regulatory, economic, and cultural environments in which the aircraft will be operating. Interaction with the passengers becomes more detailed and intense, requiring precise itineraries, airports desired, and information regarding country entry procedures. The scheduler and flight crew normally work as a team to ensure that all required information, clearances, handling requirements, and planning data are assembled in a timely manner. The key to a successful international flight is complete, in-depth planning that provides for contingencies. Most importantly, the passengers must be made aware of the additional measures to be taken for international trip planning and the closer tolerances required for the operational aspects of the trip.

If a foreign destination is not visited regularly, a flight planning/aircraft handling service should be employed to reduce the number of variables and unknowns associated with such a trip. International flight plans, weather briefings, preferred routings, overflight permits, regulations, fuel service and payment provisions, visas, and ground transportation requirements are best handled by those who provide the services on a daily basis. The vagaries of international travel in corporate aircraft are too unpredictable and consequential to be performed by the uninformed.

Flight crews and the scheduler must receive international operations training from an experienced individual or training vendor. Even though an international trip planner/handler is employed, flight department personnel must have certain elementary background information about international regulations and procedures to safely and predictably plan and conduct the flight. Since international rules and procedures change frequently, periodic recurrent training on international subjects is also a must.

Administration

Administrative practice provides the foundation on which management is accomplished. Without structure, management becomes a hit-or-miss proposition because the elements of predictability, standard procedure, and control are not present. Moreover, these practices regulate and direct the flow of work within an organization. Without some measure of administrative process, paperwork becomes unmanageable, reporting structure is muddied, feedback concerning essential operations becomes nonexistent, and chaos sets in.

Administrative practices within the flight department depend heavily on established administrative procedures employed by the principal or company. While this is not to say that the department has no flexibility in devising its own procedures, the need to fit into an existing administrative structure is essential. Moreover, following established elements of administrative practice will facilitate the department startup process. Requesting assistance of the parent's administrative personnel will help new operations and provide valuable contacts within that structure for establishing operations.

Each of these sidebar issues should have some procedure associated with it. Most established organizations have procedures in place that will accommodate most of the elements listed. Internal departmental procedures will have to be established addressing the remaining elements to produce a good fit with the parent administrative procedures.

Administrative Practice

Policy, process, and procedure (the three Ps) are the means by which actions are accomplished within any organization; they are the

---

Flight Department Administrative Elements

- Communications
  - Methods/technology
  - Process
  - Content
- Control
  - Processes
  - Overall records management
  - Reports
- Personnel
  - Qualifications
  - Staffing requirements
  - Hiring
  - Salary actions
  - Evaluating
  - Terminating
- Information management
  - Production
  - Storage
  - Control
- Financial
  - Purchasing
  - Billing
  - Bill paying
  - Budgets/variances

lubricants and channels by which work is accomplished. Administration is the umbrella under which the three *P*s combine and interact. Many organizations maintain massive volumes delineating administrative practice, some to good effect and some counterproductive. If the stated procedure is obscure, overly complicated, or in conflict with a related procedure, not only will it will not work properly, but it also will create confusion and frustration for the employees attempting to use it. The same is true for the administrative practices of the flight department. If a procedure does not make sense, is cumbersome, or conflicts with another, it will create confusion and discord.

While most procedures should be written and recorded in an easy-to-retrieve format to facilitate use and avoid misunderstanding, some simple procedures may exist as a body of unrecorded common procedures known to all within the department. Regardless of the level of utility or formalization, all procedures should be reviewed for applicability, relevance, and effectiveness on a regular basis.

Administrative procedure must serve

Fuel Receipt/Invoice Process

- Purchase fuel (pilot)
  - Receive receipt
  - Attach receipt to flight log
  - File flight logs and trip receipts
- Receive receipt (administrator)
  - Verify trip/fuel purchase information
  - Enter fuel data into database
  - File receipt
- Receive monthly invoice (administrator)
  - Locate applicable receipts
  - Verify invoice charge with receipt
  - Authorize invoice payment
  - Send invoice to accounting for payment
  - File copy of invoice
  - Discard receipt

some purpose and contribute to the efficiency and effectiveness of an operation or task. If the reason for a procedure is, "That's the way we have always done it," "I don't know, but accounting wants it that way," or "The boss wants it that way," the procedure is probably a burden rather than a useful procedure. The biggest question to ask about administrative practices is, "What will the procedure do to enhance the operation?" If a valid answer results, then the procedure probably has some utility. Administrative procedures that are simple, understandable, and easy to accomplish will be better received and practiced than those that are complicated, obscure, and cumbersome.

Finally, the fewer procedures, the better. Integrated or linked procedures accomplishing more than one goal at a time are better because they are more user friendly, and the commonality of data put into the procedure will ensure greater accuracy when compared at some point downstream.

Developing Practices

If a relevant procedure exists within the parent organization, it should be used rather than developing an ad hoc one. However, if the procedure is inappropriate or cumbersome for use by the flight department, steps should be taken to change it. The best way to accomplish this is to approach the person responsible for the procedure (sometimes difficult; be persistent) with *questions* regarding the purpose and impact of the procedure. Only after the true purpose of the procedure or practice has been determined, can it be criticized objectively and suggestions made for its improvement.

Original or department-unique procedures should be developed with the assistance of those who will have to carry them out. Consensus regarding a practice is desirable but not

Typical Administrative Procedures

- Operations
  - Scheduling—passenger, aircraft, crew
  - Postflight recordkeeping—aircraft, crew, passengers
  - Training
- Maintenance
  - Scheduled maintenance planning
  - Discrepancy control
  - Maintenance recordkeeping
  - Parts inventory and control
- Accounting
  - Expenses
  - Purchasing
  - Budgeting and variances
- General
  - Personnel
  - Facilities
  - Crisis management

necessary, however. The manager may have to hear all arguments regarding competing procedures and make a decision regarding the most desirable.

Once a practice or policy has been devised, it should be revisited from time to time to evaluate its continued applicability and usefulness. Organizations are littered with practices that once had meaning and relevance but have since become obsolete. Purging these administrative dinosaurs will streamline the operation and please the employees.

The actual design of the administrative practice should be done using actual, not theoretical, processes. Too often the procedure is designed around the ideal method rather than an in-place, realistic one. Once an actual process has been traced throughout its life cycle, enhancements may be added. A useful method of dissecting a process is by using bulleted lists or flowcharting techniques. This affords the designer a graphic view of the entire process, allowing a full picture of the critical and not-so-critical events. The sidebar on the previous page provides an example of a fuel purchase receipt/invoice handling process.

Note that each step in the process requires an individual to complete the action. It should be clearly understood who in the department is responsible for each action, and preferably, a backup person should be designated to ensure that the action is completed regardless of the status of the principal individual.

The processes listed in the "Typical Administrative Procedures" sidebar should be devised by the individuals expected to accomplish them and recorded in department administrative policy and procedure manuals so that all personnel understand and appreciate the process.

Talking with the Folks Downtown

Wars are fought over it. Marriages are ruined because of it. Governments are toppled as a consequence. Differences in cultures and cultural expectations are perhaps the strongest split known to humanity because a gulf, nay, a chasm, exists between the backgrounds and very birthright of the parties. Thus it is with many aviation departments and *all others* within the parent organization.

The very nature of the flight department sets it apart from the administrative, financial, marketing, manufacturing, and human resources types at headquarters. The location of the flight department at an airport and its unique tasking tend to make the rest of the company view it as unique, sometimes mysterious, and always different. These are the people who fly the principal, the president, and senior staff to "important meetings" in Bangor,

The Downtowners

Essential	Good to Know
The chairman/president's office	Planning
Accounting/finance	Disaster preparedness
Budget	Environmental
Purchasing	Industrial safety
Payroll	Marketing/sales
Accounts payable	
Human resources	
Administration	
Property management	
Risk management	
Legal	
Security	

Los Angeles, and Paris. These are the people who are always "on a trip" when needed to answer questions about expense reports, budget variances, or annual reviews. Just who are these people, anyhow?

Don't laugh; the flight department is an enigma to most people within the organizations it serves and destined to remain so if the department manager does not make a real effort to dispel these myths and become an integral part of the company team. More importantly, the flight department must start working with these folks downtown if it is to realize its full potential or even survive as a part of the larger organization.

Organizational Conflict

The difficulty lies in both the aviation personnel and *all others* understanding and appreciating one another—accounting versus aviation, human resources versus the pilots, lawyers versus everybody. Each discipline comes with its own set of values, expectations, and arcane language. The whole process of understanding between parties is more an exercise in foreign relations than organizational development. However, aviation may as well be the one to make the first step because it has the most to gain. Yes, flight departments come and go, but finance, administration, and human resources go on forever.

Everyone has had misunderstandings with the aforementioned departments.

- Accounting wants to know why the flight department has exceeded its monthly maintenance budget by 300 percent, and you cannot make them understand what the APU airworthiness directive is all about.

- Legal will not sign off on the hangar lease because of the airport reversion clause.

- Human resources wants all the annual appraisal form blocks filled in even though the mechanics do not do half the stuff listed for a "refinery technician" (which is how they are classified) and environmental does not know what to do with the used MEK.

The natural tendency is to throw up your hands, say *they* do not understand, and ignore the problem. Unfortunately, this will go on only for so long. Even the chairperson cannot protect his or her air force forever from an irate comptroller.

Basically, all these warring entities are on the same team: the company. They are supposedly a part of a team that is working for a common set of goals and objectives. Granted, this fact is difficult to discern much of the time. How to break the impasse? Simple, talk it through.

Every department within an organization wants to do well and to look good. Accounting wants to get its variance reports out on time, human resources wants to get the retirement plan questionnaires back and compiled, and purchasing wants all outstanding purchase orders reconciled. And you want to get your flights out and back on time. These are all very different functions, but they all work toward a common goal—to make the company achieve its operating targets and place of prominence in the industry. The problem is that everyone is too narrowly focused on his or her own interests and goals. What we have is a failure to empathize and communicate.

Get Them on Your Side

They do not understand aviation, and you probably do not understand much of what they are trying to accomplish. It is like transitioning to a new aircraft or flying to Europe for the first time—all of what is encountered is strange and does not make much sense. After a while, though, you begin to understand the new system, its language, and its culture. You may not agree with it or want to do it for the rest of your life, but you can understand it and can work within its confines. You may even like or at least admire some of what you see.

Since the flight department has the most to lose, the first move belongs to the flight department manager. Approach the appropriate people in the departments that mean something to you and ask for their assistance. Yes, ask for help. People normally respond well to requests for assistance in a field they know something about. Show an interest in their work, empathize with their problems, and offer to help where you can. Pursue the relationship if it appears fruitful; find another contact if not. Learn their language, find out what is important to them, start thinking in their terms, and cultivate a mentor or at least a coach. Invite them out to visit the hangar, show them around, take them to lunch, and ask for their help again. Tell them what is important to you and the flight department; talk in terms of supporting the corporate mission, of common goals. Build a mutual admiration society; bridge the cultural differences.

Before you start, think of what the other people want, of what motivates them. At the most basic level, we all want to feel secure—secure at home and at work. Beyond this elementary need, people need to feel that they are a part of a group, an integral part of a team working toward some worthy common goal. Moving on to people's higher needs, they want to feel that they are respected and that their efforts are appreciated, not so much that they receive a special award but merely to have somebody say, "Good job." The final step in the motivation chain must come from the individual himself or herself. Self-actualization or self-fulfillment represents the employee's personal concept of high value for the work in which he or she is engaged. These items constitute Maslow's hierarchy of needs (see "Behavioral Approaches" in Chap. 5), which has stood the test of time in the workplace. It is important to note that for self-fulfillment to be possible, *all* the lower needs must first be met.

Helping your target audience satisfy their needs will take time. However, the investment will be well worth it. Enlist the aid of other flight department members, especially your secretaries, administrative assistants, and schedulers, because they work closely with many sections of the company. Assign them tasks and goals to establish contact and build relationships with significant players within the parent hierarchy. Encourage mutual visiting and sharing of resources, where possible. Do not overlook the pilots and technicians as sources of contact with the downtown crowd either. This is an essential element in their professional development and must be pursued by them early in their careers. And if they have been assigned collateral administrative duties within the department, these "outreach" activities are a natural for them.

It may be desirable to create situations where a group of the downtowners are invited to the hangar for a formal presentation concerning the flight department, its aircraft, mission, and operations. A tour of the aircraft and facility, a presentation about your mission, and a frank discussion of your needs and aspirations will go a long way toward establishing firm communications links and bridging the cultural gaps that exist between the flight department and "them." And provide a catered lunch, too.

Coming together is a beginning; keeping together is progress; working together is success.
—HENRY FORD

Making House Calls

However, suppose that "downtown" is 300 miles away? Headquarters may be at some location remote from your airport, forcing you to do all your business via remote control. This complicates your task of establishing good relations with your "support staff" but is by no means impossible. If opportunities do not present themselves to allow face-to-face visits with the people who count, they must be manufactured. Request opportunities to discuss long-range plans, budgets, department restructuring, plans for the new hangar, or salary administration with the appropriate people at headquarters. Getting the right people to visit the hangar may be more difficult to accomplish but will be more effective than your visit to them. These are legitimate activities for the flight department manager and should be honored. Granted, this means an investment of the department manager's time, but much is at stake.

In your effort to become aviation's good will ambassador, do not forget the helpers on the other end either. The secretaries and administrative assistants who schedule the aircraft for their bosses can make or break the flight department. Without their understanding and support, their bosses will have a difficult time in thinking well of the department; secretaries are true gatekeepers and opinion molders. Why not a special day at the hangar for just the secretaries?

On your way to establishing good relations with those who can help the cause of the flight department, you also will be selling the concept of on-demand air transportation. This is something that all members of the department always must be aware of and engaged in. The position of the flight department is tenuous enough; you need all the support you can get in making the company believe that aviation is a real support function. These relationships will further that cause.

In the largest sense, you are building a support staff within the greater organization. Instead of having those departments make demands on your department, enlist their aid in helping you to achieve your departmental goals. This is possible, but remember, there is a cost. It takes an investment of time and effort to show that you understand their needs and are acting as an integral part of the larger group. The payoff is too good to pass up: Wouldn't it be great to get an annual budget or personnel request through on the first try?

Cultural differences can be bridged. The creation of your very own people-to-people program within the company definitely will help to get the flight schedule accomplished. And don't forget to say "Thank you" often.

An In-House Operation Pays Off

A financial services company that specialized in providing initial stock offering services to customers had regularly used aircraft charter services during the busy first days of an offering. The charter aircraft were used to transport company executives and clients to visit investment firms in an effort to generate interest in the new offering. In doing so, the aircraft was used to good advantage, often visiting three cities in one day and ten cities in a week.

After a year of working to find a charter operator that could be flexible enough to accommodate the last-minute and often-changeable nature of its requirements, the firm retained a consultant to determine the best course of action. In a short period of time, the firm was operating its own Lear 45 aircraft, often for its own travel needs rather than those of its clients. Not only was it able to control the aircraft's schedule, but also direct operating costs became less expensive once the company passed the 250-hour annual usage point.

Finance and Accounting

The aviation department manager probably spends the second largest amount of his or her nonaviation time on financial matters—the largest amount of time spent is, or should be, on personnel matters. Financial matters equate to resources, resources that make the difference between a struggling, barely-able-to-get-by department and one that achieves excellence with a measure of style. Therefore, the manager spends much time on financial issues just to stay ahead of the other cost centers within the organization that vie for available resources.

Each of the items shown in the financial processes sidebar are equally important and must be kept in balance if the department's goals and objectives are to be met.

The knowledge of basic accounting and finance practices and processes is essential to managing these affairs within the department. Without this knowledge, the manager is at the mercy of those within the system who do know and appreciate financial terms and practices. Perhaps the best way to learn or to brush up on this field is through an individual working in the larger organization's accounting department, preferably one who will be involved directly with the flight department accounts. In this manner, there should be some mutual exchange of information about the participant's specialties, thereby enhancing the communications process between the two departments. This exchange should not stop with the flight department manager either. Other members of the department with direct financial interests, notably the director of maintenance, parts handlers, and secretary, should establish their own relationships with accounting personnel.

A key issue in working with expenses, budgets, and accounting in general is that the accounting department must be able to process the purchase orders, invoices, bills,

Basic Financial Processes

- Planning
- Budgeting
- Recording
- Controlling/tracking
- Justifying

and credits and provide operating divisions with reports in a timely manner if management is to be able to properly control financial matters. If invoices are held and not processed and paid for weeks at a time and/or if journal and variance reports are not released until two weeks into a new month, then managers are always far behind real time and never sure of where they stand with their vendors or the budget. The accounting process is a two-way street; if the accounting department expects adherence to financial policies, then it must cooperate with operating divisions by providing accurate and timely information.

Basic Accounting Practices

- Purchase supplies and equipment
- Pay invoices
- Reimburse personnel expenses
- Verify invoices paid in a timely manner
- Receive consolidated reports of purchases and expenses (journal)
- Receive reports comparing planned with actual expenses (budget variance)

Taxes

Taxes, be they federal, state, or local, can present even the most astute accountant or tax attorney with a bewildering thicket of arcane statutes, conflicting regulations, and tax law precedents. If tax laws are difficult for the experts, then the flight department manager has almost no hope of comprehending these confounding rules and regulations. However, department managers should be aware of the rudiments of the laws to protect themselves and the parent organization.

Basically, the flight department must keep a record of all flights made, providing at least the date, departure and arrival points, names of passengers, and whether the passengers were employees *and* whether they had a business purpose for the flight. If the passenger is not an employee, or is an employee without a business purpose, information will be needed regarding the reason for the flight or the relationship to an employee or the company. These are just the basics for the flight department and the company; a more comprehensive treatment of the subject may be found at www.nbaa.org/admin/taxes. At a minimum, managers should be aware that there are significant implications to the issues shown in the sidebar.

Financial Planning

Planning as a stand-alone subject is covered later in this chapter, but since planning is an integral part of the budget cycle, it will be mentioned briefly here as it applies to the financial process. Some flight departments consider financial planning for budget purposes to be the process of adding anticipated annual inflation to last year's budget to arrive at the numbers to be submitted for next year's budget. This so-called incremental budgeting

Tax Issues

- Nonbusiness use of employer-provided aircraft
- Affiliated company group use of aircraft
- Carriage of elected officials
- Interchange agreements
- Spousal travel
- Joint ownership of aircraft
- Flight department companies
- Time-sharing companies
- Personal use of company aircraft

robs the manager of what can be a valuable tool to be used in achieving departmental renewal and progress. This "lazy man's" way of performing an unpleasant task will not allow the department to move forward, to incorporate new features, or to get rid of obsolete or inefficient processes.

Corporate strategic and short-term plans should be studied to determine the direction and goals of the company as a whole. If the departmental goals do not fit into and support the organization's plans, the budget process will be difficult. By supporting the stated desires of the parent group, the department can show that it is a member of the team, willing to subordinate its desires to the greater good and be a full contributor to the joint effort.

In helping the company or principal achieve short-term goals, however, the concept of reaching both the parent and departmental long-range goals also must be considered. For instance, if an immediate company goal is to reduce capital expenses while supporting a long-term goal of expansion into a foreign country, the department must find a means to begin flights to the target country even though the aircraft currently available are inadequate to the task. A potential answer may be the short-term lease of a long-range aircraft to support the short-term goal while achieving the long-term goal as well.

The planning process should include as many people within the department as possible. Their particular view of the overall operation should be valuable because of their experience and expertise. Moreover, engaging them in the process will make them feel a part of the team. However, this planning process should be carefully controlled and not merely a far-reaching brainstorming session to see how the department can be expanded regardless of need. Rather, the planning process should begin with a careful look at existing operations to determine desirable and undesirable elements with an eye toward improvement. This introspective look at operations should include an investigation of whether the departmental goals are being met with regard to both effectiveness and efficiency. Only after this examination of current operations should the discussion of future needs be opened.

The object of the planning process is to develop realistic goals that can be met within a specified period of time. The normal tendency is to focus principally on short-term goals to the detriment of the long view. A well-integrated set of long- and short-term goals that fit into the corporate view of the future should form the basis for a realistic and easily justified budget. To be of real value, goals must be quantifiable so that progress toward them may be measured. Merely saying that the department will have a goal of improved customer service probably will not be achieved because no performance factors have been assigned. However, a goal stating that the semiannual passenger satisfaction survey will yield 10 percent more passengers describing service as good or excellent is something that can be quantified and progress measured toward reaching it.

Budgets

A budget is an extension of the planning process that restates short-term goals in financial terms. The budget takes an abstract concept such as providing on-demand air transportation and gives it the detailed resources necessary to bring the concept to reality. Unfortunately, most people do not view the subject in such lofty terms. Rather, they look at the budget-building process as a tedious, no-win game played for the benefit of the accounting department and the budget itself as a tattle-tale device that can only show the boss that they are not performing well. These negative stereotypes have arisen in large part from a lack of understanding of the budget process and what it can mean to the department in terms of resources to be gained. Figure 4.4 illustrates the importance and utility of the budget as a means of converting plans into action.

PLANS INTO ACTION

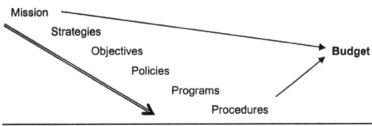

FIGURE 4.4   Budget concept.

If viewed and understood properly, the budget can be used to the advantage of the department—to help it obtain the resources it needs and look good in the process. However, the basic language of finance and accounting must be understood and appreciated first. As stated previously, this is best accomplished by finding a member of the accounting department who is willing to take on a student—you—in the pursuit of practical accounting and finance knowledge.

There are two types of budgets that the flight department will use: operations and capital. The operations budget sets forth expenses and revenues (if any), providing a breakdown by detailed accounting categories. Budgets are usually made up for a 12-month period designated the *fiscal year*. This may not necessarily conform to the calendar year but may fit a convenient period from an accounting and taxation standpoint. A simple flight department operations budget is shown in Fig. 4.5.

Note that the expenditures are broken down by variable and fixed costs. Variable costs are incurred only when the aircraft flies, e.g., fuel, maintenance, reserve for maintenance items (e.g., engine, auxiliary power unit, and thrust reverser overhauls), and crew expenses. Fixed costs are incurred whether the aircraft flies or not, e.g., salaries, hangar rent, training, etc. Where possible, the budget should provide separate variable costs for each aircraft assigned to better account for the operating costs for each aircraft.

Depreciation.   Depreciation expense, sometimes included in flight department operating budgets (not shown in this example), recognizes the fact that from an accounting standpoint, a capital asset (property or device possessing value) has a finite life span and, therefore, loses value over its life span. When its useful life has expired, only its residual or resale value remains. The decreasing value of the asset is recognized over its accounting life as a fixed expense. Flight department personnel must be aware of depreciation provisions to more fully appreciate its effects on budgets and other accounting features.

It is important to note that there are two types of depreciation: tax and book. *Tax depreciation* refers to the tax consequences of reducing the taxable value of the asset over a minimum period of time specified by tax authorities—as few as 5 years for a corporate aircraft. At the end of the taxable depreciation period, only a small residual value, often as little as 10 percent, remains. *Book depreciation* refers to the actual decrease or increase (appreciation) in value of the asset during the intended period of use by the company. In practice, the book value of a popular aircraft after 5 or even 10 years of use may actually be greater than its purchase price.

		Flight Hours per Year		
	Cost Item	250	500	750
Variable/Direct Costs				
Fuel (price × gph)	$1,200	300,000	600,000	900,000
Maintenance				
Labor (hrs/flt hr × rate)	$70	17,500	35,000	42,500
Parts	$80	20,000	40,000	60,000
Engine reserves				
Restoration allowance	$350	87,500	131,000	218,500
Landing/handling fees	$22	5,500	11,000	16,500
Crew expenses	$95	23,750	47,500	71,250
Catering/supplies	$50	12,500	18,000	37,500
Total Variable Costs	**$1,867**	**$466,750**	**$933,500**	**$1,400,250**
Fixed/indirect Costs				
Crew salaries/benefits	$200,000			
Hangar	$40,000			
Aircraft insurance	$30,000			
Training	$28,000			
Other fixed expenses	$7,000			
Total Fixed Costs	**$305,000**	**$305,000**	**$305,000**	**$305,000**
Total Costs		**$771,750**	**$1,238,500**	**$1,705,250**
Total Cost per Hour		**$3,087**	**$2,477**	**$2,274**

FIGURE 4.5   Annual aircraft operating budget example.

Practical Budgets. All of this effort requires the cooperation of many people within the company, beginning with the CEO and ending with just about every employee who has a goal to reach. It is the budget-cycle process that links the diverse organizational groups in their competition for the resources that will enable the company to complete its mission and, hopefully, make a profit. However, the means by which the cycle is completed, the financial accounting system, must be capable of completing all the stated tasks in a timely and accurate manner. Without a sufficiently fast and accurate system, the budget cycle will not function efficiently enough to be of value as a forecasting and measuring tool. For example, if the accounting department cannot enter information concerning obligated expenses in a timely manner to enable monthly reports to be sent to the operating departments by the tenth of the following month, the operating managers have no up-to-date information regarding their financial performance and progress toward their goals.

If this process cannot be accomplished in a reasonable period of time, the flight department may be forced to keep its own books or maintain a running total of expenditures by accounting category to ensure that it does not overspend the allocated amount. This double bookkeeping inevitably leads to discrepancies between the department's and accounting's systems and is a time-consuming, extra task that robs the department of productivity. However, this may have to be done as a defensive measure to ensure compliance with the budget.

Are Budgets Necessary?   A significant number of flight departments, mostly small ones, are not required to submit budgets. This is largely due to the CEO or principal absorbing the operating expenses within the general and administrative budget, a segment traditionally having sufficient elasticity to absorb large variances.

Flight departments operating under this system may consider that they have an open-checkbook fiscal system and are not fully responsible for what is spent. While most managers/chief pilots do not abuse their stewardship, this form of fiscal nonmanagement can lead to both abuse and poor management. Without being accountable to a budget, it is also difficult, if not impossible, to measure efficiency and productivity.

The real downside to this method is that the munificent CEO will not be there forever, and when the new person takes over, there may be some uncomfortable questions asked about flight department expenditures and the relevance of the department in assisting the company meet its goals.

Thus, for the sake of good management practice and anticipating the need to one day show value, building and living within a budget is a good idea, regardless of whether the boss wants one or not.

Building the Budget

Each year companies begin the budget cycle by setting forth general budget goals that flow from their long- and short-range plans. These goals may be expressed as a desire to increase sales or profits by a certain percentage or to decrease expenses by absolute amounts or percentages of individual budget line items or the entire budget. By the time these budget goals (actually mandates) have reached the flight department, several layers of additional detail may have been added to the original broad goals, thereby adding further constraints to an already constrained budget.

A budget worksheet is usually provided that shows the previous year's budgeted and actual expenditures. Spaces are provided for the budget request for the upcoming fiscal year by budget line item, as designated in the accounting department's chart of accounts. It is important to have the proper level of detail in the chart of accounts that supports the actual types of expenditures used by the flight department. Accounting departments sometimes force flight departments to use codes that have existed within the company as a consequence of its line of business, including truck heavy maintenance, small craft repairs, machinery maintenance/overhaul, and sales training. If these classifications are used, it will be difficult to determine where the department's funds are expended. Therefore, flight departments should insist that their own unique terms be placed in the chart of accounts. Table 4.1 shows a sample flight department chart of accounts.

Variable/Direct Expenses	**Fixed/Indirect Expenses**
Fuel	Salaries and benefits
Maintenance	Pilot
Airframe	Maintenance
Avionics	Administrative
Engine	Part-time employees
Parts/material	Facilities maintenance
Contract services	Support equipment maintenance
Engine overhaul reserves	Contract services
Catering	Training
Flight crew expenses	Flight planning/weather services
Fees—landing, parking, etc.	Utilities
	Administrative/office
	Insurance
	Aircraft
	Hangar
	Taxes

TABLE 4.1 Sample Chart of Accounts

Note that this is a very generalized statement of expenses. When building the budget, actual categories should be used to estimate expenditures and then subtotaled to match the broader classifications listed in the worksheet. A more detailed chart of accounts for flight departments may be found in the *NBAA Management Guide*.

The major factor driving budget estimates for the upcoming year is the estimate of annual usage of the aircraft, expressed as flight hours. While flight hours will not directly affect the fixed-cost portion of the budget, it will do so indirectly. For example, if aircraft utilization is expected to increase appreciably, additional pilots and aircraft maintenance technicians (AMTs), training, and maintenance facilities may be required, thereby increasing both variable and fixed expenses. Figure 4.5 incorporates this relationship in the budget worksheet, showing management that increased utilization reduces unit costs by significant amounts.

It is tempting to just add the value of inflation to each worksheet category and be done with it. However, such an expedient might prove harmful to the department if upcoming but obscure expenses are overlooked in the process. Moreover, if justification for an expense category is requested, it will be difficult to defend the inflation method of budget estimation. Assumptions for continuing and new expense items are very important, particularly when it comes to budget justifications. Therefore, a budget analysis method for each significant expense category should be devised and used to give the preparer a basis for justifying either existing or new requests. Such methodology is invaluable when the inevitable budget revisions are requested. A sample of one such method is shown in Fig. 4.6.

While it is not necessary to provide an analysis for each category of expense, it will provide a useful record of the assumptions and factors used in reaching a decision regarding a

Account_____   Account Number_____   Date_____

Budget Period _____

Previous Level _____

_____

Assumptions

_____

_____

_____

Analysis

Month	Event	Impact
____	_____	_____
____	_____	_____
____	_____	_____
____	_____	_____
____	_____	_____
____	_____	_____

Overall Impact

_____

_____

Justification

_____

_____

_____

FIGURE 4.6   Sample budget analysis worksheet.

particular budget request item. This is particularly true when the reason for an estimate is needed for future reference.

When building a budget, the standard methods of budget estimation and justification used by the parent organization should be followed carefully. Failure to understand the process probably will require resubmission of the budget request, bringing unwanted attention to the department. Further, a contact within the accounting department will prove very useful when questions concerning budget policy or procedures arise. Moreover, discussing budget philosophy and corporate strategy concerning the upcoming fiscal year with the flight department reporting senior should yield valuable insights.

Capital Budgets

This type of budget recognizes that certain purchases have long-term value or the ability to create value for the company. Machinery, buildings, furniture, computers, and aircraft fall into this category. Because of the ability to create or enhance value for the company, the object is assigned a finite life over which it is expensed. Therefore, if an aircraft is purchased new, it may be assigned a tax life of as little as 5 years and may be expensed over that period, leaving just a residual value at the end of the period (see "Depreciation," on page 149).

Capital budgets are important for flight departments because they are not considered a part of the annual operations budget. In essence, the capital budget is a separate budget used for buying permanent, major assets. It is a way of obtaining a new hangar, ground support equipment, aircraft equipment upgrades and interior refurbishments, support vehicles, and furniture essentially off-budget—the normal operations budget, that is. However, the capital items must be justified and put into competition with other capital item proposals made by the rest of the company. And if the company is to realize some value from the money it invests, it must be capable of producing some measurable benefit for the company. Sample justifications are shown in a subsequent section.

Any major purchase or expense should be reviewed by accounting for possible inclusion in the capital budget. Removing major items, especially unanticipated items, from the operations budget will make the department appear more in compliance with the planned expenditures and enable it to gain needed items.

Any action that increases or extends the value of a capital asset should be capitalized. The examples mentioned earlier easily meet those criteria; some maintenance actions also meet the criteria, although they may not be readily apparent. Engine, thrust reverser, and APU overhauls fall into this category because service bulletins and other improvements are incorporated during the overhaul process that may extend or improve the value of the asset. Make sure accounting knows about these features; it may be able to recapitalize the asset.

Controlling/Tracking Budgets

The budget is just a plan for spending money. It contains no method of ensuring that the plan is carried out or how actual expenses compare with the original estimate. Therefore, a method of comparing monthly expenditures with budgeted estimates, known as *variance reporting*, is used by most companies requiring budgets. An example of a segment of a variance report is shown in Fig. 4.7.

Note that the effect of a higher or lower expenditure for a single month may be quite different from the effect on the year-to-date section. For example, in Fig. 4.7 the salaries expense for the month of April was 2 percent less than the budgeted amount, but because of some over-budget expense in the previous 3 months, the net effect on the year-to-date section was 10 percent more than budget. And the large over-budget maintenance expense in April does not look as bad when spread over the 4-month period.

A method used to level out large unbudgeted expenses for departments that occasionally have unexpected expenses is the *moving-average variance*. This adds the current-month expense with the previous 11 months and divides by 12 to show the effect of the expense over a long time period. This is particularly helpful for the maintenance section, which may have large, unexpected expenses relating to nonscheduled maintenance items. The purpose

Division: Administrative   Department: Aviation   Month: April 2013

	April				Year-to-Date			
	Budget	Actual	Variance	%	Budget	Actual	Variance	%
Salaries	56,250	55,000	(1,250)	-2	250,000	275,000	25,000	+10
Maintenance	25,000	140,000	115,000	+460*	100,000	195,000	95,000	+95
(12 mo. av.	25,000	32,500	7,500	+30)				

* Note: $105,000 one-time engine repair on N123 due to foreign object damage

FIGURE 4.7   Budget variance report example.

of this method is to show the effect of spreading the current-month expense over a long period of time.

Note that the reason for the large monthly variance has been footnoted on the variance report. This is important to preclude questions concerning the variance. Preferably, the reason for a potential variance should be made known to the flight department manager, his or her reporting senior, and the accounting department as soon as it is known. Nobody likes surprises, especially large monetary surprises.

We must consult our means rather than our wishes.   —GEORGE WASHINGTON

Budget Justification

When budgets are being built, large variances are anticipated or received, or new budgeted items are contemplated, a justification statement may be required to accompany the other budget documents. This is merely a detailed description of the item or service to be purchased along with an explanation of its purpose and utility. Why do you want it? How strongly do you feel about it?

The justification should contain, at a minimum, the following items:

- Background
- Prior expenses
- Current situation
- Assumptions
- Proposal
- *Detailed* justification
- Benefits

An example of a detailed justification is shown in Fig. 4.8. This level of justification may not be necessary; ask the boss what level of justification is appropriate.

Division: Administrative   Department: Aviation   Date: 5-12-13

Item: Purchase One Stellar Model 8346 Spectrum Analyzer

Cost: $57,500 New

Useful life: 5 years

Application: Repair aircraft electronic equipment.

Justification: Aviation Department must send unserviceable aircraft electronic equipment to vendors that may require 2 day shipping/processing time. This causes the department to maintain much rotable (spare) equipment, now valued at $235,000.

The spectrum analyzer will complete the electronics repair suite at the Aviation Department's facility. This completed suite will allow in-house technicians to save an estimated $11,000 per year in repair costs and reduce the rotable equipment by an estimated $110,000.

Alternatives:
1. Continue to use the vendor services at $32,000 per year to repair spares.
2. Purchase a used analyzer @ approximately $23,000. This option is undesirable due to the lack of needed features contained in the new equipment.

Figure 4.8   Sample budget justification.

Personnel

The members of the flight department are its most important asset. Their expertise and willingness to perform are its real treasure. Without those attributes, the department is, at best, a mediocre group of aviation personnel. Their willingness to perform, or motivation, must be properly channeled and shaped if a high-quality product is to result. The channeling or directing function leads to the concept of teamwork—equally important as motivation. If the people all act independently and without regard for one another, however well they perform, the result may be chaos rather than a smoothly running organization.

The art form for the aviation department manager is to successfully direct and channel the efforts of assigned personnel to form an effective and efficient team of professionals that will provide high-quality air transportation services to the company. This art form is called *leadership*.

It is impossible to cover all aspects of personnel management in a book of this nature; therefore, the aviation department manager should have some formal or practical supervision and/or management experience and training prior to accepting the job. Many supervisors/managers have learned their craft on the job, but the mistakes made in the process are largely avoidable with some planning and forethought. Some issues are more important than others, however. The first step is to hire the right people.

Hiring

There is normally an abundance of well-qualified pilots and AMTs in the job market at any given time. In times of expansion and prosperity for corporate aviation, the best qualified

Hiring Process

- Know what you want: detailed job description, experience levels, ability to perform collateral duties, etc.
- Work closely with the company human resources department—their experience and insight is essential.
- Verify former employment and education.
- Check with the FAA Information Management Section (AFS-624, Box 25082, Oklahoma City, OK 73125) for possible airman certificate violations.
- Let the candidate interview with at least three (preferably five) people, including one or two managers outside the department.
- Fly with or observe the candidate perform maintenance functions.
- If possible, employ the candidate on a part-time basis prior to making the permanent hire decision.
- Hire for the long term. Is this person 20-year material?
- Make sure the candidate fully understands the job being offered.

people are not as apparent or available, but they are there. However, a wallet full of Federal Aviation Administration (FAA) certificates and years of experience do not ensure that any individual will work well in your operation.

The culture of the company and the department dictate the required personality of the prospective employee, not the proper certification and experience. Perhaps the greatest difficulty in finding the right employee is to determine the corporate and departmental culture and then find a match for those characteristics. If the company is mature and well structured, a less formal, free-spirit personality may not fit well. Conversely, if the company is in its entrepreneurial stage, this type of personality may prove a good fit.

Take the time to fully evaluate candidates for employment. Consider using the items listed in the above sidebar.

Motivation

Perhaps more than any other single personnel management skill, the leader's ability to motivate is the single factor that will get the most and best work out of people. Motivation creates an internal incentive within a person to accomplish a task at hand. Instilling and maintaining this incentive should be a primary job of the manager. Another issue is understanding what factors contribute to job satisfaction (see the sidebar on the next page).

Motivation comes from realizing that the job being done is essential and has value. Instilling motivation requires that the employee understand

- the mission of the company and the flight department,
- where he or she fits in the organization, and
- the importance of their job.

Positive Job Factors

- Respect shown by management and peers

- Good pay (not necessarily high pay)

- Opportunity for self-improvement

- Perceiving that the job is important

- Opportunity to do interesting work

- Latitude to accomplish the job without excessive direction

While there are no guarantees, if these factors are considered as guiding principles, the chances of having a productive, happy workforce are high. It may not always be possible to approach each task with all these tenets in mind; some jobs just are not fun. However, if the concept of motivation is kept in mind constantly, employee performance should be high.

Good managers and leaders show respect for the individual. If the individual is made to feel that he or she is an important part of the overall operation, that his or her opinion counts, and that his or her welfare is considered in every action concerning his or her employment, then that person will be a relatively happy member of the team. When management practices respect for the individual, other components of good personnel management should follow. The concept of respect will tend to promote fairness in scheduling and compensation, opportunity for advancement, positive reinforcement for a job well done, and some slack when the employee makes a mistake.

The quality of a person's life is in direct proportion to their commitment to excellence, regardless of their chosen field of endeavor. —VINCE LOMBARDI

Communicating Expectations

Nobody likes surprises, particularly on the job. If complete and detailed expectations of job performance are not provided (and updated regularly), a worker will be constantly surprised by the tasks assigned. Job descriptions, task briefings with completion standards included, and *regular feedback* concerning performance are essential items to be communicated to every employee.

If an organization's employees are its most valuable asset, then they must be maintained and upgraded so that the organization continues to realize a return on those assets. Getting the new person to a point of full productivity may take 6 months or more; this is a major task that should pay handsome dividends in the long term. With this concept firmly in mind, develop an initial *asset investment plan* for the new hire.

Most managers have figured out that while technical competence is an important aspect of a person's job qualifications, in the long run it is the person's ability to work toward common goals, to interact effectively with people, and to provide value to the organization that count the most. (These are the real reasons you hired the person, aren't they?) To get the new person started in the right direction, he or she must know what the company's values are, what product or service it offers, and what characteristics in people the organization prizes most. These elements constitute the organization's culture, that elusive but essential descriptor of what makes it unique. If the new hire has a good understanding of what constitutes the corporate culture, he or she also will understand what is expected of him or her.

Many companies hold periodic new employee orientation sessions covering the contents of the employee manual, what the company produces, and how to get around within the corporate structure. Flight department employees should be required to attend these sessions; they provide essential information and begin to provide the basis for the culture they are joining. Additionally, it is up to the employee's supervisor to continually reinforce cultural norms and customs—this is important.

Everybody wants to know if they are doing their job well or not; allowing either excellent or poor job performance to go unnoticed constitutes poor supervision. Waiting until the annual employee performance appraisal to praise or chastise is a waste of employee resources. A frequent word to correct, redirect, or praise a person's work is all that is normally needed. This is just one more manifestation of the important feedback function of management.

New Hire Checklist

- Discuss expectations (write them down)
- Introduce/reinforce organizational culture
- Provide department orientation
- Devise a training plan
- Articulate policies
- Set standards
- Describe the evaluation process/set goals
- Constantly evaluate and comment on performance
- Introduce to peers and their work
- Train to proficiency

Communication is a two-way street, too. The employee must be encouraged to provide feedback regarding the task and related expectations. Without his or her input, communication is incomplete.

Performance Evaluation

There was a time when a pilot showed up, flew the trip, did not spill the boss' coffee, and went home to wait for the next go 'round. As long as he or she did not bend metal or egos, the pilot's job was secure, and the cost-of-living increases continued. The same was true for the maintenance technicians: Keep 'em flying with a minimum of fuss, and the job "spoke for itself."

During these halcyon days, the concept of a periodic employee evaluation for aviation department personnel was considered at least superfluous, and at best, mildly demeaning. These were professionals—they were above something so crass as an attempt to evaluate their highly developed skills.

Alas, the folks in human resources now insist on at least an annual evaluation of each employee's performance, be it the custodian or the executive vice president; even the chairman and president receive one (at a board meeting). However, how do you evaluate something so strange as a pilot? Maintenance technicians are easy; they are the same as any other mechanic, right?

Experience has shown that pilots do not quite fit into the manager category, nor do they entirely fit into the technical professional group. And aviation maintenance technicians (AMTs) are different enough from steam plant workers to warrant another look at performance measurements.

Employee Evaluation Factors

- Judgment (the right decision at the right time)

- Efficiency (saving the company money)

- Resourcefulness (getting the job done in difficult circumstances)

- Potential for increased responsibility (ability to manage)

- Interpersonal skills (keep the customer happy)

- Team orientation (working well with others)

- Communications skills (*listening,* speaking, writing)

- Willingness to go the extra mile

Pilots and technicians are well compensated not only to demonstrate a high level of skill in the aeronautical environment but also to stay within the legal framework of the many rules and guidelines provided. Should they be evaluated on a level of technical performance that is at once a standard and excellent as well? If they do not measure up to the high standards of technical competence required of them, they probably should not have the job. So what is left to evaluate?

Most of us want to be more than just a success, but an *artistic success* to boot. Evaluations, therefore, should look at the margins of excellence, at the increments of performance that set one apart from one's peers. There is always a better way to accomplish a task, to please the customer, to add icing to a well-prepared cake.

The real value of communicating performance information is to give the person being evaluated a means to improve and continue to contribute to the success of the flight department. This means deciding on mutually acceptable improvement goals and developing a way to accomplish them. And most people like to walk away from an evaluation with a sense of direction and beginning anew.

Do not forget to cut human resources in on these great ideas.

Human Resources

Flight and human resources departments often endure strained relationships due to the isolation and uniqueness of the company aviation branch. As far as human resources is concerned, the flight department works out of a nontraditional, often distant location and communicates precious little information regarding its activities. The flight department may be heard from only when something is wrong with an employee's vacation balance or pension contribution. From the flight department's point of view, human resources always wants something: a performance evaluation, a pension election verification, or attendance at a boring lecture.

The uniqueness of aviation personnel makes job classification and salary administration a problem for human resources. Pilots are neither technicians nor managers, yet they share certain attributes with these occupations. Aviation maintenance technicians seem like automobile mechanics or machinery workers, yet they too do not quite fit existing job classifications. The receptionist/secretary/scheduler/accounting clerk composite person present in many departments defies all comparison. All this makes fair compensation for department personnel difficult and sometimes contentious.

The answer to all these problems is better communications. If both parties understand what the other's needs are, then most problems will become manageable if not nonexistent. The flight department manager should become quite familiar with the director of human resources and visit his or her office regularly. Additionally, a single person within the department should be assigned the duty of handling all human resources issues so that a mutual understanding of various situations can be fostered.

Detailed job descriptions will assist with both job classification and compensation issues. Human resources must understand the high degree of training and experience and mandatory certification required of aviation personnel; this is an education process to be performed by the flight department manager. Human resources also must be challenged to conduct pay comparability surveys for flight department personnel or to participate in industry salary surveys.

Human resources and the flight department may never fully understand one another, but open communications and a good-faith effort by both groups will work wonders.

Career Development

Most flight departments devote a majority of their training and personnel development time and money to technical issues. Initial, upgrade, transition, supplementary, and recurrent training on specific equipment or other aviation issues is purchased to ensure that the department is current, competent, and safe in producing its primary product, on-demand air transportation for the parent organization. Many managers consider this training an investment in safety and efficiency for the department.

Beyond these basics, however, many pilots, technicians, dispatchers, and schedulers may have a greater potential value to the department and company in nonaviation pursuits. Early development of latent organizational, administrative, leadership, and management skills in people will pay dividends later in the individual's tenure with the larger organization. And most people respond well to new opportunities and challenges if they are delivered in a considered, nonthreatening manner. This is particularly important for flight departments where upward mobility is limited due to the small size and specialized nature of the work.

Most career development occurs by providing an employee with some degree of initial training and then letting the person put theory into practice on the job. With sufficient nurturing, coaching, and observation, most adequately motivated people will master a new task within a reasonable period of time. Motivation is a key concept in performance; if an employee cannot see the value of the new task, he or she seldom will perform well.

It takes some effort on the manager's part to link what needs to be done with who can do it. Sometimes the employee volunteers; sometimes a "volunteer" must be found. The discovery of a desire and talent in an employee belongs in part to the manager and his or her people-reading skills and in part to the employee himself or herself; the link is the art form. However, the process is usually a rewarding one for both the manager and the employee.

A series of different jobs that stretch the employee's capabilities in increasingly responsible positions should lead to a person who is able to lead and manage when the opportunity to fill new positions comes available. In large measure, this sequence of development is really self-development. Development should be encouraged in all employees. It may only be chosen by a few, but all should be afforded the opportunity. See "Collateral Duties" in Chap. 3.

Small departments have a problem in growing their personnel, but the successive assignment of collateral duties, increasing levels of responsibility along the way, has the same effect as a formal development program. Pilots or technicians who are given the opportunity to plan, budget, handle vendors, work with human resources, and assist in the training of others are well on their way to being ready for a more responsible job when it becomes available. The next job does not necessarily have to be within the flight department either. A small number of talented individuals "go to corporate" when the opportunity presents itself. Has the flight department lost a member, or has the company gained a manager who fully appreciates the value of corporate aviation?

Ideally, each manager always should be training a replacement. This should not be viewed as a threat but as a means to fully exploit the talents of every member of the department. Here are a few ways of going about it:

- Get to know each employee beyond his or her job description; all employees have hidden talents.
- Seek opportunities to delegate tasks to subordinates that will give them a chance to succeed.
- Provide management, administrative, and leadership training for the most promising employees.
- Let your boss know who you are developing and why—the boss can help.
- Only give an employee as much as he or she can handle, yet everyone needs to be challenged.
- If a department member does not handle an assigned task well, find out why and redirect his or her efforts and talents.

There are many ways of investing in employees; career development may present the best return on the investment. The NBAA Professional Development Program (PDP) is an excellent means of preparing employees for new fields of work and promoting personal growth. The NBAA Corporate Aviation Manager (CAM) qualification creates a worthy goal in developing a knowledgeable and effective employee.

Leadership

All the items shown in the sidebar below may be true and have stood the test of time, but leadership may be the single most important characteristic possessed by the head of the department. This is true even if the department consists of just one airplane and two people. The leader is the person with the foresight, vision, and responsibility to carry out that vision for the organization.

Notice that the terms *manager* and *leader* are not necessarily interchangeable. The "Leaders do the right things; managers do things right" statement tells the story. Leaders motivate people; managers channel the motivation. The military found out the hard way that the management training for its officers that it invested in during the 1960s and 1970s did not necessarily translate to the needs of the battlefield. Managers thrived in Washington; they did less well in team-oriented, goal-directed stressful situations that required a *leader*.

Every organization needs a leader, someone with the vision to see beyond the horizon and to get people to march toward that unseen goal. Once the team is started in the direction of the goal, managerial qualities are needed to make sure that everybody gets there with some degree of style and efficiency.

Leadership

- Leadership is the glue that binds an organization.
- Leaders do the right things; managers do things right.
- Leaders are those with the ability to get other people to do what they do not want to do and like it.
- Leaders maintain the organization's vision.

How does one learn to be a leader (few are born with the requisite qualities)? Fledgling leaders traditionally are tested in the crucible of increasing responsibility and team tasks. They are asked to accomplish things that involve *people.* Accomplishing goals through people is the mark of a leader.

Is leadership necessary in the flight department? Definitely! Try accomplishing goals without it.

Flight Department Performance

One accurate measurement is worth a thousand expert opinions.
—GRACE MURRAY HOPPER, REAR ADMIRAL, U.S. NAVY

A popular maxim taught in business schools states that if a process or product cannot be measured, it cannot be controlled or managed. Thus accurate measurement of business processes relies heavily on a series of yardsticks designed to provide management with information that can be used to control the business. Corporate flight does not escape this need.

Whenever chief pilots and aviation department managers congregate, the subject of operating costs is usually a staple of conversation. It seems that their reporting seniors want to know how well their flight department stacks up in relation to other departments. Compensation, maintenance, fuel, and overhaul costs are all subjects of interest. Everyone wants to know how well they are doing compared with their peers. However, these measures may not tell the real story of performance in the flight department. The real question should be, "Am I operating efficiently and effectively?"

Most departments dutifully provide monthly compilations of raw data to headquarters that describe their operations. Typical measures provided are shown in the sidebar.

Flight Department Monthly Statistics
- Flight hours
- Miles, trips, and legs flown
- Passengers carried
- Load factor
- Direct costs
- Cost per hour/mile/seat-mile

While these numbers provide a measure of activity within the department, they say little about how efficiently or effectively these feats have been accomplished; they are merely statistical measures of the flight department's response to demands levied on it. They do not describe how well the department used its resources.

Efficiency versus Effectiveness

Efficiency is usually measured as a ratio of output to input for a given activity. Efficiency asks the question, "Is a given action being performed with minimum effort/resources?" compared with the concept of effectiveness, which asks the question, "Is this product or service fulfilling the organization's mission adequately?" The two concepts are different but complementary. You may be providing aircraft services very inexpensively, but in so doing, the flights may be late or delayed—efficient but not effective. Conversely, the flights could be always on time but done so at great expense—effective but not efficient. Thus the two concepts must be considered together to provide a complete picture. Some judgment must be used in applying these values:

- How much value does the boss place on efficiency as opposed to effectiveness?
- What is the corporate culture regarding this issue?
- What means should be used to measure these issues?

Ratios

Most organizations examine efficiency and effectiveness in both financial and productivity terms. Ratio analysis is a well-developed and respected method of measuring performance. This methodology reports key factors as fractions or percentages of one another. For example, simply looking at profits after taxes is a single data point that gives no indication of how profitable the company is. However, if profits after taxes are divided by assets, a figure known as *return on investment* (ROI) is generated. If a company had $1 million in assets and made a $150,000 profit after taxes, it would have an ROI of 15 percent. This number either could be compared with some other period in the company's history or could be compared with an industry average. If last year's ROI was 10 percent but the industry average is 20 percent, things are improving but perhaps not enough. ROI is just one of many financial indicators used to determine corporate health.

Similarly, the production department compares hourly output of goods produced with the number of employees required to accomplish it to arrive at an *employee productivity ratio.* Service providers may determine how many customers an employee can serve per hour. Whatever the measure, the key is a comparison between two related variables.

The most rudimentary ratio-analysis tool used by flight departments is aircraft costs per hour, mile, or seat-mile. This figure is derived by taking total operating cost for a particular aircraft and dividing it by either total hours, miles, or seat-miles flown over the same period. This should be further broken down into direct and indirect operating costs to provide a more detailed appraisal of operating efficiency.

A number of other measures of either efficiency or effectiveness are available to flight department managers (see Table 4.2). Carefully consider which measures are relevant for your department and organizational culture. Certainly, they all do not have be tracked or portrayed to headquarters, but one or two key measures of *relative performance* should be used to show management that the flight department is concerned with efficiency and effectiveness. For example, if the maintenance hours per flight hour have increased by 30 percent over the previous 12-month period, there may be some hidden factor that should be investigated further. At the most elementary level, the aircraft could just be getting older, or the quality of replacement parts could be substandard. More telling may be that the personnel turnover in the maintenance section has been so high that the department has been left with only minimally experienced technicians who may require more time for troubleshooting and understanding the systems. Without these numbers and trend information, it will be difficult to justify a replacement aircraft or higher salaries for the technicians to improve their retention.

Tracking It

A means of tracking and recording the necessary data to generate these ratios must be devised. While it is possible to generate some of these numbers with paper records, a

---

Efficiency
- Cost per flight hour/mile/seat-mile
- Fuel used per mile/hour
- Maintenance cost per hour
- Maintenance personnel hours expended per flight hour
- Scheduled flight time versus actual flight time
- Budgeted versus actual expenditures
- Crew productivity (trips/hours per crew member)
- Passenger versus ferry/maintenance flights

Effectiveness
- On-time performance versus delays
- Dispatch rate (trips flown versus trips scheduled)
- Trip denials (trips accommodated versus schedule attempts)
- Load factor (passengers per leg/segment)
- Time saved for passengers—total/ratios
- Aircraft/crew availability

TABLE 4.2   Flight Department Performance Measures

calculator, and a stubby pencil, the easiest, and in some cases, the only way to accomplish this is via a computer. Some departments have generated relatively sophisticated tracking and recordkeeping programs in-house, but most companies use the report writing capability of off-the-shelf proprietary computer software designed specifically for the scheduling and operational data records of on-demand flight department operations. Choosing the proper software for your operation is an important choice; suffice it to say that the software capabilities should be evaluated carefully to ensure a good fit for your operation. Many of these programs have features that may not be necessary or serve your operation well. Some software may not be able to generate the data and ratios you want. Ask for an extensive demonstration and ask other flight departments who use the software for their appraisal before you buy.

Comparisons

Once the desired ratios are generated, they must be compared with other reference points to be meaningful. A common method used is to compare this information with that of a previous time period. Comparison with the previous month or quarter may not be realistic, given seasonal anomalies or aircraft availability. However, comparison with the same month or quarter in the previous year may be more realistic. For the long view, comparison of the current period ratio with a moving 12- or 24-month average of that ratio is probably most useful. This method smooths out the short-term perturbations associated with periodic data and presents the values with greater perspective.

The other means of comparison is by using values generated by other, similar flight operations. Accurate and up-to-date information of this nature is often difficult to obtain, however. Some of it is available from the *NBAA Compensation and Benchmarking Surveys,* aircraft manufacturers data, and consultants' proprietary information.

Presenting the Information

What do you do with the information once it has been generated? The results of these calculations should be shared with both department personnel and corporate management, although the same information may not be appropriate for both groups. Let management know the "big picture" items, such as cost per hour, load factor, cancellation rate, etc. There is some risk of providing them with too much information, i.e., more information than they can understand or appreciate. For example, does management really want to know how productive the flight crews are, i.e., how many trips/hours/days/days away from home the crews fly? Probably not until it comes time to reengineer, resize, or downsize. It may be helpful to track these numbers internally for just such an eventuality.

Again, absolute values without a means for comparison will have little meaning for the nonaviation manager viewing the reports. Such a manager must be led to understand the significance of the information through displayed trends and comparisons with other flight departments and aircraft types and even airline travel. Graphic displays are also impressive and helpful in understanding performance information. Some companies have developed sophisticated measures of executive time saved per trip, nights away from home averted, and additional conference and work time afforded by use of a private aircraft setting versus a less conducive airliner. Any efforts in this area may be useful in keeping utilization high and your job secure.

Information versus Data

Note that the term *information* has been used instead of the term *data*. This usage is intentional to denote the difference between a large number of unrelated facts (data) and facts that have been collated, analyzed, and presented in a fashion that will prove useful and are designed to *inform*. Merely providing line after line of trip detail and passenger manifests will do little to help the nonaviation manager assess performance. This detail may be provided to selected users to be used as a permanent record for their files. Accounting may need this level of detail for tax purposes, or the president may want to scan the previous month's flights to determine who is using the aircraft and to correlate these data with other pertinent business details. Whatever the reason, there should be an explicit, stated purpose for providing high levels of detail to recipients outside the flight department.

The members of the flight department should be privy to just about all measures of performance and levels of detail. The performance ratios and other measures will give them a means with which to measure their individual and team performance. This information, in turn, can be used to motivate them through goal setting. The judicious use of information always provides real value to you and your department.

Perhaps the best yardstick for measuring performance is to use comparisons with previous time periods. However, cost alone does not tell the full story. A set of standard performance measures should be devised for your operation based on the corporate culture and standard internal policies.

No one number can tell you how well the department is performing. It is the rational assessment of a combination of measures valued by your company that tells this story.

Reports

What does the boss do with this prized information? How is it used? Does it tell him or her how well the flight department is accomplishing its mission? Does it provide information on how well the department uses its resources or its efficiency?

Unfortunately, many organizations do not ask these questions. As a consequence, flight department reports become self-perpetuating monuments to poor information management. The usual answer I get when I ask why a monthly report is sent to headquarters is, "We've always done it."

Regarding the sidebar, which one does your boss want? The second and third bullets seem to be the ones bosses *should* wish to see. Alas, bullet number four seems to take the prize in most cases.

In reality, the boss should want to see how well the department is *performing*. On-time departures, dispatch rate, number of denied trips, load factor, and cost per mile are all performance measures that may warrant management's attention, including management *within* the department. Ideally, these performance measures are tied to goals and objectives set at the beginning of the year that reflect accomplishment of the department's mission.

Why a Report Is Needed

- Management information—quick data used to make decisions

- Trends—system values are tracked for significant change or straying outside preset limits

- Exception reporting—something unusual happened, good or bad

- Idle curiosity—What's happening?

- Prediction—What will happen? (really a forecast or plan)

The point is that merely flying hours, reaching destinations, and carrying passengers do not necessarily mark achievement of the department's mission, which should be to support the on-demand air transportation needs of the company (hopefully defined somewhere).

Suppose that the company really just wants to have an airplane available to go to random destinations on short notice and does not really care about *efficiency?* Obviously, the company cares about *effectiveness* (dispatch rate, on-time departures, completed missions) but not necessarily such items as hourly direct costs, crew productivity, or maintenance hours per flight hour. OK, then just tell the company *how effective you are* using the appropriate measures.

If management wants the traditional hours/passengers/gallons measures, at least give them data from previous measurement periods so that they can see a trend. This information ideally can be displayed in a graph or series of graphs to provide quick visual recognition of possible trends.

Trends can be powerful indicators when tied to relevant issues. Suppose that flight hours have been decreasing steadily over the past six months. What does this indicate? It simply may hide the fact that the aircraft has endured a series of major unscheduled maintenance episodes and that a number of frequent flyers have been transferred to another division—be careful, the data may belie true causes.

Ask the boss what he or she wants, what measures of performance make a difference. Better yet, ask the boss how he or she uses current reports; what *information* does he or she receive from them. If the boss is honest, you can then help define what he or she believes really makes a difference within the flight department. Then report it. Figure 4.9 presents a sample report.

Do not forget your own needs within the department. Specialized reports regarding crew duty days, expenses, and maintenance personnel hour reporting may be important within the department and be of little use to the remainder of the company. Remember, in order to manage the operation effectively, ensure that you have the information you need in a form

	April 2013	April 2012	Difference
Hours	45.5	37.0	+23%
12 months	552	441	+25%
Trips	19	15	+27%
12 months	260	190	+37%
Passengers	41	44	−7%
12 months	506	588	−14%
Load Factor	2.2	2.9	−24%
12 months	2.0	3.1	−35%
Dispatch rate	100%	94%	+6%
12 months	98%	96%	+2%
Demand met	91%	100%	−9%
12 months	94%	98%	−4%
Cost per hour (12 months average)	$1455	$1510	−4%

Highlights
- 300-hour inspection conducted over third weekend
- Windshield half replaced due to in-flight fracture @ $43,000
- Joe Weil completed recurrent pilot training
- Safety audit completed—report due end of month

Upcoming Events
- All hands CPR/BBP training in May
- APU changeout in October—two-day down time
- Technician recurrent training in June for Harry Smart
- Passengers' secretaries/admin. assistants familiarization in June

FIGURE 4.9 Sample flight department monthly report.

that is of use to you. And do not forget the rest of the department—let them know how the department is performing by sharing some or all the reports generated with them. Once they see what's good and not so good about the operation, they can help make it better.

Flight Department Evaluation

Never mistake motion for action.   —ERNEST HEMINGWAY

The priorities for most flight department managers usually come down to two big ones: safety and keeping the boss happy. Having achieved these on a daily basis (you hope), the same majority of operators turn to the more mundane, unglamorous tasks of ensuring that the fundamentals of the department work well enough to support its principal goals. It is these details that determine how well the department performs.

These details assume primary importance when a review of the flight department is conducted either internally or by a third-party contractor. Flight department self-audits tend to review policies and procedures that ensure compliance with national regulations, maintenance programs, and company directives. A more comprehensive view of the department—an outside audit—examines management, organizational, customer service, and efficiency issues. The end result of both types of investigations answers the question, "Is the department safe, and does it comply with applicable regulations and policies?" Ideally, an outside audit goes one step further and yields an answer to the question, "How well-managed and efficient is the operation?"

In answering these questions, it is important to know what issues will be investigated. If the audit is done internally, department personnel simply may generate a checklist that looks at just the regulations. Table 4.3 shows such a list for the U.S. Federal Aviation Regulations (FAR). Others expand this to include standard procedures, administrative or operational, as well. When a company's internal auditors get involved, administrative and financial issues are added to the checklist to prove compliance with company procedures.

Contract auditors who specialize in flight department operations probably will use a detailed list of operational, administrative, and procedural items that encompass the normal day-to-day operational environment. The level of detail will vary with the contractor selected, but the list of items to be investigated usually will be lengthy (see Table 4.4).

The question on most aviation department managers' minds is, "Given that an audit is scheduled, how do I prepare for it?" In reality, very little can be done because these investigations rely on examining existing policies, procedures, and the records of accomplishment to determine the department's fitness. The time to prepare for an audit is sometime before it is scheduled—perhaps years before—and not after the date is set.

The International Standard for Business Aircraft Operations (IS-BAO) developed by the International Business Aviation Council (IBAC) has rapidly become the industry standard for evaluations. The primary feature that ties the audit protocols together for this method is a well-crafted safety management system (SMS). More regarding IS-BAO can be found in Chap. 6. However, the basics for any flight department evaluation follow.

The Basics.   An audit will explore the following features of a flight department:

- Standards and procedures used
- Adherence to those standards and procedures
- Methods of determining compliance with the standards and procedures

Taken individually, these elements provide the foundations for a good flight department. When all are brought together under a corporation that truly believes in on-demand aviation and a department management that uses good leadership principles, excellence is very likely.

Standards.   Standards come in all shapes and sizes: regulations, aircraft manufacturer's continuing airworthiness instructions, company policies and procedures, and best industry practices. The regulations should be as well known as the contents of the manufacturer's maintenance manual. These basic rules are the foundations of any flight operation; these come first and are the principal determinants of compliance and safety.

Company policies include such diverse items as the care and feeding of passengers, crew duty times, takeoff weather minimums, recording aircraft discrepancies, and scheduling. To be communicated effectively to corporate management, passengers, and the flight

The Essentials*
Certification: Pilots and Flight Instructors
61.3 Requirement for certificates, ratings, and authorizations
.23 Duration of medical certificates
.51 Pilot logbooks
.55 Second-in-command qualifications
.57 Recent flight experience: pilot in command
.58 Pilot in command proficiency check
.63 Additional aircraft ratings
General Operating and Flight Rules
91.7 Civil aircraft airworthiness
.9 Civil aircraft flight manual, marking, and placard requirements
.103 Preflight action
.203 Civil aircraft: certifications required
.213 Inoperative instruments and equipment
.403 General (maintenance)
.405 Maintenance required
.407 Operation after maintenance, preventive maintenance, rebuilding, or alteration
.409 Inspections
.417 Maintenance records
.503 Flying equipment and operating information
.519 Passenger briefing
.703 Operations of civil aircraft of U.S. registry outside of the United States.
Maintenance, Preventive Maintenance, Rebuilding, and Alteration
43.3 Persons authorized to perform maintenance, preventive maintenance, rebuilding, and alterations.
.5 Approval for return to service after maintenance, . . .
.7 Persons authorized to perform maintenance, . . .
.9 Content, form, and disposition of maintenance...records
.11 Content, form, and disposition of records for inspections
.13 Performance rules (general)
*A number of other regulations may apply depending on the operation.

TABLE 4.3   Regulations with Which to Comply

department, all this information should be in writing and published, normally in the form of a flight operations manual. This manual should be comprehensive and serve as a central reference work for all concerned with corporate flight operations (see the *NBAA Management Guide*).

The final category of standards looks at best industry practices, i.e., policies and procedures shown to be successful when used by similar flight operations. While comparison with unwritten standards may seem unfair at first, most flight department managers want to know how they stack up against similar departments. Being introduced to new ideas and

Administration
  Facility appearance and condition
  Passenger handling/facilities
  Flight planning procedures
  Catering
  Certificates, manuals, op spec*
  Passenger scheduling
  Financial status*
  Crew duty time management
  Dispatch and flight following
  Emergency/incident procedures
  Insurance coverage/certificates*
  Facility/aircraft security

Flight operations
  PIC/SIC qualifications
  Pilot training program/records
  Cabin attendant training/records
  Flight operations manual
  Standard operating procedures
  Operational limitations
  Flight operations records
  Pilot proficiency monitoring
  Flight information publications
  Operational standards
  International operations

Flight evaluation
  Flight planning
  Aircraft release/inspection
  Passenger handling/briefing
  Use of checklists
  Compliance with FARs/SOP
  Flying technique
  Crew resource management
  Flight records

  *Applies to charter operators only.

Safety management
  Safety program contents
  Designated safety personnel
  Personnel qualifications
  Safety awareness
  Previous incident/accidents
  Passenger safety
  Industrial safety/OSHA compliance

Aircraft maintenance
  Manager/technician qualifications
  Technician training program/records
  Maintenance manual/procedures
  Scheduled maintenance control
  Inspection program approval
  Quality control program/records
  Aircraft discrepancy control
  Deferred maintenance procedures
  Aircraft maintenance records
  Computerized records tracking
  Controlled manual procedures
  Parts inventory, inspection, control
  Vendor selection procedures
  Fueling procedures

Maintenance processes
  Work flow/job assignment
  Turnover procedures/logs
  Compliance with FARs/SOP
  Use of reference documents
  Parts control
  Inspection process
  Records procedures/control
  Cleanliness/organization
  Tool control

TABLE 4.4 Flight Department Audit Checklist

concepts in this manner is one way to be exposed to other ways of running an operation and to compare one's operation with others considered to be "best in class."

Doing It. Having all the processes and procedures on paper is one thing; doing them is another. This phase of the audit looks at how well run the department is and how well members comply with their own rules and methods. Confidential interviews with all department personnel are normally employed to determine levels of compliance, management style, and morale; an experienced interviewer can soon tell whether what is on the books is actually happening within the department. Observation by the auditors of daily work operations and on jumpseat rides in the aircraft complete the process of determining whether a

Flight Department Records

- Flight schedules
- Completed flight logs and passenger manifests
- Flight crew duty time
- Flight crew currency
- Training received
- Aircraft maintenance records
- Personnel qualification and employment
- Financial

department is really "walking the talk" and not merely "talking the talk." Incidentally, jumpseat rides are normally an optional feature of audits and are designed to see whether crews follow their own procedures and not to assess how a crew flies the aircraft.

Recording It.   Audits are really about records. The only way an investigator can really determine whether department directives are being accomplished is through written records.

The records shown in the sidebar will be examined in detail for compliance with standards, inconsistencies, and obvious "pencil whipping" (fraudulent entries). If present, more subtle record manipulation will be revealed during interviews with department personnel—people love to tell virtually everything they know when given the opportunity.

Good, well-kept records support one another, forming an interlocking story of the department's activity. Poor records lack this interdependence, displaying gaps and inconsistencies that are difficult to explain. Most important is a bullet-proof system of recordkeeping procedures that uses parallel inputs to edit and substantiate other entries.

While records may make the department and its managers appear to be doing a great job for their companies, the real reason to maintain good records is to prove compliance

Who Asked for It?

An increasing number of flight departments integrate audits into their normal management process. That is, they conduct an audit of their operation on a periodic basis, normally every 2 or 3 years. This provides management with trend information to determine performance and ensures that management remains abreast of best industry practices through the use of a third-party auditor.

Perhaps the most frequent cause for initiating an audit is a situation such as adding or exchanging an aircraft, expanding the department, or an external influence causes the flight department manager or chief pilot to become curious as to how his or her operation compares with the rest of the pack. Inherent in such a move is a desire to improve the performance of the department.

Occasionally, the flight department corporate reporting senior or other senior corporate officer will request an audit. This may stem from a desire to see how the department compares with other corporate operations or to determine whether the department is safe and in compliance with the regulations. Less frequently, a safety-related event or internal dispute within the department will cause a corporate officer to seek an outside view of the operation.

with standards in case of an accident or incident. The FAA, the National Transportation Safety Board (NTSB), insurance companies, and lawyers will pour over operational and administrative records in minute detail to determine whether the operation has missed an important mandatory compliance item. Good records provide an excellent form of insurance in case of a mishap.

Putting It Together. As noted earlier, if all the basic elements of the flight department come together under enlightened managers, excellence can happen. This excellence will be readily apparent to the auditor/evaluator soon after arrival, too. Flight department personnel normally exhibit a high level of confidence and pride in their operation. However, if poor teamwork or morale problems exist, these features will become readily apparent to the evaluator during the personnel interview process. These intangibles are an important part of the audit for the investigator. While they may not be manifested by letter-perfect recordkeeping or high standards that have been set, they are an indication of the real performance of the department.

Thus, is it possible to prepare your department for an audit? Yes, but the preparations should have started years ago. However, this should not delay initiating an audit; valuable lessons can be learned because the report received will provide a means of charting a path to a better operation. More importantly, resolve to profit from the experience by getting all hands involved in working to improve the department once they have read and absorbed the report.

Saving Money

While many flight departments are not required to submit or live under a budget or otherwise attempt to economize on any aspect of their operation, doing so is good management. Even if the boss does not care, the flight department manager should care about saving a buck just because it is a subset of being efficient, doing more with less. Granted, those who do not have to may not squeeze a nickel until it screams, but there is virtue and a sense of personal accomplishment when you strike a good deal. Whether it is saving $50,000 on a 12-year inspection or 15 cents on a gallon of jet fuel, doing so makes one feel like something good has been achieved.

The boss may not ask for the most efficient operation, but efficiency and cost savings

Money-Saving Ways

Fuel:

- Build your own fuel farm
- Buy bulk fuel and store it at the FBO's facility
- Join all fuel savings/purchase programs
- Always call ahead to ask for a discount at FBOs
- Use cruise control and track optimization techniques

Maintenance:

- Use in-house technicians to perform most maintenance/inspections
- Initiate long-term agreements with maintenance vendors
- Join a maintenance/parts insurance program

Compensation:

- Conduct/subscribe to compensation surveys
- Provide bonuses/profit sharing to keep annual salary increases to a minimum
- Use contract personnel during times of high demand

are time-honored means of measuring success of any operation. Why not see how efficient you can be?

There are three major expense categories in flight departments that may yield the greatest savings: fuel, maintenance, and personnel compensation. Concentrating on these areas makes the most sense if significant savings are to be realized. See the sidebar on the previous page for ways other flight departments save money in these areas.

Marketing the Flight Department

It is the perception of value provided that will cement the position of the flight department within the company.

The real object of marketing is to sell something, a product or service. But first, you have to let the buyers know that the service or product exists, that it has potential value for them, and that it will make their life easier and more pleasant, or contribute to their success. More on marketing later; first, the real object: sales.

Regardless of your daily activity, you are probably attempting to sell something to someone. Be it introducing the concept of a new scheduling format to the scheduler, getting the Challenger to the maintenance chief 10 hours early for its inspection, or getting a pilot to go to training a week early, you, in fact, are a salesperson. If you're not, you may be either a dictator or a do-nothing—equally distasteful options.

Why Sell?

But why sell the company on the flight department? Aren't they already sold? Utilization is good, money is no problem, and the chairman is talking about a big cabin airplane—what's to worry?

Unfortunately, times change. Chairpersons, presidents, and CEOs come and go. More importantly, CFOs and other corporate officers who may have designs on your budget come and go. What you are doing today is really history in the making; it's what you can sell tomorrow that determines your ultimate job security.

And the quarterly financial results may have a surprisingly large amount of control over the flight department's destiny. One bad quarter, no problem. Two bad quarters, they are looking for quick cuts. Three bad quarters and the entire company becomes fiscally paranoid, waiting for the budget axe to fall.

Or, an upstart commuter airline begins service between your most popular city-pairs; the drop in demand can be dramatic in the wake of these entrepreneurial inroads.

Without a broad base of support within the company, the vagaries of the business world may make the corporate aviation function tenuous at best. Service must be considered in terms of the entire company, not just the few senior executives afforded the privilege of on-demand air transportation. It is the ability of the flight department to actively and significantly support the core mission and goals of the company that produces value. Without the concept of the flight department producing value planted in the minds of a majority of the corporate decision makers, the days of the employees who work at the airport may be numbered.

The aim of marketing is to know and understand the customer so well the product or service fits him and sells itself.
      —PETER DRUCKER

The Name of the Game

This brings us to the real reason for marketing the flight department's services within the company: the flight department is an integral part of the company and, therefore, should be contributing to return on investment/return on equity (ROI/ROE) as much as the manufacturing or other service functions of the company. While these returns may be difficult to prove in hard numbers, it is the perception of value provided that will cement the position of the flight department within the company.

Does this mean that the flight department manager must employ smoke-and-mirrors sales techniques to make the company believe it's getting a good deal from its aircraft? Absolutely not. Such deceptive tactics will most likely backfire; only solid, honest means of explaining to people how corporate aircraft can help them will carry the day.

Rather, the manager must fully know and understand the capabilities and limitations of the flight department and the needs of the company to the point where he/she becomes, in effect, a highly skilled translator of needs and desires. Understanding these two elements is key to successful flight department selling and marketing.

Marketing versus Selling

While the terms "marketing" and "selling" are sometimes used interchangeably, they are quite different. Marketing is the act of promoting a product or service, while selling involves the actual act of transfer of goods or services. Therefore, both must be accomplished if a business is to be successful.

Marketing is important, since few people in the company are aware of the existence of the flight department or know of its capabilities. The aircraft are normally used by a small fraction of company personnel, and knowledge of this activity is known only to a few others. Without marketing, no one besides the same few will be aware of the activity, and no further "sales" will occur. But first, a word about corporate politics.

Marketing the Flight Department

- Create multistop trip packages for sales and marketing that minimize their time away from home and save money too.

- Hold a field day at the airport for administrative assistants and secretaries of executives to acquaint them with the operation and scheduling procedures—throw in lunch, too.

- Work with the communications department to provide a story in the company newsletter about the flight department and its people.

- Apprise key departments about the rapid-response capabilities of the aircraft in supporting critical-failure scenarios.

- Let finance know about your cost-saving efforts in parts procurement, maintenance, training, and fuel purchases.

- Create and deliver a transportation efficiency and effectiveness presentation for junior executive training courses.

- Provide aerial observation flights for site selection and survey teams.

Political Realities

If the chairperson, president, and executive vice president use the single aircraft 450 hours per year, there is little chance that they will allow other employees to use it. Further, they may not want any employees tagging along on their flights, either. So, where does this leave your marketing options?

First, do some research:

- Has the aircraft ever been used to carry customers or prospective customers to visit your corporate sites? Is this possible?

- Is the aircraft used to transport board members to the board meetings? Is this possible?

- Are essential personnel transported to sites where crises affecting the company have occurred? Is this possible?

- Does it make sense for a marketing team to hit four cities in three days using the aircraft?

Senior corporate executives may never have thought of these options. Do some research and present your proposal in a succinct and persuasive manner. One of the manager's jobs is to create value for the company; providing options to senior management is part of that task.

Company Operations

How close are you to current information regarding company operations and future plans? Do you have a good idea where the company plans to concentrate its resources within the next 12 months? How well does your boss keep you informed about where the company is headed? If you can answer two out of these three questions positively, you are probably informed well enough to begin the marketing/selling effort. If not, get plugged into the high level corporate grapevine. In order to do so, you must build a network of the movers and shakers within the company. While knowing secretaries is good, knowing administrative assistants and junior executives well enough to discuss company directions is better.

Once you have a feel for what's going on and where the company is going, begin a marketing plan.

The Plan

Marketing plans are not difficult. In their ideal form they present a new way of doing things, a way that provides value for the personnel to whom you eventually attempt to sell your idea. The key is knowing what makes sense to the other person and presenting your idea in *their terms*—you must think like them if you are to capture their interest and the eventual sale.

If your research has been done correctly, you will find more than enough reasons for more people to use the aircraft. The art form is determining which ones want it enough to support you with the powers that be. So, you must work with the potential beneficiaries to build a credible marketing and sales program. Without their help, your job will be made more difficult, if not impossible.

Pick your corporate function and let them know what's possible:

- Marketing departments want to make sales pitches to customers (just like you).

- Manufacturing personnel want to tour their facilities.

- R & D people want to visit other research facilities.

- Quality control experts must inspect their suppliers.
- Real estate personnel must visit several possible sites.

Marketing is the art and science of persuasive communication.   — DAVE KERPEN

The Tools

Perhaps the best tool in this venture is a well-motivated marketeer/salesperson. If the desire to promote your product is there, it will get done. But, to help you demonstrate value, you must provide concrete examples of how your way of doing things is superior. Ask your corporate marketing staff—they know how. But first, know what your are selling.

NBAA's *Travel$ense* travel analysis software provides you with the ability to show actual value and competitive advantage of corporate aircraft over air charter or the airlines. This computer program steps you through the many variables involved in travel analysis and presents solutions in a graphic and easy to understand manner. This software may be the best sales tool available to you because it presents practical, realistic examples of transportation solutions in business terms. Without it, you will have to do a lot of research and calculator punching.

The Presentation

Once you have your research complete and have your corporate beneficiary on board, practice your sales pitch to someone below the decision-maker's level but also within the chain of command leading to that person. Once you have their blessing (and support), take your show to the decision-maker.

Presentations are all the same. They

- define a problem,
- discuss alternative solutions, and
- provide the best possible solution (yours).

Selling 101

Selling is a universal technique. That is, one technique isn't used for selling cars and another for encyclopedias. Certain techniques common to most great salespeople have been cataloged and made the subject of innumerable books and seminars. Here are a few:

- Know your product—what do you have to offer?
- Know your prospect—what do they do; what do they need?
- Be a good listener—you'll never know their needs if you are talking.
- Speak in terms of the prospect's success—you can solve their problems.
- Ask for the order—nothing happens until something is sold.
- Be persistent—Rome wasn't sold in a day.
- Practice, practice, practice—try your spouse—there may be the greatest challenge.

What to Sell

Are you really selling seat-miles? Or air transportation? Well, yes, but in doing so, it is like selling the shoe sole (practical) instead of the entire shoe (style and protection). In reality, you are selling

- Solutions
- Opportunities
- Access to new markets
- Efficiency
- Success
- A time machine
- Assistance
- *Possibilities...*

They always speak in terms of the person being sold, never in the presenter's. They present *potential value*, not savings or profit or glory. They create a picture of success; they create possibilities.

The Sale

After you have convinced the subject of the value of the product, you must also "ask for the order." Without their OK to proceed with your idea, nothing changes. Therefore, you must say, "I would like to schedule the marketing team's trip for the second week in February, on a not-to-interfere basis with priority executive trips." Without this request, nothing will happen.

Marketing and selling are a continuous process, a never-ending attempt to promote the value of your operation. And, this is a job not just for the flight department manager but for all members of the department. The potential benefits of the department can be communicated from the scheduler to the CEO's secretary, from the maintenance manager to the person who handles the flight department's accounts and from the chief pilot to the human resources supervisor.

Without an active marketing program for the flight department it may be in jeopardy of falling victim to the budget axe or the whim of a new chief executive. Times change; the flight department must be ready to go with the flow and adapt to survive and thrive.

Customer Relationship Management, the Other CRM

> *A CRM system seeks to know and understand the customer's needs, desires, and operations as completely as possible.*

Sounds like a new crew resource management training course, doesn't it?

Actually, this CRM is in some ways more important to the life and longevity of a flight department than good flight crew communication, coordination, and problem solving. The other CRM is sweeping the business world as one of the hottest techniques to obtain, understand, serve, and keep customers.

Customer Relationship Management has been used by consumer-centered organizations for years as both a strategy and a technique to learn more about customer needs and behaviors in order to develop stronger relationships with them.

The strategic part of the concept depends on a customer oriented business philosophy and culture to support effective marketing, sales, and service processes. The technique part is developing policies and procedures that bring the strategy into being.

In big business, CRM is built around an information gathering, analyzing, and dissemination system that captures every aspect of a customer's life that may be germane to their business. A CRM system seeks to know and understand the customer's needs, desires, and operations as completely as possible. Yet, while the multimillion-dollar information processing system is important to the effort, the core of CRM is a philosophy and, eventually, a way of life that makes the customer king.

Customer as King

While many businesses *say* the customer is the most important part of their operation, that without them they would be nothing, few really follow through and *live* the customer-first life. Unfortunately, this is the case in many flight departments.

All too often flight departments fall into the trap of becoming order-takers—answer the phone, schedule a trip, fly the trip, and enter the chargeback fee. There is a ready cadre of upper level executives who regularly use the aircraft, so the concept of attracting, serving, and pleasing *customers* is lost in the daily effort to make the flight schedule work. Most ominous in this process is the tendency to believe that, with just minimal effort on the part of the flight department, the passengers will keep coming back because the alternative, the airlines, is too awful for them to even contemplate. This is the *captive audience* fallacy that creeps into many flight departments.

Make no mistake, company travelers have alternatives: outsourcing to management companies, charter, fractional, and, yes, even the dreaded airlines may serve their needs if they perceive that they are not receiving high-quality, on-demand air transportation services. Therefore, as our community discovered, good *operational* CRM was essential up front, and *marketing* CRM is essential in back, too; without passengers we might still be flight instructors, freight dogs, or check haulers. Passengers are our life; let's treat them like they are.

> *If you work just for money, you'll never make it, but if you love what you're doing and you always put the customer first, success will be yours.*   —RAY KROC

Who Are Your Passengers?

CRM attempts to define customers in terms of needs, desires, preferences, habits, likes, dislikes, and future requests. Moreover, it attempts to create a story about the ongoing *relationship* between the company and the customer. It becomes a central repository for *every* interaction between company and customer. In doing so, it tries to make every contact point between the customer and company aware of the latest episode in the relationship. In other words, CRM systems chronicle and probe the continuing saga between company and customer. Done correctly it may even provide interesting reading.

But, who are they, these individuals who hold your future in their hands? Obviously, your frequent flyers meet this definition handily, as do occasional users. Perhaps as important are those *prospects* who may some day avail themselves of the company aircraft. While it is correct to lavish your attention on frequent flyers, you should also focus on the next generation of leaders within the company. By cultivating future executives, you have a better chance of assuring the future of the flight department.

Identification of the next generation of frequent flyers can be done only by the management of the flight department getting close to executives focused on corporate strategy and progress. You are looking for the movers and shakers, the rising stars within the company who will be tomorrow's vice presidents and regional managers. The best way to find these individuals is to observe them in action. This can be accomplished only by remaining connected to the planning and management processes within the company.

Regular visits to headquarters, ongoing contact with key players, participation in management teams and projects, and maintaining an active network at all company levels will keep you in the know. These techniques will cement the flight department's relationship with both frequent flyers and the next generation of flyers. Casual chats with your frequent flyers at the airport and in the aircraft may provide an entrée to the company's inner workings, but your trips to *their turf* will provide detailed information and show that you are interested in the company's future.

Passenger Preferences to Know About

- Cabin setup—lights, temperature, etc.
- Ground transportation and handling
- Cabin routine
- Communications
- Entertainment
- Food and drink
- Reading material
- Fears and apprehensions
- Privacy
- Security

Knowing Your Passengers

What do you need to know about passengers? Most flight departments know that the boss wants specific items like *The Wall Street Journal*, Pepsi, and honey-roasted peanuts, but surprisingly few know about more-subtle preferences regarding magazines to have available and his preference for French Roast coffee and Perrier. It's attention to details like these that makes the difference between merely good and truly exceptional service. Further, the boss hates to have Wemac air outlets blowing in his face, likes Beck's Lite with rice crackers, and wants to read *Sports Illustrated* on the way home. Does he have any anxieties about flying, such as operating near thunderstorms or moderate chop/turbulence? These are the obvious elements of attentive service for the chosen few, but what about the rest? How far down the food chain do you attempt to cater to the company's executives? Perhaps the right answer is: as far as necessary to ensure the reputation of the flight department. That means *all* of them.

The list of passenger preferences shown in the sidebar is just a start. The greater level of detail you can provide will lead to improved service.

The Database

Keeping preferences straight is easy if there are only four or five passengers who regularly use the aircraft. But when the number gets greater, keeping the various permutations and combinations of preferences straight becomes an increasingly large problem. Therefore, some system of recording, disseminating, and updating customer information is necessary. Fortunately, many types of scheduling software contain a passenger preference section that covers a wide variety of items you should know about. If you don't have this capability, contact databases, such as Microsoft Outlook®, will provide the information needed.

The art form in contact database management is to keep it up-to-date and to ensure that the right people know about the changes. This is why so many companies have emphasized CRM software as the basis for their customer-centric philosophy. Virtually every contact with a customer or potential customer is recorded in detail in the CRM database. Importantly, before, during, and after any customer communication, the database is used to ensure that current information is available to the person making the contact, be it a salesman, customer service representative, production manager, or the president herself. Demonstrating an in-depth knowledge of clients is impressive to the client; it shows that the seller really cares about them. Moreover, it helps to create a relationship with the customer, a feature valued by all.

Flight department personnel—pilots, schedulers, technicians, cleaners, i.e., everybody—must be a part of the CRM process. As stated earlier, an essential element is the care, feeding, and access to the database. Virtually every interaction with the passenger, administrative assistant, or other direct report should be recorded in the database. Then, flight department personnel should regularly access the database, normally triggered by the individual's appearance on the flight schedule, to learn what's new about the passenger. Then, each contact with the passenger is an opportunity to show a sensitivity to and knowledge of them as a valued part of the flight department's life.

Cultivating Passengers

The database should also be used to trigger interactions with the passenger, potential passengers, and their assistants. For instance, do these people understand what corporate aircraft bring them in terms of time saved and value produced? Simple tutorials, such as NBAA's *Travel$ense* software, may be used to demonstrate these features. In fact, this software will produce individualized reports that portray aggregates of these values, which are powerful marketing tools. Similarly, potential passengers can be shown the value of the company aircraft compared to alternative forms of transportation.

Your best customers should be interviewed at least annually regarding their opinions about the service they receive from the flight department and what their needs for future air transportation will be. This should be a face-to-face interview conducted by the department manager or chief pilot to demonstrate how important the passenger's patronage really is (skip the written questionnaire).

Cultivating Passengers

- Provide uniformly excellent service
- Seek new passengers
- Present value tutorials
- Annual interview to discuss flight department performance
- Contact with administrative assistants
- Newsletter
- Be a business aviation advocate
- Show an interest in helping them succeed
- Cater to the company's customers

Since the majority of contacts with the flight department are via the passenger's assistant, take care to cultivate these individuals, too. Invite them out to the hangar, singly or as a group, on a regular basis to familiarize them with your operation and to show them that their assistance is appreciated (throw in lunch during the visit!).

Relationships

The object of all this it to cultivate, promote, and maintain a high quality relationship with the customer, one in which they will feel a genuine sense of loyalty to your operation. Until the customer feels a sense of attachment to the flight department, any relationship will be tenuous at best. Relationships are born and bred through a series of satisfying customer experiences. That is, each and every interaction with the flight department should be a good one, one in which the passenger feels that he/she has received some *value*.

A sense of value is created each time a friendly, helpful, knowledgeable flight department employee comes into contact with a passenger. Value is created every time a passenger is made to feel welcome, important, and *valued*. We are talking emotions here, feelings within the passenger that make them want to repeat the experience.

You provide a valuable service to passengers that saves time, offers convenience, and provides a safe and secure operating environment. But, without the warmth of human interaction and uniformly good service, the experience is diminished.

The best customer service is if the customer doesn't need to call you, doesn't need to talk to you. It just works.   —JEFF BEZOS

Preparing to Provide Good Service

Most flight departments are well standardized through the policies and procedures specified in their operations manuals. Yet, few designate passenger service standards to ensure that *all* passengers receive the same level of service *every* time. While most flight department employees make an effort to provide good service to their passengers, some are better at it than others. Therefore, why not *standardize* the level of service provided to passengers? You do it for administrative and operational reasons, why not for the reason of the flight department's continued existence?

Start simply by stating things like each passenger

- will be warmly greeted by a flight department employee,
- will have their luggage taken to the aircraft,
- will be personally escorted to the aircraft,
- will be shown to a seat within the aircraft,
- ...(and much more).

Customer Focus

- Make all personnel aware on a continuing basis that the department is customer oriented.
- Conduct a written or in-person survey of all of your regular passengers to determine their desires and perceptions of the service provided by flight department. Include the passengers' secretaries and administrative assistants in the survey, too.
- Provide customer service training for key flight department personnel.
- Institute a standardized flight crew flight briefing and debriefing procedure for passengers.
- Provide incentives for flight department personnel to find ways to improve service.
- Discuss customer service at all departmental meetings.
- Evaluate department personnel on their level of customer service.
- Provide several briefings each year regarding the flight department's mission, operations, and scheduling procedures for passengers and their assistants.
- Encourage informal contact between flight department and corporate support personnel to find out what the company thinks about your service and to communicate your service orientation to them.
- Find ways to measure passenger satisfaction on a regular basis.

CRM is as important to flight departments as having high quality maintenance, good insurance coverage, and annual recurrent training in assigned aircraft. In fact, the ability of the department to survive and thrive depends upon it. Make it a priority.

The Small Flight Department

The week started out typically frantic: Test fly the aircraft before a "revenue" trip Monday afternoon; budget variance meeting on Tuesday; renegotiate the hangar lease; find a solution to the continuing scheduling screwups in the president's office; get a replacement body for the trusty copilot's vacation; lobby for a secretary; and get the landlord to work on the balky hangar doors and get rid of the rats. And it all has to be done by the one, the only, department manager, chief pilot, head bottle washer, namely, you. Thus the joys of the small flight department.

Approximately four-fifths of NBAA members have just one aircraft. This means that there are a lot of small operations that work with between one and three full-time individuals and contract out the majority of their services. These departments must be efficient, flexible, and creative just to survive; they must have their priorities in order and run a tight organization.

The key to this organization is its head. Whether the title is aviation department manager, chief pilot, or grand vizier, this person must have a Swiss army knife collection of management, people, technical, and organizational skills. This individual should be comfortable with independence and unilateral decision making. Moreover, he or she probably prefers this organization to a more structured and bureaucratic one; self-reliance is a prized characteristic.

The manager sits at the crossroads between the parent organization and the flight department. This is an important role because the expectations of both organizations may be unrealistic or biased. If the passengers think that their one-airplane airline should be available 24 hours a day, 7 days a week, it is up to the flight department manager to help set more realistic expectations regarding availability. Conversely, if the department thinks the pilots should be treated differently by the finance department for processing travel claims, then this impression should be tested by communicating with the person in charge of travel claims. More importantly, the flight department manager becomes the aviation expert and spokesperson for the parent organization and is charged with the role of conveying the proper expectations about the department to users.

A number of the topics covered in this section have been covered in a different manner in previous parts of this book. However, the unique structure of the small flight department often demands a slightly different treatment of some aspects of the job. Thus, this section will attempt to provide a different perspective on some of the more critical issues associated with operating a flight department.

Communications

Getting the word out (and back) correctly becomes a major role of the department manager, whether it is aircraft capabilities and limitations, the daily flight schedule, hangar rent, recurrent training, pilot hiring, or equipment upgrades. The skill with which the manager presents the case for aviation and negotiates for the items needed to run the department well in large part determines the degree of success the department enjoys. If the parent organization never hears from the department or hears from it only when it needs something, it will suffer. If the manager makes an effort to become an integral part of the larger organization, the chances of the department prospering increase markedly. Do not forget to cultivate

Personal Preference Drives Purchase

A dot-com company founder made it big, sold the company at the right time, and now "works" at helping promising inventors, philanthropy, and enjoying his family. Early on he realized that the airlines could not serve his diverse and changing needs. He bought a quarter share of a Hawker 800 aircraft under a fractional program and was pleased with the result.

Over time, he realized that the 200 hours per year available to him under the fractional contract was insufficient for his needs. Further, he wanted to see the same faces in the cockpit for every flight and to have greater control over all aspects of the operation. After 2 years with the fractional company, he bought his own Falcon 50, hired a flight crew and an aviation maintenance technician of his choosing, and is building a hangar for the aircraft. It seems to be working.

the communications habit among all department personnel; managers always need a communications backup and all the salespeople they can get.

Typically, the small department manager reports directly to a high-level executive within the company or the principal and takes marching orders only from that person. An administrative assistant or executive secretary usually takes care of day-to-day scheduling issues, leaving the flight department to operate unfettered by a range of conflicting demands made by a variety of potential users of the aircraft. If the flight department is fortunate, it will have either a full- or part-time secretary/scheduler/all-purpose person to help with the administrative workload.

Support Staff

The other usual player in a small department is typically a copilot or combination copilot/ mechanic. This person also has to have a broad-based bag of skills to complement that of his or her boss. And since this may be the only other person in the department, he or she is often given "everything else" by default; there is nobody else to give "it" to. But is there?

Successful small departments have developed a support staff within the parent organization to help them with their many tasks. Instead of remaining the remote and mysterious people at the airport who fly the boss around, the clever departments have recruited company personnel to do their work for them while they fly. These departments have done so by communicating well with the principal's office, human resources, finance, and administration departments to make them appreciate their unique role within the company. By asking questions, learning to appreciate the other's roles within the company, and establishing a friendly relationship with the important departments, the flight department has gained allies and a support staff to assist it with its functions.

The relationship of the flight department manager to the senior executives and principal's supporters may prove advantageous in securing support from the larger organization. These executives can open doors for the flight department that normally may take long hours of effort to open through normal channels. However, realize that potential support staff may resent the high-level attention provided to the mysterious folks at the airport. The solution to this is usually to make a special effort not to appear special or untouchable but to identify with the overall goals and objectives of the company. If your potential supporters see you as team members instead of prima donnas, your access to them is much easier.

However, even with a great support staff, there is still much for the flight department to accomplish. Maintenance providers, hangar landlords, fuel salespeople, aircraft cleaners, parts vendors, catering suppliers, recurrent training companies, and janitorial services all make demands on the flight department's scarce time resources. This is where good organizational and prioritization skills are essential. There is often too much to do for the given amount of time in which to accomplish it; something has to give. If the support staff cannot handle it, then it is up to the flight department to do it.

Again, the smart flight operators get the vendors and suppliers to help them. Instead of figuring out what the scheduled maintenance plan, costs, and alternatives will be for next year, ask potential maintenance vendors to do it for you as a part of the bid process. Get the caterer to analyze the previous year's consumption and provide a quote for a structured catering plan for the next year, including periodic deliveries and restocking catering supply cabinets. Arrange with the landlord or janitorial service to conduct a monthly inspection of the hangar to take care of minor maintenance items before they get to be major items. These and other suppliers of goods and services should be providing *full service* and not just face-value service; make them work for you and become involved in your operation.

An invaluable means of support may lie within that personal computer sitting on your desk. The flight department management software available today can tie scheduling, trip planning, budgeting, invoicing, maintenance, and personnel files together into one tidy package. Communications features available in this software tie you to the boss's office, accounting, and the flight planning/weather information provider, too. Sure, it takes some getting used to, but once you become accustomed to its idiosyncrasies, you may never go back to paper and the mail. Seriously, the computer revolution has taken the corporate aviation world by storm, proving particularly beneficial to the resource-strapped small department. The personal computer may become your most useful "staff member." And let the software vendor set up your desired program for you.

Networking

Since the experience and knowledge base for the small department is relatively small, a network of other flight departments is essential to finding the best deals and for assistance in time of need. If another flight department on the field can recommend a maintenance or parts vendor from its own experience, this is valuable information. Word-of-mouth recommendations made by someone who has a similar operation will save you much time and effort in searching for a vendor on your own. And if you need a part, a special tool, advice, a copilot, or merely someone to commiserate with, your trusted network probably will pay immediate and bountiful dividends. Building the network may not be easy, particularly if you operate a relatively rare aircraft or are located at a small, remote airport with few corporate operators. Working with the aircraft manufacturer's customer service organization or aviation association should soon yield suitable networkers. Operators at airports that are normal destinations for your department are candidates for your network, too. If you get to those facilities more than once every 2 weeks or so, they may be able to provide mutual services and aid.

Networks may be formalized into time-share operations to create a mutual assistance pact when aircraft availability or heavy trip requirements become a factor. The ability to call a nearby operator and have him or her cover or back up an important trip not only keeps the stress level under control, but it also makes your organization more valuable and credible to management within your company. Naturally, time-share agreements are not to be entered into lightly, and they do require the involvement of lawyers, accountants, and senior executives (your support staff). However, do explore the possibilities of these mutually advantageous arrangements.

Cyber networks are becoming popular among all flight departments, but particularly among small ones. Bulletin boards posted on business aviation websites provide a ready outlet for questions, opinions, and insights regarding everything to do with running the operation, including finding employees, vendors, foreign handlers, parts, and local hotels. The NBAA Airmail System is especially well used.

Think Backup

The one-airplane airline is out of business if its single aeronautical asset becomes unserviceable due to either scheduled or unscheduled maintenance. However, this need not lead to disastrous consequences. Scheduled maintenance should be scheduled to coincide with known slow travel periods for the normal users of the operation and well publicized in advance to preclude possible disappointments. Unscheduled maintenance is a bit more tricky but manageable. An arrangement with a local charter operator to provide short-notice backup for your flights is a good solution to the specter of an unserviceable aircraft. Operators on the field are specially helpful because of their proximity and potential for short response times. Interchange and time-share arrangements (previously mentioned) are ideal ways to obtain a backup for your operations. However, creating a mind-set among the regular passengers that an occasional trip may have to be delayed or canceled due to maintenance or weather is an essential part of the department manager's role, too.

Planning

One of the most important functions for the small operator should be that of planning for the future. Whether it be preparation for the heavy transportation requirements for the annual board retreat, equipment upgrades, augmenting the department with additional personnel, or planning for the purchase of the next aircraft, foresight and planning always should be in the mind of the manager. If not, events surely will overtake the operation, and the value of the department will be reduced and credibility lost. Credibility of the operation is probably the most important asset of the flight department; it must be protected. The manager must allocate a sufficient amount of time to look to the future, long and short term, to ensure the continued health of the department.

This planning must be linked to action through an individual who sets priorities properly and communicates well with all. The narrow range of assets available to the manager of a small flight department makes the ability to view the future in clear terms with priorities well formed absolutely essential. Obviously, the manager must not do this in a vacuum but must use all the assets available to him or her from the CEO down through the janitor to accomplish the mission.

The manager of the small flight department must be single-minded in dedication to the department's mission: Providing the best possible air transportation for the parent organization. If the manager really believes in this concept and is at least an adequate performer, the rest is relatively easy. Well, maybe not easy, but clear and well-defined. However, this is no different for a flight department manager of a department of any size, is it?

The greater thing in this world is not so much where we stand as in what direction we are going.
—OLIVER WENDELL HOLMES

CHAPTER 5

Flight Department Management

While most flight department personnel have a good feel for what it takes to fly and maintain aircraft, they often have little appreciation for how to integrate those operations with other elements that go into running a flight department. Scheduling, personnel, reports, budgets, planning, communications, and control all have a significant impact on flight department operations but may be foreign concepts to the people responsible for running the department. In many ways, aviation personnel lag behind their corporate counterparts in knowledge of these essential elements because while the aviation people were focusing on the narrow disciplines of flight and maintenance at the beginning of their careers, the folks downtown were wrestling with budgets, human resources, planning meetings, and the like. The business folks *grew* into their management positions, whereas the aviation people were *thrust* into the world of management at some point in their careers. In other words, it is catch-up ball for the people out at the hangar.

Many people in corporate aviation have been exposed to the world of management through college courses, company training, military service, and jobs taken to feed their aviation habit. However, unless they were given the opportunity to put the principles they learned into practice soon after receiving their training, they probably soon forgot them. There also seems to be a common misconception that sound business and management principles do not necessarily apply to the world of aviation. Aviation personnel tend to get caught up in the technical and romantic aspects of flying and think that "all the rest" will take care of itself. Perhaps it is a matter of priorities, but management often receives short shrift in the aviator's world.

In many cases, the main reason that many flight department personnel are not ready for management positions is that they have not been tasked with other than aviation duties on their way up through the ranks. Flying and maintenance tasks may be all they have been expected to perform during their tenure. They may never have been given the opportunity to confront budgets, training plans, long-range maintenance planning, performance reviews, or aircraft replacement justifications. Moreover, because they were never given the opportunity to *learn* through these essential functions, they are unprepared for the giant step to management. Remember this as a need for your subordinates once you attain the lofty perch of manager.

The best executive is the one who has sense enough to pick men to do what he wants done, and self-restraint enough to keep from meddling with them while they do it. —THEODORE ROOSEVELT

Management 101

Whether you have an MBA or are merely an ATP or senior aviation maintenance technician, it is a good idea occasionally to review the basic principles of management. By doing so, you will be providing yourself with a useful perspective for viewing your daily tasks in terms of a theoretical framework, one that may yield valuable solutions.

Manage

- To direct or control the use of.
- To direct or administer (a business, for example).
- To direct, supervise or carry on business or other affairs.

The Basics

The dictionary definitions of *manage* in the sidebar are really quite complete; if they were followed, all management functions would be accomplished. It is interesting to note the Latin root of *manage* is *manus,* meaning "hand." The derivation comes from the use of one's hand to direct and shape events and perhaps even to coerce. The implication is one of control—of things, resources, processes, and people. The underlying thought is that one must be "in control" if management is to occur.

Most management writers list the following functions of management:

- *Planning.*   What does the organization need to thrive?
- *Organizing.*   How do you order available resources to achieve the goals?
- *Resource allocation.*   People, money, and time
- *Leading.*   How do I get it all going in the same direction?
- *Controlling.*   Am I really making progress, and if not, how can I change?

However, this can be shortened to just:

- *Planning.*   Goal setting
- *Execution.*   Making it happen
- *Feedback.*   How are we doing? (See Fig. 5.1.)

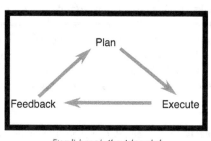

Don't break the triangle!

Figure 5.1   Essential management tasks.

Obviously, organizing, staffing, and leading must take place, but they really are subsets of execution. Regardless of how many elements are included in the management process, it is important to dwell on the point that it is not only a process but a *continuing process.* It is really a cycle of events that never ends. Projects, people, and other resources may come and go, but the process never ends.

Planning

Planning looks for a destination and a way to get there. If it is done correctly, planning involves looking for several ways of getting to a destination, for detours or new roads may be encountered on the way. This is the essence of strategic planning. Although planning was covered in Chap. 4, a slightly different view may be helpful.

Strategic planning looks at the long view, the big picture. It asks, "Where are we going as an organization?" and "What are the alternative means of achieving our mission?" The articulation of the ultimate organizational goal is known as the *mission statement*—"To make uncommonly good cookies and crackers" (Keebler Foods), "We rent cars" (Avis), or "We move money" (Brinks, Inc.). Organizational elements of the company should have a mission statement that supports the overall mission: "We provide safe and efficient on-demand air transportation to the XYZ Company."

The next level down in the process establishes goals or objectives needed to achieve the mission statement. For example:

- Achieve a minimum 99 percent dispatch rate for all passenger-carrying flights.
- Receive a least a 95 percent approval rating from passenger surveys.
- Acquire or build a new hangar by December 2005.

Note that these objectives are measurable but not necessarily concrete, real tasks. Note also that they do not state exactly how the objective will be reached.

Operational planning looks at the short term but still attempts to support the mission and objectives statements. For example:

- Replace the interior on the Gulfstream at the next 72-month inspection, scheduled for mid-2004.
- A 2-year training plan
- The monthly flight schedule

Note that some of these are long range and some short range, but they are all operational plans because they are concrete and measurable. In contrast with strategic plans, they address ways and means of achieving these shorter-range goals.

This planning must be done with full knowledge of what the parent company is doing about the future. If corporate headquarters is moving to another city within the next 2 years, it makes little sense to look for a new hangar at the current airport. Corporate plans and objectives are sometimes difficult to obtain, and it is frequently even more difficult to know what the latest plans are. Therefore, your network within the company is a vital source of information about what the latest plan is.

Execution

Once a plan is developed, the more practical aspect of making it a reality comes next. The elements of execution include

- *Problem definition.*   What are we really trying to accomplish?
- *Standards development.*   How do we know we are doing it right?
- *Resource allocation.*   Where do I place available resources?

The problem definition stage is, arguably, the most important aspect of project execution. Without complete knowledge of the task at hand, many surprises are possible. Take some time on this one. Some action items include the following:

- Develop a *complete* task statement.
  - Is it a complete interior refurbishment or just upholstery?
  - Is this a good time to do the APU STC?
- List available resources.
  - Do I have to send a technician to oversee the refurbishment of the aircraft while it is at the service center?
  - Can I use my contingency fund, too?
- Define the project schedule or time to process.
  - Can we do the landing gear trunnion replacement during the repaint?
  - Will this schedule provide enough contingency time to make the board meeting schedule?
- Develop alternatives.
  - If the aircraft is not finished on time, will the boss use charter?

Standards development seems simple enough on the surface, but what standards are in question? Time, quality, airworthiness, good operating practices, usage of people and money, or company policy? And why not standards you have developed for the department? If you do not designate standards and communicate them to your people, vendors, the company hierarchy, and *yourself*, it will be difficult to determine the quality of performance.

Examples:

- Materials used in the modification will be specified in writing prior to commencing the project.
- The project will be completed in four phases in accordance with the following schedule.
- Personnel completing the course will do so with a score of 5 or better.

Resource allocation simply means, "Am I using my available money, time, and staff to best advantage?" How you place these scarce commodities often makes the difference between success and failure. For instance, should you use your lead technician to babysit the Challenger during a major inspection at the service center or use him to train the new technician on the King Air? Or should you wait until after the European board meeting to send your best international captain to management school? Perhaps the best way to accomplish this well is to simply list the pros and cons side by side on a sheet of paper to make sure you consider all alternatives. And don't forget to include on the same sheet other projects that vie for the same scarce resources.

Feedback

Feedback and its natural partner, control, are the features that make the entire process work on a continuing basis. The ability to obtain timely, accurate, and germane input regarding a project or process provides a means to correct it if it is not going according to plan or schedule.

During the planning phase, feedback mechanisms should be developed to provide necessary information concerning the progress of the project. The most important aspect of feedback development is a means to measure progress, be it time, quality, quantity, or resources allocation. For instance, if no standards are set regarding dispatch rate for trips or costs per hour or passenger satisfaction, then performance cannot be measured. Similarly, if quality, time, and price standards for a TCAS installation are not defined for a vendor, success or failure can be questioned only with difficulty. Take the time to define standards so that feedback is possible.

Feedback must be timely and complete if it is to be of use in managing an ongoing process or project. And the time spent in developing the feedback should not be out of proportion to the payoff expected from the information received. Report systems design is a science in itself that will not be discussed here, but it is an integral part of the success of the feedback loop. (For more information, see Chap. 4.)

Once feedback is received, it must be evaluated and acted on if it is to be of value to the management process. If the 600-hour inspection is 2 days behind schedule or the training budget is overspent by 50 percent and it is only June, something must be done to get these or any other deficient projects back on track. This is when you should return to the planning and execution phases to see what alternatives are available. Can you hire additional part-time technicians to complete the inspection prior to the Aspen trip? Should you ask for additional training funds or delay the international operations training until next year?

It is this constant process of planning, execution, and feedback that spells the difference between a successful manager and one less so. It is attention to detail while not losing sight of the big picture that separates the good from the not so good. There is another dimension of the management process that cannot be overlooked—that of people doing the managing and work.

Management is tasks. Management is a discipline. But management is also people. Every achievement of management is the achievement of a manager. Every failure is a failure of a manager. People manage rather than "forces" or "facts." The vision, dedication and integrity of managers determine whether there is management or mismanagement.   —PETER DRUCKER

No Shortcuts

The greatest problem managers seem to have with the management process is shortcutting the system. Figure 5.1 uses a closed triangle to demonstrate the three-step process. However, there is a tendency to break the integrity of the triangular process once feedback has been received.

Let's say that a hangar is being built for the flight department. The foundation has been poured and structural steel support fittings anchored. A delay in receipt of certain steel components has opened the option of modifying the hangar structure to accommodate the shortage. If the manager makes a decision based on the need for expeditious completion, the integrity of the triangle has been violated. Any feedback should be considered first as a potential modification to the initial plan and not simply used summarily to modify the execution process. Doing so thwarts the original intent of the project and jeopardizes its ultimate utility.

The plan should be considered the driving force for any project or ongoing process. This is most evident when people disregard their organization's mission in favor of a short-term, visible goal that seems to make sense. For instance, if an operator opts to keep its 25-year-old aircraft despite its failing reliability and escalating maintenance costs, it may be ignoring the "reliable" and "efficient" segments of its mission statement. More important is the failure of a department to ensure that every action taken must consider safety as a primary consideration—this should be part of the primary plan for all flight departments. Never break the integrity of the triangle!

Theories of Management

Within the framework of the basic management process, there are a number of theories that both describe the means to manage and further define management itself. These theories are numerous and sometimes conflicting; therefore, one must be selective in the use of these various theories. The theories described here are possibly the most popular that have arisen in corporate America and certainly the most enduring.

The Beginnings

As a consequence of the industrial revolution, there arose a desire to make work processes more efficient and effective. Late in the nineteenth and early twentieth centuries, the scientific approach to work and management took shape. Frederick W. Taylor's *Principles of Scientific Management* is the seminal work of the scientific management movement, which featured time and motion studies to achieve efficiency. Henry L. Gantt focused on the need for realistic work standards based on study and measurement. His work schedules, now known as *Gantt charts*, were widely adopted. Finally, Henri Fayol, a French engineer, worked on organizational issues and was the first to define management by its constituent parts (planning, organizing, commanding, coordinating, and controlling) in his 1929 book, *General and Industrial Management.*

Behavioral Approaches

As it became evident in the post-World War I period that time and motion studies were limited in achieving increased productivity, management theorists began to focus on the worker rather than on the work itself. Elton Mayo found that worker productivity is related to social and psychological factors as well as the work itself. He called in 1933 for managers to consider human relations factors, and people-management skills began the human relations movement in management. Chester Barnard concentrated on the need for effective managerial communication and motivation skills and emphasized that managers' real power comes not from their position but from acceptance of the workers. And perhaps the most famous behaviorist, Douglas McGregor, developed the concepts of theory X and theory Y in which pessimistic and optimistic views, respectively, of the worker are set forth. His 1960 book, *The Human Side of Enterprise,* introduced the corresponding theories of authoritarian and participative management.

Motivation theory has been and continues to be a major part of the behavioral approach to management. The theory states that workers will not willingly perform that which they are not sufficiently driven to accomplish. Abraham Maslow held that individual unsatisfied needs are the main source of motivation. He stated that these were hierarchical in nature from the most basic to the most mature; e.g., all lower needs must be present for the higher ones to occur. The needs are

- Physiologic/survival
- Safety
- A sense of belonging
- Ego status—self-esteem, reward systems
- Self-actualization—working for the sake of work itself

Frederick Herzberg took this theory one step further, stating that the first three needs must be present before any motivation (embodied in the two highest needs) could take place. In fact, without the first three, the employee would be dissatisfied.

Modern Theories

Theory Z management is based largely on Japanese approaches to management, which feature long-term commitment, worker-management cooperation and discussion, and decision making that relies on group consensus. The popular terms *quality of worklife* and *quality circles* are components of theory Z. This spawned total quality management (TQM), one of the hottest theories of the 1990s. TQM tenets include total quality control, continuous improvement, and quality improvement teams. All these were fathered by the quality guru, W. Edwards Deming, the man who transformed Japanese industry of the 1950s into the force it is today.

The cult of worker excellence and the ability to manage chaotic change in the workplace were popularized by Tom Peters and Robert Waterman in the 1980s. They developed a formula for determining excellence within companies and applied it to a number of companies to prove their point. Their book, *In Search of Excellence*, became a runaway best seller in management circles in 1982. Their central thesis of excellent firms involved the need for constant improvement and change. Subsequently, Peters' book, *Thriving on Chaos*, discussed the need to move from hierarchical management structures to more horizontal, fast, cross-functional, cooperative ones.

To meet the demands of the fast-changing competitive scene, we must simply learn to love change as much as we hated it in the past. —TOM PETERS

Reengineering involves front-line workers performing complete tasks without departmental boundaries or supervisory checks dividing and slowing down the operations. While not a new theory (Toyota pioneered the basis of reengineering in the 1950s), its rediscovery has enabled corporations to take advantage of new technology in manufacturing processes by freeing workers from stifling organizational paradigms.

A single individual has lived, survived, examined, discarded, and added to virtually all these theories. Peter Drucker started as an economist and became a management consultant for General Motors during World War II. Since that time, he has devoted his life to the practice of management, writing scores of books on the subject, including my personal favorite,

Management: Tasks, Responsibilities, Practices (Harper & Row, 1974). Drucker brings an enlightening mix of disciplines to the task of management, seeking to find a balance between the needs of the worker and the company.

Future Theories

Theories of management keep academics and consultants active and in business. However, their postulates serve as a useful way to view the world of work and the means of accomplishing it in an orderly and efficient manner. Each theory provides us with a different perspective on how to gain the best productivity from the worker and ourselves.

Undoubtedly new management theories will emerge that will expand our understanding of the phenomenon. In time, however, these too will yield to still newer paradigms. The point is to gain as much as possible from each theory without being totally taken in by its tenets. We have a tendency to unequivocally embrace hot new theories to the exclusion of other proven means of getting the job done. There never has been a complete and perfect management theory; there probably never will be. Use the theories as they come as aids, not ends unto themselves.

Managers do things right. Leaders do the right thing.   —WARREN BENNIS

Learning It

Obviously, this brief look at the theory and practice of management will not be sufficient to make you an ace. The best way to learn about management is to be exposed to its principles and theory on a continuing basis and then practice, practice, practice.

Management Reading

- Carnegie, Dale, and Associates. *Managing Through People.* New York: Simon and Schuster, 1978.

- Drucker, Peter F. *Management: Tasks, Responsibilities, Practices.* New York: Harper & Row, 1974.

- McGregor, Douglas. *The Human Side of Enterprise.* New York: McGraw-Hill, 1960.

- Peters, T., and Waterman, R. H. *In Search of Excellence.* New York: Harper & Row, 1982.

- Blanchard, Kenneth, and Johnson, Spencer. *The One Minute Manager.* New York: William Morrow, 1982.

- Collins, Jim. *Good to Great: Why Some Companies Make the Leap…and Others Don't.* New York: Harper Business, 2001.

- *NBAA Management Guide.* Washington: NBAA, 2012.

Note that these are not new or trendy books—for a reason.

Local colleges normally have a variety of basic management courses available at night, and many companies conduct their own supervisor and management training in-house. Either of these is good to start with, but the company training may be preferable just because it gets you closer to company people. Knowing nonaviation company personnel will help you with your network of contacts and will teach you more about the corporate culture.

At the very least, read a book on the subject of management. The bookstores are full of them, some good and some not so good. You are probably better off reading the classic management gurus rather than studying the management technique of the month, which goes away almost as soon as it came; the classics age well.

Once you have taken Management 101 or even 102 and read your book, do not stop there. Continue to take courses and read about the practice of management. The pages of *Fortune, Forbes, Fast Company,* and *Business Week* are filled with tales of managerial rights, wrongs, dos and don'ts, usually told in an entertaining fashion.

Your organization probably has its share of good managers and role models. Think about it: At least a few of your manager's must be doing something right to keep the organization afloat. Each of these good performers is a potential mentor or coach. Find the right person, and establish a relationship. They do not have to know anything about aviation; in fact, that quality is probably an advantage.

You Can't Get There from Here

A large consumer plastics products company in the Midwest employed a number of boutique manufacturers to supply it with specialty products. In many cases, these suppliers were located in rural locations not near airline-served airports. The ones that were near airline airports often received only infrequent service from commuter carriers. Design, engineering, purchasing, and quality control personnel needed regular access to the suppliers, so 30 years ago a small twin-engine piston-powered Piper Navajo was purchased to meet the need.

As the company grew, so did its need to access an increasing number of suppliers, distribution centers, and customer outlets. The single aircraft has become a fleet of four turbojets, each operating more than 400 hours per year in support of company needs. The long-time CEO is enthusiastic in his endorsement of company aircraft, saying, "For many of the places we need to go, you can't get there from here, but the company airplane can!"

Management Skills

Theories are great if nothing has to be accomplished. However, the object of management is to do exactly that: *to accomplish objectives in a productive and efficient manner.* Managers must develop and exercise certain skills to apply the theories in much the same manner as a pilot or technician does in the pursuit of his or her trade. Pilots need to understand the flight environment and have good hand-eye coordination, spatial relationships, and motor skills. Technicians must understand mechanical devices, be good troubleshooters, and have good motor skills. All these attributes become skills when applied to their trade. The same is true for the manager. Some of the skills needed include the following:

Communication.   The ability to transmit one's ideas to others in a timely, clear, and concise manner is perhaps the most important skill for a manager. Moreover, if the manager cannot understand and appreciate what subordinates are saying, then true (two-way) communication has not taken place. It is said that at least half of communication should be spent listening and the other half thinking about what to say.

Written and oral forms of communication are equally important for management. The ability to express oneself clearly and concisely in writing is a particularly useful attribute. Memoranda, letters, performance appraisals, and budget justifications are all essential written communications that will recur with disturbing regularity; be prepared and practiced. The ability to say what you mean to individuals or groups is an essential mark of a leader and a manager.

Control and dissemination of information are forms of communication that also must be addressed. Everybody does not need to know everything you know, but some need to know more than others; knowing the difference is the art form.

Decision Making.   This skill allows the manager to choose between alternative courses of actions, plans, and materials. Without it, nothing happens. Without the proper techniques of choice, the wrong things happen. This is an important skill that involves data gathering, analysis, and judgment skills as well. Only time, practice, and perhaps a good mentor will help the decision-making process.

Motivation.   The ability of managers or, perhaps more appropriately, leaders to get people willingly to do their bidding is more art than science. Yet, without this skill, they will never get the most out of their people or accomplish the desired objective with speed and efficiency. Motivation involves understanding the human psyche and knowing what stimulates people into spontaneous action, i.e., generating action without manipulation. Group motivation is somewhat different from individual motivation; both should be mastered.

Time Management.   This skill involves setting priorities and determining the proper allocation of personal resources in achieving the object of the priority. Self-discipline and the ability to see beyond a specific component of a task are also important. Work planning and delegation come into play for time management.

Conflict Resolution.   Resolution of differences with superiors, peers, and subordinates is another art form that spells success for managers. The win-win approach to conflict resolution is always best, when possible. Communication skills are an important subset of this skill.

Group Dynamics.   Small work groups, peers in meetings, and larger groups are all multiples that must be dealt with. Again, communication and motivation skills are subsets of group management that must be mastered. The interplay of people and human nature in the workplace reveals an infinite number of permutations and combinations. Understanding the nature of the individuals within the group is step one; understanding their individual motivations and ability to cope with change are important subsets of this effort.

Acquiring Skills

The appropriate skills can be encountered in the classroom, but they must be practiced to become a part of a manager's repertoire. A firm understanding of management theory is a useful prerequisite to the actual skills-acquisition process. Assignment of responsibility and authority to accomplish certain tasks are the usual means of gaining the requisite management skills. And the skills do not come overnight. Years, not months, of concentrated effort are often needed to master the rudiments of these essential skills.

The National Business Aviation Association (NBAA) has developed the Professional Development Program (PDP) to help flight department personnel learn and appreciate the intricacies of management practice. The organization acknowledged the excellent job being done in flight departments to remain technically and operationally proficient but identified a need to strengthen knowledge and skills in management and leadership areas. Concentration on these areas was deemed necessary if sufficiently talented individuals were to be available to manage and lead the flight departments of the future. The NBAA board encouraged industry and an academic task force to develop a curriculum to be available through a variety of delivery methods and targeted to the needs of current and future aviation managers.

A survey of NBAA members indicated that management education is an unfulfilled need in the corporate aviation community. Survey results also were used to focus on four broad subject areas and to determine the overview, goals, and topics (performance objectives) under each of the areas. The NBAA-recommended skills represent essential areas of knowledge for aviation professionals. They were developed through consensus among educators and aviation managers. Here's what they came up with:

- *Business Management*—Management issues related to financial matters of the flight department, as well as contracting, confidentiality, and community relations.

- *Human Resources*—Management issues related to staffing, training, performance, compensation, human factors considerations, and compliance with human resource requirements.

- *Leadership*—Management issues related to vision, strategic planning, team building, decision making, communication, and currency with professional knowledge.

- *Operations*—Management issues related to the operation of aircraft to ensure safe and efficient service.

- *Technical and Facilities Services*—Management issues related to aircraft, facilities, and ground support equipment.

These issues are expanded in Table 5.1 to provide greater detail.

A number of training organizations and academic organizations have been authorized by NBAA to offer PDP courses. For information on PDP and the following program, see the NBAA website.

To capitalize on a systemized study of the above subject areas, the NBAA has developed the Corporate Aviation Manager (CAM) program. This program identifies qualified professionals to lead flight departments in companies that use business aircraft. Through a testing and certification program individuals prove their level of expertise and commitment to the aviation industry, enabling then to achieve individual professional and organizational goals. The certification program concentrates on the PDP areas shown above.

NBAA and the business aviation community acknowledge CAMs as professionals who have demonstrated an exemplary level of industry knowledge and management skills. It is an excellent career-building tool.

Business Management
• Communicate the value of aviation resources by interacting with key people about aviation department capabilities that will help the company achieve its goals.
• Implement the budgeting process through analysis of revenues, expenses, and forecasted needs to prepare the budget, monitor for deviations and report the results.
• Manage aviation department processes in line with strategic and tactical goals, consistent with data gathered within and outside the company.
• Manage the aviation department finances in accordance with generally accepted accounting principles to achieve department goals.
• Manage risk in the aviation department through appropriate protection of assets.
• Identify, recruit, and coordinate a team of qualified individuals within the company with the highest level of expertise, which will contribute to achieving department goals.
• Implement procedures using company policies to safeguard information and physical assets of the company.
• Lead the aviation department in a manner that is responsive to and involved in community concerns with the goal of maintaining good relationships and increased understanding of environmental issues.
Human Resources
• Determine the level of staffing needed for the aviation department.
• Develop knowledge and understanding of job-related training for all personnel within the aviation department to ensure competence in each prescribed discipline.
• Promote personal and professional growth through training and education by providing opportunities, financial support and scheduling flexibility for career development.
• Prepare to fill key positions by identifying internal or external candidates and providing the necessary training and growth.
• Maximize employee performance by aligning individual and company goals.
• Evaluate compensation levels in the aviation department using accepted tools to attract and retain employees.
• Develop the ability to respond to employee performance by rewarding or disciplining as appropriate.
• Ensure compliance with human resource regulatory requirements and company policies.
• Recognize and understand how human performance is influenced by human factors in all work environments.
Leadership
• Develop and/or maintain the aviation department vision, mission, and values in accordance with the company business plan to ensure understanding and commitment among stakeholders.
• Develop goals and strategies consistent with the aviation department vision, mission, and values. Analyze current status and trends relative to future needs to align the aviation department with the company's business objectives.
• Exercise leadership by being a role model, empowering personnel, building effective teams, promoting ethical behavior, encouraging the exchange of information, and making sound decisions to achieve aviation department goals and company objectives.
• Promote effective verbal and nonverbal communication strategies to engage personnel to enhance performance and understanding at appropriate levels of the aviation department and the company.
• Enhance professional knowledge using industry resources for personal effectiveness as an aviation department manager.
• Continually evaluate the effects human factors have on the efficiency, safety, and vulnerability of the aviation department.
• Understand, develop, and implement a safety management system (SMS).

TABLE 5.1 NBAA PDP Performance Objectives

Operations
• Develop and maintain a system of standard operating procedures (SOPs) for the aviation department using manufacturer's specifications, regulations, and accepted industry practices.
• Ensure that essential scheduling and dispatch procedures are in place to conduct safe and efficient flight.
• Develop the knowledge and understanding to establish a record-keeping system to document regulatory compliance and initiate appropriate action.
• Understand the safety component of the International Standard for Business Aircraft Operations (IS-BAO). Plan and develop and implement aviation emergency plans with consideration for organizational and community procedures.

Technical and Facilities Services
• Implement a system of standard operating procedures for ground operations in accordance with accepted industry standards and practices to ensure safety for passengers and employees.
• Maintain aircraft and installed components in accordance with manufacturer's specifications and pertinent regulations to provide safe, secure and efficient transportation of passengers and products.
• Maintain emergency equipment in accordance with manufacturer's specifications and pertinent regulations to ensure reliability, effective service, and responsiveness in an emergency.
• Maintain cabin information systems and passenger service items in accordance with manufacturer's specifications and pertinent regulations to ensure reliability, comfort, and effective service.
• Maintain aircraft spares, supplies, and other inventories by following appropriate regulations and industry practices required to provide safe, available, reliable, and efficient service.
• Apply security procedures in accordance with regulations, airport requirements, and company policies to provide a secure environment for passengers, employees, and assets.
• Contract with qualified vendors and service providers, using accepted business practices to procure needed services, equipment, and supplies for the department.
• Standardize operating procedures by requiring uniform practices and procedures for quality maintenance service.

TABLE 5.1   NBAA PDP Performance Objectives (*Continued*)

The Flight Department Manager as a Business Executive

Nobody in business aviation ever said, "When I grow up, I want to be a business executive." Executives do dull, boring things that fathers do all day: shuffle papers, talk on the telephone, and go to meetings—yuck! "When I grow up, I want to be a pilot and fly all over the world" or "I want to tear airplanes apart and put them back together."

And sure enough, this is how most of us got to where we are today, with a simple wish and the single-minded determination to achieve the goal of flying or maintaining airplanes. We chose the profession; it did not choose us. However, a select few of us not only did well enough at the aviation job, but we also were chosen for a position of leadership, authority, and responsibility and were given a position in management. And if we stuck with it long enough, a very few rose to the heights of flight department manager. In many companies, this means a director-level position—a business executive. From Cessna 152s to the big time in 20 years or less; this is no mean achievement. However, what does it really mean to be an executive?

The aforementioned leadership, authority, and responsibility are not bad descriptors for the occupation (see sidebar). It is usually a step above middle management and a step below corporate officer. It means increased prestige and perquisites and a bigger salary. And yes, executives do talk on the telephone, shuffle papers, and go to meetings. It means preparing budgets, planning for next year, hiring, firing, coordinating, defending the aviation function, and being called back from vacation when something goes wrong. Most of all, it means entering a foreign territory, learning the new language of corporate America, and crossing swords with MBAs and accountants.

The new flight department manager is thrust into a strange new arena, one that has little to do with airplanes, flight schedules, and making a slot time at National Airport. Not only is the entire weight of operational matters placed squarely on the department manager's shoulders, but administrative, financial, human resources, public affairs, and marketing are also heaped on for good measure. Yet these are just the details of the job. In a greater sense, the new executive must acquire or develop a larger view of the department's job. Moreover, he or she must gain an appreciation for the goals and objectives of the company itself.

No longer can such a person be interested in just making airplanes fly. His or her perspective becomes one of transportation, support, and helping to meet company objectives. Not only must he or she begin to talk to his or her fellow executives in these terms, he or she must *think* in these terms as well. We are talking transformation, metamorphosis, big change. However, just how does a person trained to fly or fix airplanes adapt and adjust to this strange new world?

If the system worked the way it was intended, the fledgling executive would have been trained to assume the mantle of executive ever since he or she started with the corporate aviation world. Management training courses, night school, collateral duties, increasing amounts of responsibility, and assisting the flight department manager should have all happened to provide corporate management with an opportunity to observe the candidate manager under fire and to have provided the candidate with the necessary tools and experience to prepare himself or herself for the role. While this sounds good and very textbookish, reality is seldom so. The new manager is usually thrust into the new position suddenly, somewhat unexpectedly, with little notion of the real substance of the job.

If senior management has chosen well, the fresh-caught manager will have adequate management skills to survive a transition period and, more important, will have the potential to grow into executive material. If management has chosen badly, the department probably has lost its best line captain or head technician and gained a mediocre manager. Even with the proper background, there is always a chance that the new person will not make it; there are a lot of unknowns in any personnel move.

Flight Department Manager's Ideal Traits

- Curiosity
- Vision
- Commitment
- Innovation
- Marketing
- Appreciation of people power

The new manager can increase his or her chances of success as an executive by developing and nurturing a few simple personality traits. Those listed (and treated more fully) in Chap. 3 and the sidebar have stood the test of time. The best way to come up to speed in these traits is through a coach or mentor within the corporate hierarchy. Cultivate these individuals early in your corporate tenure. They can be your key to success.

There is a separate classification of flight department manager that has been gaining popularity in recent years, that of the nonflying or non-aviation-experienced manager.

This type of manager is not flooding the ranks of department managers but is becoming a significant force within the community. Presumably, these individuals are chosen for their management and executive characteristics, so they have a head start in these areas prior to showing up at the hangar. However, just as the experienced pilot or maintenance technician must learn the management and executive ropes, the nonaviation type must learn the complex and arcane world of the airplane. Moreover, he or she must learn the unique culture and personal characteristics of aviation personnel.

Apart from the possible up-front resentment of the pretenders to the throne of department manager, the nonaviator must understand and appreciate the characteristics and limitations of aircraft and the standard procedures that attend a sophisticated flight department. The rules, regulations, and procedures that reside in a flight department can be daunting to the novice. And department personnel may not be terribly honest or forthcoming regarding the facts when the new person asks. One such new kid on the block likened his first few months on the job to an extended snipe hunt; reality and fantasy were difficult to separate.

Strangely, the naiveté of the nonaviator presents him or her with a great opportunity to test the mettle of his or her new charges by having them explain to him or her the significance and importance of all aspects of the aviation world. Sure, he or she will receive some bad information, but in so doing, he or she will soon find out who the bad and good guys are. The lack of bias and preconception the new manager brings to the job may just make up for his or her lack of aeronautical knowledge. A fresh viewpoint and outlook may be what the department needs to renew and improve itself. (Come to think of it, an experienced new aviation manager may want to take the clean-slate, challenge-existing-assumptions approach, too.)

The nonaviator must immediately rely on his or her lieutenants, the chief pilot and director of maintenance, if he or she is to survive. And this reliance should continue long after he or she has been in the job. By empowering these senior aviation personnel, managers free themselves to concentrate on the larger issues required of the executive. And because the aviation personnel are freed of the larger nontechnical tasks, they can concentrate on their areas of expertise. And this brings us back to the experienced aviator who has been designated the flight department manager.

This individual is at a disadvantage because he or she has to split himself or herself between the demands of remaining current in an aircraft or staying abreast of maintenance issues and learning the larger view of management within a large company. At best, the manager will be unable to keep his or her position as a full-time pilot on the flight schedule or to be involved in all maintenance planning decisions. At worst, he or she may have to give up these familiar pleasures entirely in favor of the executive life. The demands of running a complex, costly aviation operation may not allow much more than token participation in the aeronautical arena, and this may prove disconcerting to the individual or even unsafe because of a lack of proficiency. There are lots of flying and maintaining department managers in corporate aviation today, but each one of those individuals has had to make a conscious decision regarding the proper mix of aviation and management he or she chooses to undertake.

There will be times when the flight department manager longs for the simple life of 5-day road trips, night circling approaches, and a phase check that is days behind schedule. However, many have made the transition to the life of making an executive transportation system work in support of a growing corporation. The aeronautical and the administrative tasks do not necessarily have to be mutually exclusive in the manager's life, but for the company and department to benefit, the *executive* must choose the job with the vision and larger purpose. Certainly this is something we would not have picked when we started down this road, but it is something we can grow into.

An empowered organization is one in which individuals have the knowledge, skill, desire, and opportunity to personally succeed in a way that leads to collective organizational success.
—STEPHEN R. COVEY

What the Boss Should Know

The boss deserves and needs to know what makes the flight department tick, why it does what it does and how well it is performing.

"Just keep on doing what you're doing." This is the answer offered by many corporate flight department reporting seniors when asked what they want to know about flight department operations. The question is often asked by the flight department manager/chief pilot in an effort to keep the boss informed about various aspects of the company aircraft operations. Unfortunately, this hands-off approach from the company may do more harm than good.

The truth is that the reporting senior has a day job as the company CEO, CFO, VP of human resources or security manager and has lots on his or her plate. The flight department has probably been presented as an additional "gift" to overload an otherwise jam-packed existence. The saving grace of the flight department is that its manager usually does a good enough job to deserve a large amount of autonomy and independence. Thus, if the flight department stays within budget, keeps the top executives happy, and doesn't generate any incidents, the reporting senior is more than willing to permit the aviation operation to operate under a largely laissez-faire form of oversight and management.

Black Art at the Airport

Let's face it, aviation is a complex and highly detailed operation. Compared to the business world, it has an entirely different language, set of regulations and highly developed skills; the nature of aviation is closer to nuclear power generation than it is to general business practice. Strangely, many company reporting seniors choose to ignore the details of their flight department and rely on the "let me know when something goes wrong" theory of management.

This theory overlooks a fundamental feature of the flight department: it is one of several essential support functions within the company that need to be managed. Human resources, accounting, risk management, and information technology are all integral parts of the company and are carefully managed to derive optimum performance; why not the flight department, too?

I can hear the groans from flight department managers now: "Leave us alone! We're doing just fine." Maybe yes, maybe no. How do you know unless the flight department operates to a set of standards and has its performance measured against those standards? Accountability keeps us honest and creates high performance levels.

Flight departments often suffer from their company's benign neglect, which inevitably leads to poor communication, missed signals, unhealthy surprises and, ultimately, lack of respect for the department. The flight department should be considered an integral part of the business enterprise, subject to normal performance and management controls.

Two-Part Challenge

The reporting senior needs to have knowledge of the flight department at two levels: big picture and ongoing performance. The big picture provides essential aviation background elements, i.e., what it takes to run the department successfully, legal implications for the company, and the effect of department standards and limits on the operation. The ongoing picture provides the boss with a periodic *howgozit?* that speaks to operating statistics, performance measures, and a view of things to come.

Both views of the department are needed, first to bring the boss up to speed about the realities of corporate aviation and then to offer a view of how well the flight department continues to serve the company. Each view is necessary both to build a foundation of knowledge about our world for a person who knows little of it and to apprise the reporting senior of the challenges, opportunities, and successes of the department.

Basic Education

Everything the flight department does supports its mission of providing safe, reliable, and efficient on-demand air transportation to the company and its customers—might as well set the scene using your prime directive. Describe the mission by dwelling on the elements of safety, reliability, and efficiency, with heavy emphasis on safety.

The department operates to higher standards than those required by the FAA in order to make the mission possible. Discuss the essential higher standards listed in the flight operations manual in terms of why they make the operation safer. An important subset of this is that these standards help protect the company from excessive liability exposure. (See the sidebar Points for the Boss' Safety Education.)

These higher standards may restrict the department's ability to conduct certain operations. "We can't go into the 3500-foot airport near the chairman's resort home, but we can meet him at a more suitable airport and drive him there in 40 minutes to ensure his safety" is an example of this. Point out the potential

Background for the Boss

- Flight depart mission
- Higher operating standards for safety
- Standards' effect on operations
- Resources required
- Keep each other informed
- People information
- Good communications essential

Points for the Boss' Safety Education

- Corporate aviation safety record (*NBAA Fact Book*)
- Airport limitations
- Range/payload trade-off
- Takeoff/landing weather minimums
- Fuel reserves
- Duty/flight time limitations
- Required maintenance intervals
- Severe weather restrictions
- Pilot/technician qualifications/ currency requirements
- Minimum equipment requirements
- Authority of the PIC
- Initial and recurrent training requirements
- Selecting high quality vendors for maintenance and training

negatives that go with the limitations, but provide an acceptable alternative as well. Severe weather, lack of approaches and airport facilities, and duty time are prime candidates for this discussion.

All of this requires resources. Number and quality of people, equipment, training, parts, contract maintenance and backup air transportation are all essential topics in this discussion. An integral part of this is that the department will attempt to choose the services and products offering the best value to the company, not necessarily the least expensive. The rationale for this always comes back to safety and reliability. A word of caution: don't overplay the safety card too often; you don't want to be found out as the "boy who cried wolf." A budget helps create expectations that will reduce surprises.

The flight department provides services to the company's highest-level executives and customers; the department manager should keep the reporting senior informed about its successes and failures. Nobody likes unpleasant surprises, least of all your boss when informed about some failing of the flight department. These tales should come from you, not from others, and not at the end of the month; the sooner the better. Similarly, let the boss know when something good happens, too. Importantly, he/she should be willing to reciprocate and share the good and not so good with you.

The flight department's people are its most important asset; you should keep the boss informed about the good and the bad here, too. 'Nuff said.

You should meet regularly with the boss to keep the lines of communication open. Monthly meetings are ideal, preferably brief, and should always have an agenda. These meetings show that you care and provide a platform for you to tell the good news about the department. These huddles also offer a valuable information source for you regarding the internal working of the company and its future direction. Additionally, a regular report should be provided to keep the boss informed; again, brief is better.

Ongoing Information for the Boss

- Operational statistics
- Performance measures
- Good news
- Future events

While all of this information and philosophy is designed for background information, it all should be revisited regularly. Just because you said it once doesn't mean that the concept is fully understood or imbedded for life. Because the boss has other concerns and is an amateur observer of the aviation scene means that repetition and reinforcement are essential.

Regular Updates

How well is the flight department performing? If that question were asked about the company's core business product, performance measures, such as widgets per hour, return on investment, or percent of customers served within a period of time, would be offered. Each of these measures speaks to the concept of *efficiency*, the ratio of input to output; this is the language of performance in business. Sales volume, customers served, and miles traveled may also be mentioned but these are merely measures of demand and a function of the marketing and sales efforts. Most businesses look at both: effectiveness and efficiency. But, the real measure of success devolves to efficiency.

Most reporting seniors receive monthly reports detailing how many flights were flown, passengers carried, and gallons burned. Few, if any request measures of efficiency: cost per flight hour, trips completed vs. requested, dispatch reliability, budget variance, and aircraft availability. Yet, these same managers hold their other direct reports to strict performance goals in their area of expertise; why?

Many companies consider their corporate aircraft sunk assets that were acquired to facilitate travel for top-level executives. If the aircraft are fulfilling that mission to the subjective satisfaction of the frequent flyers, then all is well. The fault in this philosophy is that the aircraft could be performing additional services for the company, flying more people, facilitating more missions, and expanding company business, i.e., the aircraft assets are not being maximized to the point of peak efficiency. This is a hard sell for the flight department manager and reporting senior when the top five executives are personally pleased with flight department performance for their needs. Advocacy for corporate air transportation is an essential job for the flight department manager, which should be actively supported by the reporting senior; knowing flight department capacity and performance will help both do this job.

Basic monthly flight department statistics, such as hours, miles, and passengers, should be provided with a 12-month look-back in graphic format to identify trends. Additionally, measures of efficiency may be displayed to include cost per hour, reliability, availability, and trips flown vs. schedule requests, also with some method of comparing current numbers with some past time. Many of these reports are readily available from flight scheduling programs or may be generated using imbedded report writers. Other more esoteric measures of performance are possible, depending on company culture and preference of the boss.

Don't neglect the opportunity to blow the flight department's horn by *briefly* telling the good news about how a critical flight was saved, money was saved on contract maintenance or a special service was provided to a passenger. Similarly, provide a brief apprisal of what the flight department will be doing in future months, such as scheduled downtime for the aircraft, telephone/computer system upgrade, or personnel changes.

These reports should be generated prior to the monthly meeting with the boss to form a basis for discussion, should it be necessary. An added bonus from these reports is that they enable flight department managers to measure their own performance and provide flight department personnel with a picture of their operation and its performance.

How Much Is Enough?

How much does the boss really need to know? Are there things best left unsaid?

Some departments punch the "all reports" button on their scheduling programs and ship a forest-worth of reports to the boss each month. Suffice it to say that he/she will read only a small percentage of the data received and, more importantly, may be presented with data/information that are little understood or appreciated. In doing so, you may be providing information that could possibly harm the department because it is not fully understood or able to be placed in perspective by a naive or uninformed individual. Does the boss really need to know the average number of flight crew duty days or number of maintenance staff overhead hours expended last month? Perhaps for special one-time purposes, but not on a regular basis unless the concept has been fully explored with the report recipient.

Don't intentionally withhold essential or relevant information, but don't provide irrelevant or nongermane data, either.

Communicate

The boss deserves and needs to know what makes the flight department tick, why it does what it does, and how well it is performing. Regular interchanges with the reporting senior will ensure that these essential communications occur. Remember, the boss can be an advocate or merely a place holder for the flight department at headquarters; give him/ her the information needed to be the flight department's cheerleader and advocate within the company.

Growing the Next Generation—Succession Planning

There are many ways of investing in employees; career development may present the best return on an investment that will yield the managers of tomorrow.

It's taken eight months to find the "right" person to expand the department's cadre of pilots. While many of the candidates had the right flying stuff, they often lacked the more desirable non-technical attributes such as interpersonal skills, ability to communicate, and administrative/management potential. These characteristics were important when considering the concept that each employee should bring added value to the small, but growing, flight department.

The selection process was long and involved; hundreds of resumes, scores of telephone interviews, face time with six candidates, simulator evaluations, team interviews with flight department and HR personnel, reference checks, and finally, presenting the chosen one to the CEO for the final blessing. This one feels right!

Only time will tell whether this pilot was a good hire. But, it is not too early to start thinking of the long-term future for this employee, given that he/she potentially has a long and illustrious career ahead. It's time to starting writing this individual's company career plan.

Investing in Employees

Each employee should be considered a valuable asset, one that can help the flight department and, ultimately, the company achieve their mission and goals. As such, each employee requires an ongoing investment to ensure that the company receives an adequate return on that investment. While this may seem a cold and calculating way of describing the value of an employee and how that value is to be realized, this view is meant to encourage you to consider each person in the department a unique resource to be used wisely.

Most people approach jobs with an expectation of mutual benefit. More to the point, a job forms a contract between employee and employer to create a mutual support group in which the employee provides specified services and the employer provides compensation and the opportunity for personal satisfaction within the employee. Without this last part, it can't be much of a job, since monetary compensation alone has proved to be a poor motivator. To ensure the contract is viable and producing for both parties, each must invest something into the process.

Investing in employees takes many forms. Training, education, increasingly responsible positions, diversity in job opportunities, and the freedom to seek their level of competence and excellence are all means of investing in the individual. Note that training and education are just two methods of investing: the other methods are not as well-defined or as easy as sending a person to Flight Safety school for initial aircraft training or to an NBAA Professional Development Program course. No, investment requires goal setting, analysis, market research, and commitment, just like your personal investments do.

But, remember why you invest in the first place: to realize some return on the investment. While it takes money to make money in the marketplace, it also requires insight, perception, constant monitoring, and rebalancing for employees. While these actions may not be easy, if done correctly, they should yield significant returns in terms of employee satisfaction, increased skills, expertise in diverse fields, greater performance for the department and, perhaps, a successor worthy of your own excellent performance.

First Things

Step one should be to get the new employee acquainted with the standards, policies, and procedures used within the department. This sets the scene for the individual, provides a frame of reference, and establishes a foundation from which to begin. If the standards and procedures are mature and current, this step should go a long way toward getting the unpolished gem off on the right foot.

Employee Growth Hormones

- Development plan
- Training
- Challenge
- Increased responsibility
- Evaluation/follow-up
- Team environment
- Recognition
- Advancement

Shortly after this step, the individual's supervisor should sit down with the individual and carefully outline what is expected in terms of job performance and expectations. Qualities such as promptness, safety culture, customer orientation, adherence to standards and personal grooming standards should be mentioned. This should be followed with a discussion of mutual expectations regarding job performance, opportunities for advancement/upgrade and educational opportunities. This early supervisor/employee discussion should create a set of expectations for both parties, expectations that they can refer to and build on during the life the relationship. This discussion should set the scene for the future growth of the employee.

Many companies provide an indoctrination session for new employees to introduce them to the mission, business, and culture of the company. If done correctly, this session is a valuable introduction to the reason for the existence of the company and how the new employee fits into that picture. This is especially important for flight department personnel, who lead their lives removed from the main business of the company.

The Road Ahead

Once the new employee has completed the indoctrination and can find the restroom, the real work begins. It is important for them to gain a sense of comfort and confidence about their job before starting any major investment moves. This will probably require several months, during which time the supervisor can observe the individual's strengths and weaknesses. Knowing what is OK and what is needed will form the basis of the employee's initial development program. Normally this is accomplished through a company personnel evaluation program.

These programs usually identify an employee's strengths and weaknesses, outlining methods to capitalize on the strengths and improve the weaknesses. Even if the company does not require such an evaluation system, something similar should be instituted to enable measurement and continuation of the individual's progress toward goals. Regular checks on the mutually determined goals should be made to determine whether progress is adequate or not; informal quarterly revisits to the subject seem to work well. If progress is poor or artificially restricted by other events, try a plan B; if progress is good or completed, set new goals.

Challenges

Many psychologists subscribe to the theory that humans thrive and grow via personal challenges and meaningful experience. That is, without challenge, personal growth either does not occur or it does so very slowly. Without stimulation and challenge, the organism tends to vegetate or regress.

Whether or not you subscribe to this theory is irrelevant, but common sense dictates that those who extend themselves and take on new jobs and tasks usually tend to succeed and progress faster through the ranks. Part of this is ambition, part motivation, and part opportunity. It is the supervisor/manager's task to enable the employee to progress by providing motivation and opportunity.

Unfortunately, many individuals are not self-starters, choosing to take life as it comes. While this does not necessarily mean that they take the path of least resistance, some often need to be shown the way to a better future or to be nudged in that direction, sometimes more forcefully than others. If truth be told, any of us can identify with a teacher, boss, or mentor who had to give us a jump-start to set us on the road to a better, more satisfying life. This, then, is the challenge of all supervisors/managers: enable, demonstrate, and provide direction as means to progress down a path that will be mutually beneficial for both employee and employer.

Many challenges arise out of organizational need: A computer chronically crashes; the interface between supplier and accounting is perennially messed up; cabin snacks/supplies are never quite right for the passengers. From these humble beginnings careers are spawned. Not necessarily stellar, high-roller careers, but a start, nonetheless. Each situation provides an opportunity for the supervisor to challenge an employee and for the employee to rise to the challenge. In doing so, the supervisor can observe the individual reacting to challenges and new situations, and the individual can learn more about his/her environment and increase his/her skills. While most of these situations turn out to be win-win for all parties, if the employee stumbles or does less well, an opportunity for coaching or even a job shift is possible. Certainly, not everyone is suited to all jobs, but it is through attempting different tasks that we learn and progress.

The Plan

After a few trial challenges, both employer and employee should have a better idea of what is possible and what makes a good fit. This is a good time for both to sit down and explore possible futures for the individual.

Each employee has certain obvious talents that make them initially attractive to fill whatever position they apply for. But, they also have other, less obvious, talents that may surpass their obvious ones. Pilots with a flair for decoration, technicians who are volunteer as emergency medical technicians in their off-hours, and schedulers who are literacy tutors in their spare time—these and other talents reside in your employees, largely unknown to even their close working companions. Bringing out these unique talents is a task that each manager should work on for each employee; using them is mutually beneficial to employee and organization.

Each employee should have some idea of what their possible career paths will look like while in the flight department. Chief pilots, chief inspectors, and flight department coordinators are all possible, given the right set of motivation, training, and opportunity for beginning pilots, technicians, and schedulers. Ultimately, all have a shot at the top job, aviation department manager.

It is at this initial career path meeting that the possibilities, options, and opportunities should be laid out, realistically and openly. Many will not opt for positions of increased responsibility and power, preferring to hone their technical skills to become the best pilot, technician, or scheduler that they can be. These same people may awaken to the desirability of higher-level jobs years after joining the department. It is the responsibility of the manager to keep probing and providing information to the individual to open the best opportunities for company and employee.

Training and new job assignments will flow from these conversations to enable the candidates to realize their potential. And, just because a person doesn't want to be a manager doesn't mean that they can't grow and progress within the department. Being a better technician or a pilot who is the resident instrument procedures expert are also signs of growth and increased value for the company. Again, the task of the manager is to explore and unleash whatever potential is present.

Small departments have a problem in growing their personnel, but the successive assignment of collateral duties and increasing levels of responsibility along the way have the same effect as a formal development program. Pilots or technicians who are given the opportunity to plan, budget, handle vendors, work with human resources, and assist in training others are well on their way to being ready for a more responsible job when it becomes available. The next job does not necessarily have to be within the flight department, either. A small number of talented individuals "go to corporate" when the opportunity presents itself. Has the flight department lost a member or has the company gained a manager who fully appreciates the value of corporate aviation?

Possibilities

Ideally, each manager should always be training their replacement. This should not be viewed as a threat but as a means to fully exploit the talents of every member of the department.

Whether an employee is destined for the top is not as important as opening up the possibilities for all to achieve their full potential within their jobs. If done correctly, a high performance team will surely result.

There are many ways of investing in employees; career development may present the best return on investment.

Growing Employees

- Get to know each employee beyond their job description; they *all* have hidden talents.
- Seek opportunities to delegate tasks to subordinates that will give them a chance to succeed.
- Provide management, administrative and leadership training for the most promising employees.
- Mentoring, coaching, and occasional redirection nurture the process
- Let your boss know who you are developing and why—he/she can help.
- Only give an employee as much as they can handle, but everybody needs to be challenged.

Mission

The Nonpareil flight department provides safe, effective, and efficient air transportation services for company employees and customers.

It has been said often that one of the primary jobs of the manager or supervisor is to train his or her replacement. While egos sometimes interfere with this process, this practice often happens by default rather than design. Preparing employees for leadership and management positions is an essential element of job progression for a department, an element best not left to chance. But, the process should not stop with management; regardless of position, the job of each department member is to ensure that the next generation, the freshman class, is progressing to the point where they may someday become seniors and carry on.

A manager develops people, including himself.   – PETER DRUCKER

Flight Departments in Trouble

Seeing the warning signs preceding any major event often may be difficult. Frequently the indicators are subtle and inconsequential. A little bit here, a little bit there—they all add up.

Consider a failing engine. The subtle signs of impending failure are often overlooked during day-to-day flying. The slightly delayed light-off, slower spool times, lower oil pressure, transient vibration, higher climb and cruise ITT readings, and perhaps, strange noises. The individual indications normally do not tell the whole story, but taken together, they create a pattern of indicators that may spell trouble.

Therefore, the trick is to be alert for slightly abnormal indications and to look for trends and combinations of indicators. The point is that you have to be on the lookout for the subtle signs of trouble. If you wait until the more visible signs of trouble appear, it may be too late to recover.

Normal Operations

The boss sweeps by the flight crew on the way down the airstair saying, "Good job, guys." Or while waiting for the remaining passengers to show, the senior vice president for operations says, "I don't know what I'd do without the flight department." These and other brief passenger comments often form the primary performance feedback indicators for the flight department and, owing to their general nature, provide a favorable aura that may conceal more substantive and critical comments.

However, if the flight schedule is full and the boss is happy, what more could a flight department ask for? Under normal circumstances, not much. Unfortunately, anecdotal and nonquantified forms of feedback often lead to a false sense of security. Yet many flight departments blithely charge ahead, going through the motions of scheduling trips, flying flights, working with vendors, and so on without getting any substantive feedback from those who control their destinies.

Flight departments often are lulled into a false sense of security because of their closeness to the corporate hierarchy and because of their relative remoteness from the rest of the company. These features of a flight department's existence actually may prove to be negative rather than positive features. Regardless of the pros and cons, flight departments must be constantly alert for warning signs that may have negative consequences for them in the future.

What follows are just a few of the more common signs that I have observed in my consulting practice that may spell trouble for flight departments. See if you can identify with any of these.

Warning Signs

Few Supporters.   This is the situation in which only the top three of four executives in a company regularly use the aircraft; others are sometimes invited to join the chosen few, but rarely. While this may be justified in some, especially entrepreneurial, companies, it provides the flight department with little opportunity to demonstrate its real value to the company. It would be like saying, "Human resources, accounting, and information services may only be used by the manufacturing division; research and development, administration, marketing and sales have to find the own services." This is not very practical or prudent.

The flight department needs all the supporters and cheerleaders it can get. When a downturn comes, or when new management takes over, broad-based support is essential for the continued good health of the aviation operation. And everyone in the company needs to know how valuable the flight department is, not just the chosen few permitted to use its services. Only continual interaction between flight department employees and personnel at all levels of the company will achieve this; a company network of believers is a necessity.

Decisions Being Made by Accountants.   When top-level executives must defer to accountants and/or the finance committee of the board of directors, this usually means severe turbulence ahead for the department. Basically, this indicates that stringent cost controls are in effect and that each action taken by executives is placed under a fiscal microscope, giving the bean counters veto power over most corporate decisions. If no means of demonstrating the value of on-demand air transportation are available, at a minimum, utilization is sure to decrease. Worst case: The flight department is declared a nonessential perquisite and eliminated.

The antidote to this situation is to have developed a means for demonstrating value produced by the flight department and to have developed a database showing this feature over time. The NBAA's *Travel$ense* software is an ideal means of accomplishing this.

Little Interaction with Company.   The flight department stays to itself, has little communication with the company, and relies on one or two people for information about the company. Taken by itself, this sounds like an ideal setup: Stay at the airport, fly airplanes, and leave the business stuff to the boss's secretary. However, flight departments traditionally have suffered from their isolation and lack of interaction with the company.

Basically, it's out-of-sight, out-of-mind. Only the select few passengers ever think about the aircraft and then just as a means of transportation, not as a valuable part of the company. News about company activities that could aid the flight department is lost, and the flight department rapidly becomes known as "those people at the airport." Worse, the broad base of support that sustains most departments is not developed, opening the door for the budget cutters and new management to question the value of the operation.

Too Much Negative Attention.   The flight department reporting senior only wants to hear good things about the strange group that lives at the airport. The same is true for legal, human resources, information services, and finance. If the administrative and managerial overhead required to maintain the flight department rises to an unacceptably high level, a management company, fractional ownership, and charter become increasingly attractive alternatives for the company. Personnel problems (often the biggest negative), contracting faults, vendor difficulties, and unresolved financial issues are all examples of this phenomenon.

Good management and leadership will prevent little problems from becoming big ones that only corporate headquarters can handle. And timely interaction with the experts downtown can alleviate future problems that may get out of hand.

Reactive, Defensive, Head-in-the-Sand/Detached Bunkerism.   What I am trying to get across here is that the flight department is acting as an autonomous entity with little interaction with the company. This disease sometimes infects flight department managers who believe that their "patron" in the company (chairperson, CEO, chief financial officer) is powerful enough to protect them, come what may. Alternatively, the flight department feels that it is remote enough to ride out any administrative or fiscal storm that may arise. In reality, the flight department manager may not be comfortable enough with business processes to enter the fray downtown. Therefore, he or she decides to ignore the process and continue to do what the department knows best—fly.

> **The Warning Signs**
>
> - Few supporters within the company
> - Decisions being made by accountants
> - Little interaction with the company
> - Too much negative attention
> - Reactive, defensive, head-in-the-sand department management
> - Merger/acquisition changes
> - Low utilization

Unfortunately, patrons go as rapidly as they come, and businesspeople are persistent. Surprising as it may seem, the flight department provides a service that supports the company and is subject to common business practices and metrics. The flight department must be run like a business, just like any other supporting business unit within the company. Failure to do so will result in micromanagement from above, personnel changes, or worse.

Merger/Acquisition Changes.   Merger/acquisition mania has been a fact of worldwide corporate life for years. Economies of scale, cornering markets, and tax advantages are all reasons for this phenomenon. With mergers and acquisitions come change, often in the form of downsizing and doing more with less. This may mean, at worst, the demise of the flight department or, at best, a reduction in size and possible relocation. The problem with these actions is that they may take years to accomplish, and during that period, the fate of the department goes largely unknown. Rumors are rampant, and the uncertainty level becomes an unwelcome adjunct to all operations.

Rather than hunkering down and awaiting your fate, become a part of the decision process. If you are being acquired, educate the new company regarding the advantages of corporate aviation; develop a plan that demonstrates value for the merged venture. If the new company already has a flight department, collaborate with it to provide a win-win consolidation of assets. If you are the dominant partner in the marriage, develop a plan to serve the other company as rapidly as possible. Most important, stay in touch with the movers and shakers in your company to gain information about and to influence the action.

Low Utilization.   If the airplane flies two or three times a week, everybody is happy. However, when utilization falls to one or two times per week, will management decide that a definite trend has developed? Why the lower utilization? Have passengers found another way to travel? Is the company going through a slump? What's wrong?

Regardless of the reason, lowered demand for the flight department's services should be viewed with concern. As aircraft utilization falls below 250 to 300 hours per year, alternative forms of transportation may become more attractive/economical. A subset of this is that corporate management becomes concerned about the positioning (deadhead) rate, load factor, or fitting trips together in an effort to increase efficiency. Management may be missing the point of what the actual mission is for the aircraft; high deadhead rates and low load factors actually may be dictated by desired utilization patterns.

Do not wait for low utilization or corporate meddling to reach epidemic proportions; see the downward trend early, and work to reverse it. Normally, a little marketing effort with your passengers and potential passengers will either bolster declining flight hours or reveal a valid reason for lower demand. Strive to increase the two to three times per week to four to five times.

Bottom Line

The warning signs just mentioned rarely occur in isolation; several are linked to create complex situations that are often difficult for flight department personnel to detect. Therefore, regular objective reviews from both inside and outside the department are essential to the continued well-being of the organization.

Communications, poor and lack of, are the common denominator for all the warning signs mentioned. This applies to both communications within the department and with the company. Transmitting and receiving information is really job one for management personnel in any organization, especially highly specialized groups remote from the parent organization. Communicating must come before flying, administrating, financing, and managing.

The flight department should be considered a business unit that supports the overall goals and objectives of the company. As such, it must

- Stay closely connected to the company,
- Strive to be effective and efficient,
- Maintain a broad base of support within the company,
- Actively seek feedback that can be turned into service improvements,
- Look for early performance/status warning signs,
- Be run like a business, and
- *Communicate!*

Flight department personnel must be constantly alert for warning signs that may threaten the future strength and viability of the organization. To do less may endanger not only the organization but also the career development of its occupants.

> *In every instance, we found that the best-run companies stay as close to the customers as humanly possible.*
> —TOM PETERS

Staying Connected

> *Being connected with the company permits the hangar dwellers to become an integral part of the larger organization.*

Flight departments are privileged in that they are normally located apart from corporate headquarters, enjoying their remoteness and autonomy from the head shed. No corporate politics, few administrative hassles, or prying eyes to destroy the peace and harmony of the airport environment; these characteristics mark the existence of most flight departments.

Answer the phone, schedule the trip, fly the flight, and turn in the paperwork—the process works well, why do anything more?

Why indeed? The flight department provides a well-known service in a highly predictable manner, so why rock the boat and get involved with company affairs of which you know little? Many flight department personnel think that delving into company business is not their place and may even jeopardize their existence because of their lack of knowledge of the business world. But, this has proved to be the ostrich approach to providing on-demand air transportation, which may feel warm and cozy yet exposes the operation to unknown and unwanted hazards.

Unfortunately, flight departments are not always perceived to be the most essential service organizations within the company. Consequently, budget cutting and reengineering exercises often target the flight department for reduction or elimination. Efforts to reinforce the value of the flight department should be ongoing; being well connected to the larger company is the first step in that process.

It's a Business

Flight departments provide a service to the company, a transportation service. Only incidentally do they use aircraft to fulfill their mission; otherwise they are like other internal service organizations that support the company.

Human resources, accounting, information technology, and risk management all provide services that permit the company to accomplish its primary mission. So too the flight department functions in a support role, contributing to the overall success of the company.

Even though some of these functions may be outsourced, each remains connected to the main business through primary and secondary contacts and informal networks. These connections enable the service organizations to remain in touch with company plans, trends, and needs. It would be hard to imagine any of these support groups not being in touch with the ebb and flow of company events for more than a week or so. If they did, they would rapidly distance themselves from the company's mainstream demands and desires, eventually failing to provide high quality and timely services to those who depend on them.

Flight departments provide *safe, effective, and efficient air transportation services to the company, its employees, and customers.* If done correctly, these services will become an *indispensable* resource for the company. This should be the goal for each flight department: to make themselves *indispensable.* This business service should become regarded as essential to company operations as accounting, HR, or IT; with exemplary performance, the flight department may become *more important* than other service arms of the company—something to shoot for, anyway.

Reasons to Be Connected

1. The company is expanding (or contracting) rapidly in a time of great change; what does this mean for the flight department?

2. A new CEO from outside the company will soon take control; little is known about his opinions and preferences regarding on-demand air transportation.

3. Business as usual—answer the phone, schedule the trip, fly the flight.

The first two are obvious reasons to get closer to headquarters. The uncertainty and doubt that these events bring provide good cause to collect intelligence, do research, make plans and, more importantly, gather allies in your strategy for self-preservation and the continued good health of the department. The third condition, business as usual, is the time during which you should build relationships within the company that may prevent the shock and awe created by situations one and two.

Unfortunately (or fortunately), condition three rarely lasts. Successful companies on the cutting edge can't afford to remain static, to ignore market forces, or fall behind. Business as usual is rarely good business. Insidiously, condition three probably masks incipient conditions one and two; without good connections within the company, there is little incentive for people to tell you the news in a timely manner.

Being connected within the company permits the hangar dwellers to become an integral part of the larger organization; the flight department is designed to contribute to the continuing success of the overall organization. Flight department personnel receive regular paychecks from the *big company*, are provided support and benefits, and may receive bonuses as a consequence of the company's profitability; these should be reasons enough to feel a warm and abiding attraction to the mother ship.

The greatest reason for being connected is that the flight department can't do all of what it needs to do without outside support. Employment law, OSHA, budgets, accounts payable, contracts reviews, purchasing assistance, and state and local laws are all reasons to seek expert advice and support for the flight department. Flight departments are aircraft operations specialists; if you wouldn't expect the accounting department to arrange for an aircraft charter, why should you know everything about state aircraft taxation?

But, from a purely selfish standpoint, being connected means knowing what's going on within the company. Staying at the hangar and waiting for the telephone to ring won't provide the sort of information and intelligence the department needs to survive in a fast-changing business environment. Understanding the ebb and flow of company operations, knowing about new directions, feeling a part of the overall organization, and participating in company initiatives is the glue that holds companies together; it's hard to tap into all this while warming your chair at the hangar.

Connectors

- The boss (there may be several)
- The boss' assistants
- Chief financial officer
- Accounts payable clerk
- Purchasing/contacts officers
- Human resources person
- Risk manager
- Legal counsel

Relationships

Connections with headquarters are really about relationships. It is the abiding links with individuals, which build and maintain the information channels most valuable to you. Without these relationships, you will remain isolated and largely clueless about what the company is doing or needs.

Importantly, relationships are not built on the spur of the moment or overnight. If you need information right now about a critical work-related issue, a random contact garnered from the company telephone book may not be very helpful. Rather, ongoing relationships and, hopefully, friendships are the *lubricants* that facilitate business-related functions. These connections take time and rely on mutual assistance and trust to work properly.

While relationships may work well via telephone and email, the best contacts thrive on in-person connections. So, the aviation manager, chief pilot, director of maintenance, and scheduler should go downtown regularly to establish and cultivate their relationships. Just as important is to invite your *relations* to the hangar for an occasional visit; the aircraft is a powerful relationship-building tool and training aid. And, after they visit the hangar, throw in lunch as an extra added attraction.

Since some relationships take time and effort to establish, starting sooner rather than later is good advice.

Opportunities

To most company personnel the flight department is a little known splinter group that flies the bosses around. Using this not-so-complimentary description, most employees in the larger company will not regard flight department personnel as essential, productive members of the team. Therefore, it should be the flight department's primary tasks to dispel the splinter group and other myths about itself.

The flight department's reporting senior is a key person who can open doors to the larger company. Since this person is likely to be an integral part of the headquarters staff, his/her contacts should provide a rich source of people who can help and support the department. But, first the boss has to know what you want and need; this requires regular contact, preferably face-to-face, to discuss status and needs. Agenda-driven monthly meetings with the boss provide an opportunity for the aviation manager not only to talk about challenges and opportunities but also to educate the boss about aircraft operations and safety—essential functions of the job. And, incidentally, the boss may see fit to drop a few hints about what the company is about to do.

The department's frequent flyers may be a source of contacts, too. Casual pre- and post-flight conversations with these key people can yield valuable connections and insights. Care should be taken not to abuse these relationships or to cross the line between service provider and peer.

The best contacts may be people within the company who provide essential services the same as the flight department does. Human resources representatives, accounts payable clerks, information technology technicians, and facilities managers are all there to help the company succeed—this also applies to the flight department.

Don't overlook lower level contacts, either. Some of the best information comes from the least likely places. Mail room clerks, janitors, cafeteria servers, cashiers, receptionists, and secretaries all have large networks that possess tremendous amounts of information. Much of the information will be of little use, but it only takes one valuable tidbit to make the relationship pay handsome dividends. Strangely, the workers in the trenches are often the first to know the most significant revelations regarding forthcoming events—every contact is valuable.

But, relationships must be established with these individuals so that they may become productive members of *your team*. Since every contact provides an opportunity to sing the praises of the flight department and the value it brings to the company, co-opting as many people as possible to this point of view will help fulfill your mission.

Giving Back

The old adage *it's better to give than to receive* is certainly true when building relationships. More specifically, if the flight department is perceived to be a member of the company team, to be truly helping it succeed, the better its chances are for receiving assistance. So, the flight department must venture beyond the confines of the hangar and volunteer to assist with company projects and initiatives. Long-range planning, employee benefits, customer satisfaction surveys, and even the United Way campaign are all opportunities to join the larger effort.

The real benefit of this activity is the contacts gained as a consequence of participation. Extending yourself beyond the aviation world will generate many good contacts, contacts that will provide information sources and an ever-expanding network.

Ultimately, working as a part of and for the larger enterprise will do more toward convincing the entire company that the flight department is an integral part of the entire organization than any pronouncement made from on high. Being a part of the larger team will bring the greatest rewards from staying connected. See Table 5.2.

Sources of Involvement

- Regular meetings with reporting senior
- Corporate planning sessions
- Transportation analyses
- Accounting/budget reviews
- New product/services reviews
- Environmental, health, and safety studies
- Internal audits
- Customer service reviews

Desired outcome: To sensitize flight department personnel to the needs of their customers and to develop good customer service practices within the department.

Part 1   All personnel

- The concept of customer service—what it is and isn't
- Who are our customers and what do they want?
- What business are we in?
- What impresses people; what doesn't?
- Attitude, body language, and tone of voice
- Develop a service plan
- Measure service performance
- Work as a team

Part 2   Flight crews

- The travel experience
- Appearance and body language, part 2
- Creating service expectations in the customer
- Safety and passenger apprehension
- What happens when things go wrong?
- The flight crew—passenger bond
- Role playing—dealing with the customer
- Making it better one flight at a time

TABLE 5.2   Customer Training Outline

Is It Worth It?

Operating the flight department takes a lot of effort and long hours to stay safe and keep the *customer* happy. So why go beyond the aviation world and venture into a world of which you know little and which you couldn't care less about?

Flight departments often get their greatest shocks and are most surprised by the things that they don't know. Company downsizing, plant closures, expansion plans, budget cutbacks, new management teams, and changing markets often jolt the cloistered existence of the *people at the airport*. If they had taken time to integrate with the company, to establish relationships, to build networks they might have been able to better prepare for the resulting trauma or even have headed it off before it happened.

Becoming a part of the larger company is an investment in the future health and well-being of the flight department; the dividends may be significant. It's worth the risk.

Lessons Learned

Is there a perfect flight department?...I'm still looking.

Learning from our mistakes is a valuable but often costly experience. Working for the wrong boss/company, forgetting to put the landing gear down, and failing to acknowledge a spouse's birthday are definitely valuable learning experiences, but the cost may be high indeed. Unfortunately, there are too many mistakes, some too costly, for us to experience; learning not just from our mistakes but also from other people must be considered—there just is not enough time to make them all on our own.

Perhaps the best people to ask about mistakes are those who make it a business to observe and comment on the mistakes of others. These people may be teachers, counselors, or consultants who observe errors and attempt to provide perspective and guidance for those making them. Of course, the credibility of the observer/fixer is important—we have all received bad advice from those unworthy of the teacher/counselor name.

I make my living by, among other things, evaluating flight departments and charter operators against federal and company standards and best industry practices. As a consequence, I get to see a variety of flight operations: large and small, single-task and multimission, fixed and rotary wing, and single and multiple bases. As you may imagine, these run the operational gamut from excellent to unsafe and the managerial spectrum from super to couldn't manage their way out of a paper bag.

The good news is that the great majority of operations I visit are safe and try hard to get the job done effectively and efficiently. This is a tribute to the professionalism of the pilots, technicians, schedulers/dispatchers, and receptionists who make it all happen, for these individuals are sometimes not well led within their department nor supported by corporate headquarters.

Predictably, certain patterns of performance emerge when a significant number of operations are examined. Here are some common themes that separate the really great ones from the merely good and those that need work.

A Sense of Mission

Those departments that believe that their job is to fly airplanes for passengers who may occasionally sit in back may be setting themselves up for a fall. The operative expression in this job is *customer service,* not flying. First, *service,* then *transportation,* and then *aviation.* *"Safe, reliable, efficient on-demand air transportation for company employees and customers"* is the concept that must be internalized by all concerned within the flight department. The concept of a flying club that allows people to go along for the ride is a perversion of the main mission. And mission statements hung on walls in attractive frames and not lived on a daily basis only waste picture frames. Ya gotta live it!

Standards

Not surprisingly, the best departments have a well-developed and current set of administrative and operating standards. These standards reflect the actual operating policies and procedures used by the department and are not merely window dressing for the folks downtown or the insurance broker. These standards are memorialized in a flight operations manual that is well known and used regularly by all within the department. Scheduling, duty time, crisis management, standard operating procedures, and maintenance all have prominent parts within the manual; all the parts fit together into a well-thought-out set of guidelines and rules that leave little doubt in the minds of those who live by it. However, standards, no matter how good, cannot do the job without having a means of making sure that everyone is singing from the same hymn book.

Leadership

Without this critical element, all the rules in the world will not make an organization perform well. There must be a means of transmitting a vision of what the organization is all about to get people to follow the rules. If policies and procedures are merely enforced, the operation will lack spirit and a sense of direction that mark great organizations. Notice that it is *leadership* that is required and not merely *management skill.* Inspiration, a sense of direction, counseling, and a role model are the elements that separate the leader from the manager.

Management is obviously needed, but leadership wins in the end. Leaders already may be good managers, or they can get by with minimal management skills; managers often have a harder time becoming leaders.

Rapport with the Company

The best departments have good relationships with a variety of headquarters personnel, not just the secretaries of their passengers. Knowing where the company is headed and how it plans to achieve that progress is essential information for the flight department. Without this knowledge, the flight operation will never be prepared to support the company with its desired direction. In fact, it will always be behind its needs. Being able to chat easily with the chairperson or CEO may make the department manager feel like he or she is "in the know," but it may just mean that he or she can chat easily with the heavies.

It is detailed knowledge of the direction of the company or principal's interests with all their twists, turns, and changes that will adequately prepare the department manager for the future. The CFO and heads of research and development, planning, manufacturing, facilities, and human resources are the ones who are more likely to speak substantively of enabling events and timeline. The very top executives usually speak in terms of concepts, the big picture, and a sense of direction; their lieutenants have the details and make it happen.

Close connections with the workers in human resources, accounting, facilities, risk management, and safety by a variety of flight department employees will make the operation more knowledgeable and efficient. These company people are experts in their fields and can help the flight department run itself like the business that it really is.

Teamwork

This is an often maligned and misunderstood term that is the capstone of successful flight departments. This means that everyone is rowing in the same direction, helping where they can, pitching in when the tough work falls on just a few—these are all examples of people working under a sense of mission toward a common goal. Team members do not really care who gets the credit or glory—they all benefit from it. The secret to teams? Teams are formed and fostered by leaders and coaches.

Is there a perfect flight department? I doubt it, but I am still looking. Until I find it, I will continue to push the qualities listed above. And even the best departments cannot be best all the time, only during peak periods of well-focused effort. However, they can (and do) attempt to excel in everything they do. This probably will keep them in the good to very good category most of the time—not too bad a grade on a consistent basis.

Your task is to decide if my observations are valid and that I am credible. If so, you do not have to make all the mistakes of those who came before you. And even if you do not believe everything I have told you, you will be better off than you are now. You can't beat that.

Good management consists in showing average people how to do the work of superior people.
—JOHN D. ROCKEFELLER, JR.

CHAPTER **6**

Operations

Overview

The essence of business and personal flight operations is to provide on-demand air transportation services to the principals or to a company and its guests. While this sounds simple, there are many operational considerations to be made to ensure safe, reliable, and timely service.

The key word is *service*. In reality, the flight department is a service provider, similar to human resources, information technology, and other groups that support the company's or principal's mission and goals.

Operations Perspectives

"An operator shall ensure that all employees know that they must comply with the laws, regulations, and procedures of those States in which operations are conducted." So begins ICAO Annex 6, Part II, *Operation of Aircraft—International General Aviation—Aeroplanes,* Section 3 pertaining to large and turbojet airplanes. This is a deceptively simple statement encompassing many facets of a very complex subject. It is difficult enough if this directive applies to just one country (State) but imagine how difficult it is for all 191 of the ICAO member States to comply with this statement!

Fortunately, most business and corporate aviation operations occur within one or a few States, which greatly simplifies the problem. Yet regulations packages are measured in pounds rather than pages, given their size and scope. Therefore, the knowledge of regulations is essential, and the means of complying with them is even more critical. Compliance methods, standards, procedures, processes, recordkeeping, and training requirements all go into ensuring that aviation operations featuring complex and fast aircraft are performed completely and correctly.

Getting Organized

While some operators crassly claim that "we don't need no stinkin' rules" to operate their flight department, evidence shows otherwise. Those consultants who work with corporate flight departments, myself included, will tell you that the most safe, effective, and efficient are those with a well-conceived set of policies, processes, procedures, and standards (P³S) that either meet or exceed State operating rules, manufacturer's standards/recommendations, and industry best operating practices.

Do private operators need large, general flight operations and maintenance manuals to permit safe and effective operations? Do they need exceptional recordkeeping requirements? Are their procedures exceptionally complicated and comprehensive? No is the answer to all of the previous items; but what they do have is a well thought-out set of P³S that covers all the important features of their operation and is appropriate to their size and mission.

As we go through this chapter, you will begin to see what is important and what is not. This should help you formulate ways and means of portraying what you consider to be important when it comes to operations.

Also, as you go through the chapter, think about what policies and procedures you want to specify for your flight department and how they are to be communicated. In addition, consider how you will know that each of the items (and regulations) are actually being followed and what means for recording compliance with each item will be used. You've seen it before in this book: *If you can't measure it, you can't manage it.*

Many flight departments believe that the most important element of their venture is operations—putting aircraft in the air so that passengers get to their destination safely, effectively, and efficiently. While these are essential and laudable items, the ops portion of the department must fit well with other business concerns—customer service, communication with *all* concerned at corporate headquarters, personnel management, safety, security, and numerous other concerns. All of the "very important operations considerations" must closely mesh with everything else required to ensure that the *total package* works well together.

Compliance

Airlines have extensive groups of inspectors and check airmen to ensure compliance with their own rules and those of the government. The regulatory authority also provides each airline with operations and maintenance inspectors to oversee these overseers.

The parallels to these airline measures within the flight department are often few and far between. Compliance with operations manual policies and standard operating procedures may be left up to the integrity of the individual and the leadership abilities of senior flight crewmembers and technicians. Even if these measures work fairly well, there is seldom any means to track compliance inspections or evaluations. The "good 'ole boy" and "professional colleague" networks kick in to protect the weak performers or blatant scofflaws. After all, there are *two* pilots up front, aren't there? And when was the last time we had an instance of maintenance error?

So, what keeps corporate aviation so safe if there is no surveillance mechanism to keep track of who does and who doesn't comply? The answers keep coming up as *individual integrity* and *professional pride*. To their credit, pilots, technicians, and schedulers in corporate aviation often perform safely and with excellence *in spite of* department management, policies, and procedures. The great majority of personnel in personal and company aviation want to do a good job, abide by the regulations, and make the company look good. This is acceptable for the finest 98% of the population, but what about the maverick, scofflaw, or rugged individualist? These types may not generate the most accidents or incidents, but they probably generate the most gray hairs

Perhaps the reasoning should be, "Can this department afford to rely on the absolute integrity and professionalism of every member all the time? What if somebody slips up or just has a bad day? Can we accept individual performance as insurance against mishaps, or should there be a *system* in place to provide a last line of defense against human error?" This system should serve as a check and balance against human frailty and an uncanny affinity for Murphy's Law.

The airline model contains all the right stuff: standards, limitations, surveillance mechanisms, and management systems. Flight departments opt to pick and choose among these to find what suits them. The trick is to pick enough of the higher standards to ensure safety and professionalism without jeopardizing the need for flexibility required in company and personal aviation. Unfortunately, some may choose combinations that do not have enough of the right stuff to make the system work effectively in terms of safety, standards, and service.

Realize that what you are running is, in fact, an airline. If you are to continue to attract customers and operate safely, you must operate to some standard that exceeds Part 91. So, when it comes time to reflect upon your operation, ask yourself, "Is this any way to run an airline?"

Regulatory Compliance.   The world of aviation is one of the most completely regulated activities in the world. With the possible exception of nuclear power, aviation interests have more regulations that any other industry. Even under a nation's general operating and flight rules (e.g., ICAO Annex 6, Part II and US FAR Part 91), which govern most corporate flight operations, the regulatory burden is considerable (see Table 4.3 on page 170).

Then come the other federal agencies with their restraints; in the United States the Environmental Protection Agency (EPA), Occupational Safety and Health Administration (OSHA), Federal Communications Commission (FCC), and Equal Employment Opportunity Commission (EEOC) were all once remote and nondemanding agencies, but now they are forces to be reckoned with on an increasingly frequent basis. These agencies often have state-level counterparts that impose a welter of sometimes confusing regulations that actually have little to do with aviation. State regulations often concentrate on workplace issues, particularly the protection of workers, and they must be taken seriously to avoid unpleasant involvement with these agencies.

In addition to the government regulations, the company usually imposes its own policies and procedures on both the aviation and administrative aspects of the operation. In many cases, these company-imposed rules are often more restrictive and demanding than those of the government. Airfield limitations, higher instrument approach minimums for new pilots-in-command, minimum fuel requirements, duty time limits, and weather restrictions are common examples of these restraints. And even smaller companies have layers of regulations affecting personnel, accounting, administration, health, safety, and security. Although the flight department is usually located far from corporate headquarters, it is not immune to these strictures.

Achieving regulatory compliance is difficult enough for the regulators, and when the other official constraints are added, the problem takes on extraordinary proportions. Attempting to understand and comply with these constraints on a case-by-case or situational basis seldom works well, given the constant and all-pervasive nature of the regulations. Merely knowing and understanding all of these regulations is almost impossible, unless the primary mission of the department is sacrificed to the regulatory juggernaut. Clearly, assistance is needed.

Assistance can come from several areas: the parent company or contractors, regulatory agency personnel, and your own system designed to ensure compliance. The parent company also has to comply with many of the nonaviation regulations; therefore, they should already have procedures in place to accommodate many of those regulations.

For the ones they don't track, efforts should be made through the department's network of flight departments to determine applicability and means of compliance. The agencies themselves can be of assistance, if time and persistence is available for the inquirer. The fear of investigation from regulatory personnel who provide the information may be a factor, but not knowing the rules may be even more hazardous.

Developing a system to ensure compliance with the many regulations and procedures is a must-do item to avoid running afoul of our many watchdogs—see the following sidebar.

Model Compliance System

- Determine what regulations apply to your operation (sometimes the hardest part).
- Know the applicable regulations or find/hire someone who knows them.
- Grow experts within the department who will be responsible for tracking specific regulatory areas.
- Ensure that there is a means for keeping up to date on the regulations.
- Develop a comprehensive record-keeping system to track your compliance actions.
- Devise a tickler system to highlight recurring compliance actions and reporting requirements.
- Conduct periodic audits or inspections of the areas affected by the regulations.

You won't win many points by being in compliance with the regulations, but the thought of loss of points from noncompliance should be sufficient motivation to do the right thing.

The best flight departments practice their excellent standards every day.
—JOHN SHEEHAN, AVIATION CONSULTANT

Flight Operations Manual

Most flight departments have a flight operations manual (FOM). Larger departments are the most likely to have and use a FOM, while smaller, especially single-aircraft operations, tend not to have a manual or have one but seldom use it. Why don't all departments have, let alone use, a manual?

Answers to this question run the gamut from "We only have two people, so we don't need one" to "It's just more paperwork." Variations include "Everything we need is in the regulations or aircraft flight manual" or "We used to have one but we never used it."

Some flight departments have a FOM just to meet an insurance requirement or association membership requirement. If these are the only reasons to have a manual, it will undoubtedly not be used. So, why have a manual?

The FOM is primarily a communications device, a means for getting the word out about how the flight department will be run. Policies, standards, administrative and financial issues, and procedures to run the department in a logical, measured fashion are all items that should be communicated to members of the department and also to the principal or the company as a whole. It sets standards and limits for the operation—items that all must know.

The contents of the manual should be used to set the overall tone for the operation of the department and to establish the culture in which all members of the department will work. If employees know what is expected of them regarding work performance, required paperwork, standards of conduct, and operating policies, they should be more accepting of them. Without these items in writing, the department is making up its procedures every time the members perform an act, and is thus literally winging it. Without some form of reference, policies and procedures developed on the spur of the moment will become the norm rather than the exception. With no set form of work policy, supervisors also tend to vary their pronouncements to fit each situation as it arises, and inconsistency surely will be the result.

The FOM forms a contract between company senior management and the aviation department personnel, a contract that provides the expectations of both the senior executive and the department staff. If the manual has been signed by the company president or principal, there will be little tendency for the passengers to abuse the privileges of the aircraft or to ask for something not permitted by the FOM. This is important when passengers "demand" to go when the weatherman says no or when you are asked to go beyond your 14-hour duty day just so they can sleep in their own bed.

A manual is as essential a part of a flight department's operation as is an aircraft flight manual, manufacturer's maintenance program, or company financial handbook. Developing (and updating) one is a valuable and productive experience. This raises the point of how should a FOM be developed. There are purveyors of generic, one-size-fits-all manuals that merely insert your corporate name into the appropriate places of the computer program, put a nice cover on it, and ship it to you in the desired quantities for a bargain basement price. Or, departments have been known to "borrow" another company's manual and "excerpt" large portions of it for their own. Either process will yield, at best, a marginally acceptable manual, one that will seldom be used or provide adequate guidance for department members or the larger organization. A manual developed for a multiple turbojet operation will probably not be suitable for the single turboprop operator. This is because of the order of magnitude of operations: differences in backup capability and tempo of operations are all markedly different.

There are different sets of expectations for large and small operators, differing levels of administrative requirements, and differing methods of communications required within the department. But the most important

Flight Operations Manual Preface

The XYZ Corporation operates its aircraft as a management tool to improve its efficiency and ability to compete in the marketplace by providing safe, available, and reliable on-demand air transportation.

The policies and procedures contained in this manual are consistent with management practices and philosophies of the company. The contents of this manual are to be followed in all cases unless extraordinary circumstances dictate otherwise. Any deviation from the procedures shown in this manual may be approved only by the aviation manager or me.

It is especially important for all passengers using our corporate aircraft to realize that, for safety-of-flight reasons, the operational authority of the pilot in command of a flight is absolute; this authority must not be questioned regarding flight operations. If there is some question concerning the pilot in command's actions, it should be brought to the attention of the aviation manager after the fact.
/signed/Chief Executive Officer

aspect of a FOM is that it should represent the *actual operation of its department* and reflect the detailed procedures and processes used by its members.

Building It

Here lies the key to a successful manual, one that will be used by all members of the department and will be updated on a regular basis: its authors must be the people who have to use the manual. *Department members have developed their procedures through an evolutionary process, for which they were responsible and with which they feel comfortable.* If they are suddenly presented with a neatly formatted, bound volume that purports to be a description of how they do things, they may initially be curious enough to at least look at it. But as soon as they come across something that doesn't ring quite true or, heaven forbid, is a radical departure from reality, the manual is instantly suspect or suddenly alien. If too many of these inaccuracies are discovered the manual is relegated to that part of the bookshelf reserved for the company personnel and style manuals.

But if instead the members of the department are challenged to develop a manual that truly reflects their actual day-to-day operations, they will probably respond by developing a credible, useful document. They will do so because they are, for the most part, interested in their jobs and want to find ways to improve them. In large measure, the development of the manual becomes a process of discovery, questioning, and refinement: discovery of things they were unaware that they all did; questioning of existing procedures that may not be well founded; and refinement of long-standing procedures that need a bit of polish to make them truly exceptional.

The process becomes at once the first meeting of the standardization committee and an excellent team-building tool. Pilots and mechanics like to talk about what they do, share ideas, and compete for the best way for doing things. "Building" the manual is often a very positive experience for the entire department.

While most people think of a FOM as the sole domain of the pilots, the maintenance and administrative segments of the department must have their say, as well. Like it or not, the department cannot exist without the entire team working together. Schedules and flight statistics have to be processed for the department to run properly, and without the maintenance function, the pilot's career would be short-lived, indeed. Everyone within the department must have a clear understanding of how all the segments fit together, or else chaos (even more than normal) will result.

The process of development is at least as important as having an FOM. For, without the full commitment of senior management and all department personnel, the mere fact of having a manual is of little consequence and, perhaps, a wasted effort.

How to get started? The NBAA Management Guide has some useful guidance on the subject and a detailed outline of a FOM in it. NBAA often conducts workshops on the subject of flight and maintenance operations handbooks, and there are a number of consultants that specialize in providing development and drafting services. The best way to get started is to put all the department personnel in a room and tell them to list the policies and procedures they would like to see developed and published—there will be plenty of input.

Excuses, Excuses

What about the department that says, "We have been doing things the same way since I can remember—everybody knows the drill; why put it into writing?" First, if they have been doing it the same way forever, is it still the right way? The outside world changes with amazing rapidity, so shouldn't the flight department change along with it, at least regarding some issues? Second, will the same people be in the department forever? Are contract pilots or technicians ever used, and don't they need to know what the policies and standards are? What should the folks downtown—the people at company headquarters who regularly deal with the department—know about its operation? Finally, can everybody really remember the way a certain issue was handled the last time it came up?

The preceding comments are not just for those departments that do not have a manual. They also apply to departments that have a manual but don't use it. This is probably the worst situation, to have a set of policies and standards on the books and not use them. This says, "We have directives but they are just for show and do not apply to our real operations. We must be flexible in meeting our operational requirements and can't be tied to strict procedures." Variations on this are the selective use of portions of the manual—use the rules when they suit the situation and ignore them when they are inconvenient or overly restrictive. An example of this might be: although the manual states that non-type rated pilots will not fly in the left seat, an individual pilot in command (PIC) may allow his second-in-command (SIC) to fly in the left seat on deadhead legs. Or, the manual states that once any part of landing gear actuation system integrity has been broken, another technician must inspect it; because it is 2 a.m. and you are the only technician at the hangar, you decide, "just this once I will return the aircraft to service without the second set of eyes looking at it."

I often hear the comment, "We don't use that procedure anymore because we got the new aircraft," or "The new chief pilot likes us to do it a different way." OK, use the new procedure, but *change the manual to reflect the new procedure*. Even a single instance of ignoring a standard operating procedure (SOP) or policy will lessen the power and utility of the manual. Once the first exception has been made, others will come easily.

Changes

Changes don't have to be a big production, either. Approach the chief pilot or chief of maintenance with the suggestion and see what they say. Ideally, they will refer it, formally or informally, to a number of other department employees for comment. If they like it, the change is as good as made. But the change must be put into writing, normally as an interim change, approved by the department manager, and distributed to everyone within the department. Then, once a year, all the interim changes should be incorporated into the manual along with other changes made during the annual review of the document. Yes, to keep the FOM relevant and up to date, it must be examined annually.

Annual reviews should be accomplished by a small group of individuals from a variety of disciplines within the department. It need not take long; only a discussion of items requiring change need be discussed. If the document has been actively used during the year, a list of prospective changes should have been collected and will serve as the review committee's agenda.

Active use of the manual throughout the year is the key to keeping the department aligned with its contents. In many departments, the FOM forms a centerpiece for the entire operation. When any unusual situation crops up, the first move to resolving the issue should be to the FOM. "What does the ops manual say?" should be the words heard when an operational question arises. Enforcing and reinforcing this tendency is a prime responsibility of management. Managers must be the keepers of the flame if the manual is to be respected and used. The old term "use it or lose it" applies in spades to the FOM.

Since many manuals are now created, distributed, and maintained electronically, it is a relatively easy task to make changes to its content. However, there must be change and access controls to e-manuals to protect them from unauthorized changes or access. A change control document should accompany each manual to ensure that the manual viewed is the latest version with changes highlighted to emphasize their newness or differences.

Every Situation?

What should be included in the manual? While the simple answer is "whatever you want," a more detailed answer is difficult. The manual should reflect the operational culture of a flight department and contain all items necessary to run the department safely and effectively. The usual main headings include administration, operations, and maintenance—but a lot can be included under those headings. The real answer to this question is to include whatever it takes to standardize all aspects of the flight department and to effectively communicate those policies and procedures to both department personnel and corporate management (see Table 6.1).

Suggested Topics for a Flight Operations Manual	
I. Administration	H. Crisis Planning
A. Mission/Objectives	1. Organization
B. General Policies	2. Notification
1. FAA Enforcement Actions	3. Information Gathering
2. Smoking	4. Initial Actions
3. Alcoholic Beverages/Drugs	5. Communicating the Crisis
4. Medical Fitness	6. Records
5. Interaction with Media/Press	7. Participation in the Investigation
6. Security	8. Media
7. Insurance	I. Charter Operations
C. Organization	1. Company Policy
D. Job Descriptions	2. Qualification of Operations
1. Aviation Department Manager	**II. Operations**
2. Chief Pilot	A. Safety Management System
3. Captain	B. Control of Flights
4. Copilot	C. Scheduling
5. Flight Attendant	D. Basic Operations
6. Flight Mechanic	1. Dispatch Release
7. Director of Maintenance	2. Airworthiness
8. Technician	3. Flight Crewmember Duties
9. Aircraft Handler/Cleaner	4. Flight Plans
10. Scheduler	5. Operational Reports
11. Secretary	6. Maintenance Test Flights
E. Personnel	7. Positioning Flights
1. Appearance	8. Aircraft Noise Policy
2. Records	E. Passenger Relations
3. Primary and Collateral Duties	1. Policy
4. Use of Part-Time Personnel	2. Crew Duties
F. Pilots	3. Authority of pilot in command
1. Authority of Pilot in Command	4. Schedule Changes
2. Qualifications	F. Limitations
3. Training	1. Airports
4. Contract Pilots	2. Weather
5. Pilot Currency	3. Aircraft
6. Duty Time	4. Crew
7. Physical Fitness	5. Hazardous Materials
8. Alcohol/Drug Policy	G. Standard Operation Procedures
9. Flight Evaluation	1. Policy
G. Aircraft Maintenance Technician	2. Crew Assignment
1. Qualifications	3. Basic Crew Responsibilities
2. Duty Time	4. General
3. Training	5. (Specific)
4. Physical Qualifications	

TABLE 6.1 Flight Operations Manual Topics (*NBAA*)

Suggested Topics for a Flight Operations Manual	
H. Abnormal Operations	G. Aircraft Maintenance Records
1. Emergencies	1. Responsibility
2. Terrorist Hijacking	2. Record entries
3. Accidents/Incidents	3. Computerized Tracking System
I. International Operations	H. Aircraft Parts
1. Training	1. Ordering
2. Qualification	2. Receiving
3. Planning	3. Inspection
4. Documentation	4. Inventory Control
5. Security	I. Aircraft Fueling
III. Maintenance	J. Maintenance Test Flights
A. Purpose/Mission	K. Checklists
B. Basic Policies	1. Accident/Incident
1. Airworthiness	2. International Operations
2. Inspection of Work Performed	3. Flight Crew Evaluation
3. Airworthiness Status Documentation	L. Forms
4. Controlled Publications	1. Manifest
C. Airworthiness Release	2. Flight Log
1. Postflight	3. Abnormal Operations
2. Release Document	4. Airworthiness Release
D. Scheduled Inspection and Maintenance	5. Maintenance Discrepancy
1. Maintenance Programs	6. Deferred Discrepancy
2. Work Assignment	M. References
E. Unscheduled Maintenance	1. Advisory Circulars
1. Discrepancy Reporting	2. NTSB Part 830
2. Deferred Discrepancies	N. Telephone Numbers
3. Work Assignment	1. Flight Department
4. Recording Work Performed	2. Company
5. Returning Aircraft to Service	3. Emergency
F. Vendor Selection	4. Government
1. Approval Process	5. Associations
2. Approved Vendors	
3. Major Work	
a. Request for Proposals	
b. Evaluation of Proposals	

TABLE 6.1 Flight Operations Manual Topics (*Continued*)

How long should it be? Small departments can probably get by with as few as 10 pages for starters, large departments may need as many as 100. Again, the answer is whatever it takes to get the word across clearly and concisely to those who need to know. The first edition of this masterpiece should be just the beginning of the living document that reflects the changes that occur within the department. The first edition may become a collector's item but should pass out of fashion. Does the manual have to cover every situation that may be encountered by the department? No, because it can't possibly cover all situations. But it can provide policy guidance that will form the basis for sound, safe, and consistent decisions. If the concepts of safety in all operations, high levels of service to customers, and value provided to the company are emphasized throughout the manual, what to do in unusual situations should come easier to all concerned. See Table 6.1 for a list of topics for a Flight Operations Manual suggested by NBAA.

Singing from the Same Hymnbook

Perhaps the best reason for having a customized manual is to use it as the basis for determining how well the department is performing. If you don't perform actions consistently, how can you measure performance? That is, if you do the same things differently every time, there is no standard by which to measure.

The expression "singing from the same hymnbook" comes to mind when considering the need for a viable FOM. It is hard enough to get the congregation to sing at all, let alone with the right words and notes. And if they haven't grown up with or helped to develop that hymnal, they may not sing at all. Getting the "congregation" to sing from the same hymnbook can make some beautiful and powerful music; the same is true for a flight department.

Standards

Aviation is arguably one of the most regulated activities within the United States. Perhaps nuclear power may be more regulated, but aviation regulations occupy a significant chunk of the rules bookshelf. However, only a few of the rules speak to technique or operational procedures. Certainly, the rules that apply to not flying below the glide slope in a turbojet aircraft or exceeding 250 knots below 10,000 feet brush lightly against procedure but have little to do with how the aircraft is operated in the sense of technique.

The government rightly leaves most issues of operational procedure to the operator; it is his rules that govern the detailed operation of the aircraft, planning for flights, caring for passengers, and maintaining operational records. But not all operators have specific operational procedures for their crew to follow. The air carriers, large air taxi operators, and some corporate flight departments have detailed SOPs carefully crafted and placed in their flight operations and standardization manuals, and most of them insist upon adherence to them. But some operators, particularly corporate operators, either do not specify SOPs or fail to demand compliance with them.

Sterile Cockpit

Flight crewmembers may not engage in, nor may a PIC permit, any activity during a critical phase of flight that could distract from or interfere with the proper performance of any flight crewmember's duties. Critical phases of flight includes all ground operations involving taxiing, takeoff, and landing, and all other flight operations conducted below 10,000 feet, except cruise flight.

IS-BAO Sets the Standard

Why have operational standards for business aviation? Some will say that a State's civil aviation regulations are more than adequate to ensure safe operations anywhere within that State and that the ICAO standards provide direction for flights outside those sovereign shores. So, why do most business and corporate operators around the world generally exceed the regulations of their State of registry? Safety.

Yes, safety, because over time operators have found that regulatory and international standards don't quite provide a sufficient level of direction for flight crews encountering the wide variety of operational conditions even in domestic operations.

First, operators discovered that periodic PIC checks for large aircraft operations were not adequate to properly prepare their pilots for the variety of emergency and abnormal situations that may be encountered. Therefore, recurrent training became a standard practice. Then, the challenge of getting all flight crews to perform their cockpit duties in a similar and predictable manner, especially in abnormal situations, led to standard operating procedures and standardization programs to verify compliance with the procedures. Finally, the realization that constraints on the use of aircraft, as specified by regulatory and aircraft certification restrictions, may not be adequate for safe operations led most operators to adopt additional limitations regarding critical items, such as airport types, runway lengths, types of instrument approaches, range profiles, and severe weather.

Operators devised their own operations manual filled with the items *they* considered to be important. While many of these items provided an additional buffer against unsafe operations, they were conceived and implemented in piecemeal fashion with little regard for their interactivity and actual safety effect. Each new policy, procedure, or action was usually devised in reaction to a specific incident or brainstorm of the flight department hierarchy. Often, the items within this hodgepodge of standards, procedures, limitations, restrictions, admonitions, and guidance worked against one another, often confusing members of the organization and creating potentially unsafe conditions. Sound familiar?

Enter IS-BAO

The International Business Aviation Council (IBAC) was founded in 1981 to represent, promote, and protect the interests of business aviation in international policy and regulatory forums. It began with just five national business aviation associations (including the British Business and General Aviation Association) and now comprises 15 national and regional business aviation associations. One of its tasks is to work actively with ICAO and other international organizations to ensure that their standards and regulations are in the best interests of the international business aviation community.

In the late 1990s, it became apparent that organizations like ICAO and Joint Airworthiness Requirement bs might be moving toward some sort of certification of business aircraft operations. Rather than let these initiatives evolve into the requirement for an air operator's certificate, IBAC convened a meeting of its affiliates to explore the possibility of creating a set of international standards and best practices that would serve the worldwide business aviation community. Over a two-year

IS-BAO Elements

- Safety management systems
- Organization and personnel requirements
- Training and proficiency
- Flight operations
- Operations in international airspace
- Aircraft equipment requirements
- Aircraft maintenance requirements
- Company operations manual
- Emergency response plan
- Environmental management
- Occupational health and safety
- Transportation of dangerous goods
- Security

period, virtually all IBAC affiliates participated in creating global, performance-based standards and best operating practices. The standards and recommended practices (SARPS) contained in ICAO Annex 6, Part II, *Operation of Aircraft—International General Aviation Operations—Aeroplanes* was used as the foundation for the standard, ensuring universal acceptance. Importantly, the new standards were structured so that they could easily blend their State civil aviation regulations with industry best practices.

The International Standard for Business Aircraft Operations (IS-BAO) was introduced in 2002 via IBAC member organizations. The standard comprises a full range of topics applicable to all business aviation operations (see sidebar on previous page), large and small. While operators may choose to undergo an audit based on the standards to achieve IS-BAO registration, the real purpose was to provide business and corporate flight departments with standards and guidelines that could make their organizations more safe, efficient, and effective.

Notably, this standard was written by business aviation personnel for business aviation personnel. Experienced business aviation pilots, maintenance technicians, and managers representing a wide variety of operational backgrounds and experiences from around the world contributed to the IS-BAO.

Safety Management System   The core of the standard is the safety management system (SMS). It forms the centerpiece because it links and integrates *all* standards efforts behind a single goal: safety. Conventional safety programs throw together a bit of training, a smidge of standard operating procedures, a pinch of equipment, a touch of maintenance and a *soupçon* of management in the belief that the resulting stew will ward off incidents and accidents. Unfortunately, the cook's tastes may not cover all palates or situations.

A SMS is a "systematic and comprehensive process for the proactive management of safety risks that integrates the management of operations and technical systems with financial and human resource management." If you parse this mouthful, you wind up with, *systematic, proactive, integrates, and management* as the operative concepts. In other words, an integrated system looks to the future of the operations and manages it. More directly, it systematizes the normal hodgepodge approach to safety.

The system:

1. Establishes a safety policy tailored to the organization
2. Makes people accountable for safety-related tasks
3. Applies management techniques to control risk
4. Actively looks for hazards
5. Takes action to eliminate or reduce them (mitigation)
6. Tracks the action for effectiveness

The goals of a SMS are to:

1. Manage risk to a level as low as reasonably practicable
2. Optimize safety performance in an operations and business environment

The SMS concept is not new. Nuclear power, railroads, manufacturing plants, the military, and airlines have been using forms of well-integrated safety systems for decades. The civil aviation world is a relative newcomer to the concept and has drawn heavily on the trailblazing industries that came before them. While the concept sounds complicated, it is actually logical and easy to use once implemented.

The four concepts underlying the system as articulated by ICAO are:

1. Safety policy and objectives
2. Safety risk management
3. Safety assurance
4. Safety promotion

Each concept incorporates all of the issues raised above into a well-organized and integrated system designed to *assure* safety. Moreover, the safety of the organization can be measured through a risk assessment process and validated by an internal evaluation process.

Difficult to establish and more difficult to maintain you say? IS-BAO has devised a number of establishment aids known as the Safety Management System Toolkit that comes with the standard in order to step organizations through the process. Further, most national or regional business aviation associations already or will shortly conduct SMS implementation workshops to facilitate the process. But establishing the SMS and IS-BAO is not an overnight task; most flight departments require at least six months of steady, but not difficult, work to bring the program to fruition.

The Remainder   When you look at the sidebar on page 231, you should be familiar with the elements: it should look like a well-developed flight operations manual used by quality flight departments. The point being that most good flight departments already have a majority of the IS-BAO standards in place. Some of the standards may be a bit different, since they are based on ICAO SARPS; and elements, such as the training, maintenance, and emergency response plans, may have a few more items than the normal ops manual, but these differences are relatively minor. Besides, there should be some elevation of existing standards if an organization is to aspire to those considered to be world-class.

To help with the process of implementing the system, IS-BAO provides an Acceptable Means of Compliance (AMC) section in its Standards package, which are examples of

---

An IS-BAO registrant wrote, "We achieved IS-BAO registration and have discovered several side benefits:

1. Lower insurance rates
2. Better team concepts for all employees
3. Upper management approvals
4. Empowerment for all aviation departments to make the safest decision
5. Allowance for a better presentation of the flight department

We are proud of our accomplishments. I look forward to the evolution of the standard and the benefits that continued and meaningful collaboration will provide to our operation."

policies and procedures that may be used to satisfy the standard, and a generic company operations manual (GCOM) on which you may pattern your own emerging manual.

Again, the standards for these items are mature, well developed, and in wide use around the world. Unlike theoretical and esoteric standards often attempted, these are tried and tested by some of the world's finest flight departments. You will be in good company by adopting these new standards and recommendations.

Many think that IS-BAO is just for the large flight department and is way above the capabilities of the small, single-aircraft department. Since all features of the standard are scalable, a significant number of small operators have either purchased the standard or become registered. Also, special means of compliance are provided for single-pilot operations and very light jets (VLJ).

Air Operator's Certificate (AOC) holders will benefit from safety, procedural, and high-level standards provided by IS-BAO as well. While States provide additional regulations for certificate holders, they are not often well integrated or procedurally driven; State aviation authorities are not immune to the hodgepodge effect.

If you wish to become a registered IS-BAO flight department, you may contract with a qualified, independent auditor to visit your operation and ensure that all items shown in the standard are fully complied with and that there are records, when necessary, to prove compliance. In addition to generating a feeling of pride and accomplishment within your flight department, you will have been evaluated by a person experienced in business aviation operations evaluating your operation to a set of standards, not personal opinions.

Why IS-BAO?   This is a question articulated by many flight departments, who believe that they rank fairly high among the world's business aviation organizations. Why indeed?

While an AOC will not be required world-wide for business aviation operators, ICAO, EASA, and some States are or will shortly be requiring many elements of IS-BAO. The newly revised ICAO Annex 6, Part II now requires turbojet and large aircraft (>5700 kgs MTOM) operators to have an SMS, flight operations manual and training, fatigue management, and maintenance control plans for international operations. EASA has proposed similar standards within their recent operations notice of a proposed amendment. Finally, Bermuda and the Cayman Islands have already notified their business aircraft register that it will have to either comply with that State's stringent regulations or, alternatively, become IS-BAO registered. Therefore, international operations will soon require IS-BAO-like standards; why not incorporate the best?

The following reasons for adopting IS-BAO have been provided by flight departments themselves:

- *World-Class Standards.*   Be assured that IS-BAO standards have been devised, vetted, and revised by high quality people from your aviation community, not regulators or nonaviation specialists.

- *A Well-Integrated Living System.*   All parts of the standards are designed to complement one another. Importantly, the standard is revised annually to refine and upgrade its contents.

- *Safety Culture Development.*   Excellent safety cultures are developed by involving all members of the corporation, from the CEO downward, while providing them with appropriate safety tools and communications methods. The SMS serves as an excellent methodology for elevating the flight department's culture.

- *Effectiveness and Efficiency.*   The systemization of flight department processes and procedures promotes measurement of all activities within a flight department and the ability to fine-tune them.

- *Customer Confidence.*   Saying that your department adheres to world-class safety standards sends a strong message to corporate executives and passengers. The added degree of confidence creates a bond between flight department and passengers.

- *Cost Containment.*   How costly is an accident or incident? Aircraft damage, injuries, lost time on the job, personnel liability, or lost lift capability all add up to large sums. IS-BAO provides the ultimate insurance policy. While insurance companies will not normally say that premiums will be reduced as a consequence of IS-BAO registration, consideration is certainly given.

Stabilized Approach Criteria

All flights must be stabilized by 1000 feet above airport elevation in Instrument Meteorological Conditions (IMC) and 500 feet above airport elevation in Visual Meteorological Conditions (VMC). An approach is stabilized when all of the following criteria are met:

- The aircraft is on the correct flight path.
- Only small changes in heading/pitch are necessary to maintain the correct flight path.
- The airspeed is not more than V_{REF} + 20kts indicated speed and not less than V_{REF}.
- The aircraft is in the correct landing configuration.
- The sink rate is no greater than 1000 feet per minute; if an approach requires a sink rate greater than 1000 feet per minute, a special briefing is required.
- The power setting is appropriate for the aircraft configuration.
- All briefings and checklists have been conducted.
- Specific types of approach are stabilized if they also fulfill the following:
  ○ Precision approaches must be flown within one dot of the glide slope and localizer.
  ○ During a circling approach, wings should be level on the final approach when the aircraft reaches 300 feet above airport elevation.
  ○ Unique approach conditions or abnormal conditions requiring a deviation from the above elements of a stabilized approach require a special briefing.

An approach that becomes unstabilized below 1000 feet above airport elevation in (IMC) or 500 feet above airport elevation in (VMC) *requires an immediate go-around.*

- *Pride in Accomplishment.*   Most operators adopting IS-BAO find that the process of achieving registration provides a good team-building experience for the organization and yields a group of individuals that more fully appreciate the details of running the organization.

More than 1400 organizations have purchased the standard and 600 have become IS-BAO registered. If you are interested in exploring the standard further, see www.ibac.org. Better yet, speak to a registered IS-BAO organization about what the adoption of the standard has meant to the organization—all registered organizations are listed on the website.

IS-BAO has set a new standard for business aviation, one that has elevated performance levels in all aspects of flight department life. Explore its benefits.

While you may fly with the same individual every day, SOP creates expectations between the two of you that form the foundation of preflight and cockpit communication. If you fly regularly with more than one person, the SOP is an essential part of the cockpit routine. If you fly with contract pilots, the SOP gives the temporary pilot an idea of what to expect and makes him or her a full participant in the process instead of a student to be trained.

Developing Your Own   So, you don't want to go with IS-BAO. OK, but while you are developing your own SOP, be sure to include operational limitations, too. Private operations, even in large, long-range turbojets, are not subject to many limitations other than those contained in the aircraft flight manual. Issues, such as minimum takeoff visibility, operations in the vicinity of thunderstorms, and how much safety factor you will add to the Airplane Flight Manual's (AFM) runway performance numbers, are all critical and must be addressed.

Where to start developing the SOP? Your aircraft training provider's basic introductory manuals for the type of aircraft you fly are a good start. For limitations, look to the regulations imposed on the airlines and charter operators for guidance. ICAO standards and recommended practices (SARPS) also provide good guidance. For SOP ideas, search for advisory materials in the following organizations:

- NBAA
- ICAO
- FAA
- Transport Canada
- CAA UK
- CAA Australia (CASA)

Procedures must be developed in a measured and rational manner, too. Arbitrary or whimsical procedural changes are soon seen for what they are and either winked at or ignored completely. Standards must be firmly based in reason and experience, tested, and validated.

Everyone must be singing from the same hymnbook as well. There must be a departmental SOP, one that is up to date and adhered to by *all* participants. The one that was developed for the Jetstar in 1971 may not be of much use with the 7X. The training manual you got from the contract trainer is good but it's not *yours.* You have to *own* the procedures as a department; somebody else's won't do.

Take time to carefully evaluate everything contained in the airplane flight manual, operations manual, and contractor's training manual. You may not want to use it all or may find that some things are missing.

Developing SOP is a joint venture, too. Every pilot should have a chance to comment on and discuss a draft SOP unless regulatory issues or hazardous conditions dictate otherwise. If consensus cannot be gained on a certain procedure, the chief pilot should act as the tiebreaker.

Operations

The subject of operations contains many subsets; virtually all activities having to do with the actual process of scheduling through flying the flight and delivering the passengers are a part of the operations process. While it would be tempting to address every subset, this book is not long enough, nor would every issue covered be applicable to all departments. What follows are a number of subjects that seem to be of universal interest within flight departments.

Operational Control

The fundamental concept of operational control is essential to any aircraft operation. This term defines who or what entity is in charge of a flight from beginning to end.

ICAO defines the term *operational control* as the exercise of authority over the initiation, continuation, diversion, or termination of a flight in the interest of the safety of the aircraft and the regularity and efficiency of the flight. The FAA uses basically the same definition for any type of flight operation. Under ICAO, *operator* is defined as a person, organization, or enterprise engaged in or offering to engage in an aircraft operation. Finally, *pilot in command* (PIC) is the pilot designated by the operator or the owner as being in command and charged with the safe conduct of a flight. Note that this discussion will be only for private aircraft operations; commercial operations, including air taxi and fractional ownership, operate under specific and more stringent rules because of the for-hire implications of their operations.

So, who is in charge? That is usually left to the owner/operator to decide. This entity clearly has responsibilities for operational control issues, but these responsibilities are usually delegated to someone in the flight department or are shared between the owner/operator and the flight department. It all begins when someone says, "Let's fly." In corporate/business flying, this is normally one or more individuals within the operator's organization who is authorized to initiate a flight. Then, a designated person within the flight department reviews the request for operational capability and safety issues and places the flight on the schedule. Then, unlike many commercial operators, it is up to the PIC to initiate, continue, divert, or terminate the flight. Yet the PIC is also governed by the operator's policies, procedures, and standards, which one hopes are defined in detail within the company flight operations manual.

Why is something so seemingly simple and routine so important? While the PIC in many flight departments operates as his or her own dispatch authority, it is important that all concerned understand the chain of responsibility and the need for well-defined operational practices and standards. Without this interlocking chain of command and these standards, the organization is less organized, safe, and efficient. Therefore, the operational control concept deserves a good deal of attention and refinement to ensure safety and compliance.

Mr. Fixit Travels in Style

A large manufacturer of window glass fabricating equipment introduced a radical new technology that was enthusiastically embraced by its customers. Unfortunately, operational problems with the new equipment soon had customers clamoring for technical support to solve equipment breakdowns. The manufacturer had qualified technicians available but often could not reach its customers in less than two days because both manufacturer and customer were often not located near airline airports with adequate service.

In an effort to support the company's largest customer who was having significant problems, the CEO dispatched the best technician in the company via King Air service. The technician reached the plant site hours after the call for help was received and had the machine back in operation before the end of the day.

As a result, the company aircraft is now used on a priority basis for what the CEO refers to as "911 calls."

Setting Limits

Aviation regulations, AFM, and insurance policies are replete with limitations that form a structure in which all aircraft operate. While these items may appear to be overly restrictive in some cases and certainly comprehensive, they may not cover everything necessary to ensure safe operations for your department. The FARs, AFM limits, and insurance provisions under which corporate aviation operates are designed as minimum standards and generalized limits. They do not refer to specific flight crew, weather, or operational situations that may require additional attention and, ultimately, limits.

For example, while the AFM may state that the aircraft may operate into an airport with a runway of a specific length under certain wind and weather conditions, many flight departments set minimum runway lengths (and widths) for their aircraft types to ensure an additional margin of safety. Similarly, they may restrict less experienced pilots-in-command to higher takeoff and landing weather minimums or even longer runways until they reach some level of experience in the type of aircraft to which they are assigned.

Devising additional limitations should be done carefully and in response to a real or anticipated condition that could create a hazard or administrative failure. Overly restrictive limitations or restrictions may unnecessarily reduce the operational effectiveness of the department and engender the ridicule or animosity of department personnel.

Flight Crew Scheduling

The safety, efficiency, and reputation of a flight department's operations rest in large part on the performance of the crews assigned to fly individual trips. Therefore, care should be exercised in what may at first be a process that involves only the practice of rotating crews equitably through the schedule. The morale of the crews and its effect on the passengers' perception of the department are often at stake during the scheduling process. The tendency to allow this process to become routinized should be resisted.

A scheme should be devised to both equitably and logically assign crews to trips (see sidebar on next page).

The tendency to assign crew pairings without regard to the personalities involved may sometimes create situations that either are potentially hazardous or create unnecessary friction between crewmembers. While the personalities of the crewmembers ideally should not enter into the scheduling equation, incompatible crew pairings should be examined and dealt with. In small departments where there are few possibilities to avoid certain pairings, the individuals must be counseled and the problem resolved by their supervisor to ensure safe and professional operations. In larger departments where multiple crew qualifications allow for some crew flexibility, the temptation to allow some "custom pairing" should be resisted as well; if not, the problem will inevitably worsen. Custom pairings are also unfair to the other crewmembers, since their scheduling flexibility is reduced.

Custom pairings may be justified when considering the experience levels of crewmembers, however. Pairing the newest copilot with the just-qualified captain may not be a good procedure. While both are undoubtedly qualified, the combination of two junior pilots may not be good judgment.

Standby status for flight crews is an unpleasant fact for corporate pilots, since the true value of the corporate aircraft is its on-demand feature. But there should be some rational, predictable considerations applied to this status. Two-pilot flight departments have little choice regarding their standby status, often being considered to be on duty 24 hours a day, 365 days per year, responding to a pager as the primary source of "trip planning." Some small departments have devised duty schedules with their corporate management that allow them to take half- or full-days off from responding to trip demands. More frequently, a single crewmember is given scheduled time off, and a contract pilot

Limits, Restrictions, and Standards

- Takeoff and landing weather minimums
- Airports—size, facilities, instrument approaches
- Severe weather operations
- Fuel—planned and minimum landing
- Inoperative equipment (beyond the minimum equipment list)
- Flight crew
  - Pilot currency
  - Health and medical certification
  - Low time in type and less-experienced pilots
  - Flight and duty times
  - Multiple aircraft qualifications
- Training
- Aircraft noise and noise abatement
- Passengers, baggage, and cargo to be carried

is put on standby in his or her stead. Alternatively, a charter company offering a suitably short response time can be alerted to accommodate the company's potential needs during a crew stand down. Larger flight departments often have the flexibility to provide crewmembers with scheduled days off and should integrate these into the schedule as a matter of course.

Multiple aircraft qualifications for individual pilots have generated some controversy. There is no regulatory and usually no insurance restriction limiting the number or type of

Flight Crew Scheduling Considerations

- Rest and duty time limitations
- Crew pairing—senior/junior and experience in type mix
- Senior/primary passenger preference
- Pilot currency requirements
- Seniority
- Nights away from home base—short and long term
- Experience and upgrade requirements
- Scheduled time off
- Personal preference (when possible)

aircraft in which an individual may act as pilot in command. But the complexity of today's aircraft and the recurrent training demands placed on pilots beg for a limit to the number of aircraft a pilot can safely keep fresh in his or her brain. The industry norm for the number of turbine-powered (including turboprop) aircraft a pilot can operate as PIC appears to be no more than two. Mixing aircraft categories (airplane and helicopter) becomes more complex, yet a number of companies allow pilots to operate as PIC in at least one turbojet and helicopter simultaneously. Regardless of types or categories, many companies require a certain number of hours per month or quarter for a pilot to remain qualified as PIC and sometimes semiannual recurrent training in *all* types.

Flight Crew Duty Time Limits

This topic is subject to much discussion and debate in the on-demand aviation community. The unpredictable nature of corporate aviation operations is both a blessing and a curse: a blessing for the passengers who enjoy on-demand air transportation and a curse for flight department managers and crews because of their inability to provide structure and predictability to their lives. But the controlling element in determining duty time is safety. Whatever determinants are devised, the manager should always ask himself the question, "Can the flight crew perform their mission safely, given their rest history and the nature of the current mission?"

Corporate aviation duty and rest practices vary widely because of the diverse missions within the community. Duty time limits usually fall into the range of 12 to 18 hours, with a common average being 14 hours. Type of operation, number of legs, length of legs, and availability of day rooms (for rest) are all variables that are used to alter and modify these numbers. NASA/Flight Safety Foundation studies have provided guidance to corporations seeking flight and duty time limits for their crews. Their recommendation for two-pilot flight crews is a maximum scheduled time of 14 hours of duty and 12 hours of flight time and a minimum of 10 hours of rest between duty periods.

Factors that may be considered in refining the duty time equation include total flight hours in a duty period, number of legs flown per day, and length of legs. A 14-hour duty day that includes 4 hours of scheduled flight

Altitude Alerter Procedure

On receipt of a newly assigned altitude, the

- Pilot-not-flying (PNF) will read back the altitude and set it in the altitude alerter, point to it, and repeat the altitude.
- Pilot-flying (PF) will point to the alerter and state the altitude observed in the window.
- PNF will note any discrepancy.

time is quite different from one with 8 hours of planned flight time. Some departments adjust the duty day based on the number of flight hours over 6 scheduled hours. Similarly, a duty day consisting of 7 planned legs is probably more tiring than one with just 2 legs. If flight time is to be accounted for, 30-minute legs are more demanding than 3-hour legs. Certainly, all of these factors have some bearing on the duty time issue; however, none of them have to be used, or they can all be used to modify duty time limits.

Rest time for corporate aviation usually varies between 8 and 14 hours, 10 hours being the norm. Again, the variables of consecutive duty days, circadian rhythm considerations, type of duty day, and availability of rest facilities enter into the rest time equation. The cumulative effect of three consecutive 14-hour duty days can be impossible to fix with even fully allocated 10-hour rest periods; therefore, some cumulative duty time limit should be imposed, such as no more than 55 hours in five days. Ideally, a 24-hour rest period is usually provided at least once every 7 days.

While air carrier rest and duty times are inviolate, some flexibility must be built into those times used for corporate aviation. Canceling a flight because a crewmember may run over his scheduled duty time by one-half hour may not make sense if the aircraft, weather, and crew are in good shape. Therefore, the authority to extend duty times or contract duty times once the flight has begun, by some finite amount—usually not more than two hours—may be given to the flight crew as a consensus decision. Management should be the final arbiter of duty time extensions and should make this a standard procedure that involves conscious risk assessment as well. The ability to alter these times should be used sparingly, or the standards will be undermined.

More problematic are the long, duty-day issues that arise with ultra-long-range aircraft operations, often capable of flying for 14 hours or more. Augmented crews (three or more pilots), double crewing, and en route crew changes are all means of dealing with what may become a 24-hour duty day. No clear guidelines are available for these operations at the time this is written. Suffice it to say, caution is a prime consideration when contemplating very long duty days, especially because operations in foreign airspace are usually more challenging than domestic flying.

Equally problematic is the concept of operations during the window of circadian low, normally considered to be 0200–0600 local time. Beginning, operating through, or ending during this period has been shown to interfere with a person's normal day-night/wakeful-sleep schedule, often to the point of significantly degraded human performance. Therefore, duty and flight times should be shortened when flight crews operate during these times.

In the early 2010s, the concept of fatigue risk management systems (FRMS) was introduced by ICAO and adopted by many States to more carefully analyze and customize duty, flight, and rest times for all forms of civil aviation. This approach has the advantage of modifying formerly rigid standards to accommodate the many variations occurring within the various operational profiles. Applying the principles of risk assessment and mitigation to a field as diverse as fatigue management has yielded significant progress in controlling this essential aspect of flight operations.

How Many Pilots?

While many business and corporate aircraft require two pilots to fly under the national authority's certification regulations, more than two pilots are often required to safely and effectively run the operation. Scheduled time off; flight-, duty-, and rest-time limits; and administrative and training duties all work on a continuing basis to take away from the time available to pilots in flying trips.

Too often, single-aircraft operations require their two pilots to be on call 24 hours a day, 7 days per week in anticipation of a last-minute or pop-up trip. While these pilots may have a significant number of days off each month, none may be planned; all occur just because the aircraft is not scheduled. This uncertainty regarding one's ability to have a predictable life is the principal complaint within the corporate pilot community and ranks high on the list of reasons pilots give for leaving such demanding operations.

The natural consequence of not having enough pilots is the necessity to hire a day or contract pilot to fill-in for one of the two full-time pilots who must go to training, take a vacation, or gets sick. While many contract pilots are well trained and experienced, the last-minute nature of getting a fill-in pilot does not lend itself to being highly selective. Unfortunately, the pilot chosen to fill in for a day may be only minimally qualified in type, may not be current, and may have little knowledge of the company SOP. Consequently, the company PIC spends an inordinate amount of time briefing the contract pilot and overseeing his or her performance—hardly an ideal circumstance.

The best solution for avoiding this uncomfortable situation? Hire at least one more pilot. How many pilots to hire? Use the NBAA pilot staffing formula as a baseline to determine the number of pilots required (see Table 6.2).

Some companies use 1.5 crews (three pilots) per aircraft as a rule of thumb to estimate the number of pilots required. This would apply only to aircraft flown not more than 400–500 hours annually and with an average number of overnight stays. It is essential to customize pilot requirements.

Consideration should be given to hiring part-time or temporary flight crews, especially during vacation periods and heavy travel times. However, any part-time or temporary pilot should meet experience and proficiency requirements for full-time pilots, as outlined by the company. They must be familiar with the flight department SOP, aircraft checklists, and crew callouts. Importantly, they should exhibit the same levels of professionalism and deportment required of full-time crewmembers.

Considerations Regarding the Number of Pilots Required

- Number of aircraft operated
- Company flight time and rest considerations
- Aircraft hours per year
- Number of concurrent trips
- Number of trips that keep the pilot flying for a number of consecutive days
- Number of trips overnight (RONs)
- Number of night flights
- Vacation policy
- Training policy
- Manager/pilot ability to fly a "standard" amount

Checklists

"List complete, ready to go," sayeth the copilot. You taxi into position with the ever-present question, "Are we really ready?" Not that you distrust your trusty copilot, a fellow captain and cockpit companion of many years, events, and escapades, but does the checklist contain everything you really need to be safe for this takeoff and was the copilot's recitation and review of the sacred items thorough and complete? Philosophy aside, you always execute your own private killer list—parking brake, controls, spoilers, flaps, igniters, and anti-ice.

Checklists are made to backup the all-too-fallible human memory. Certain controls must be set correctly within the aircraft to accommodate upcoming or existing conditions. All is well until we forget or misset or set the

Description	Calculation	Totals
Number of Days Pilot Needed		
Work days per pilot	52 weeks per year × 5 days per week	260
Days not available	• Vacation: 15 days	
	• Holidays: 11 days	
	• Sick leave: 5 days	
	• Training and physical examination: 10 days	
Subtotal of days not available		−41
Total days available for duty	260 days − 41 days	219
Number of crew seats (if two per aircraft)	2 aircraft × 2 pilots per aircraft	4
Number of operating days per year	7 days per week	365
Number of flight-crew days per year	4 pilots × 365 days	1460
Number of pilots required (rounded up)	1460 flight-crew days required/219 days available	7
Estimated Pilot Hours per Year		
Estimated number of flight hours per year		850
Number of flight-crew seats	1 aircraft × 2 pilots	2
Total number of pilot hours	850 flight hours × 2 pilots per flight crew	1700
Duty/flight ratio*	3 hours duty/1 flight hour	3
Number of duty hours per year	1700 pilot hours × 3 (the duty flight ratio)	5100
Total hours available for duty	219 days × 8 hours per day	1752
Number of pilots required (rounded up)	5100 duty hours per year/1752 hours available for duty	3
*This ratio assumes that a pilot spends at least three hours on duty for each hour in the air. The duty time includes preflight, postflight, and ground time. The ratio could vary for a company that often requires long periods of waiting between flights on the same day.		

TABLE **6.2**   Sample Calculation of Number of Pilots Required (*NBAA*)

wrong control. The consequences of such boo-boos may be minor—no engine anti-ice without visible moisture is not a big thing. But an attempted takeoff on a contaminated runway with the parking brake set is a disaster in the making.

Dire consequences come in many forms: neglecting to set the altimeter to a local setting on the way down may net you a violation from ATC when you level off 500 feet low. Failing to turn off the seat belt sign in smooth air during cruise may bring embarrassing queries from treasured passengers. Forgetting to set the flaps correctly prior to shut down may bring good-natured gibes from your ground crew. Not lowering the landing gear may create a settling sensation late in the flare followed by unusual noises.

Every pilot has a favorite checklist story: failing to shut down engines, not turning lights on at night, missing pressurization needs, or securing all generators at an inappropriate time all make for good war stories around the pilot lounge. Many of these stories could have led to more disastrous consequences had the circumstances or checklist item been a bit different. Many times, we fail to realize the real purpose of the checklist and how to use it during different phases of flight. More importantly, there may be a lack of understanding among crewmembers as to the appropriate use of the checklist. First, why use checklists?

We're All Human   The need for checklists is an acknowledgment of a common human failing that prevents us from remembering long or complicated sequences of actions required to consistently produce predictable outcomes. In other words, we forget. Pilots and nuclear power plant operators are not the only ones who need assistance in remembering; planning calendar and smartphone manufacturers would be out of business if all of us could remember what comes next. Aircraft operating checklists are part of the total discipline that goes into safe, predictable aircraft operations, whether the platform be a Tri-Pacer or a Citation X. We use checklists because we don't want to forget important items of aircraft operation.

Verify or Do?   There are different ways of using checklists that sometimes escape mention among crews. For instance, the "challenge-and-reply" or "challenge-response" methodology is the most popular method of checklist employment. Does this mean "challenge-*verify*-response" or "challenge-*do*-response"? In the first case, the person responding to the checklist has done all or most of the items before calling for a particular segment of the list, and in the second case, the responder is waiting for the challenge prior to completing the associated action. The verification method is most common for normal checklists, while the do-list method is used for abnormal or emergency lists to ensure, or make doubly sure, that an action is taken and it is the correct one, too. Is this clear among your crews?

What about "silent" checklists, where the PNF notes the accomplishment of each item in sequence and reports completion of that checklist segment to the PF? Is this method allowed for certain segments at *all* times? Consistency is the key—if the checklist is normally challenge-and-reply, but during a time of high workload, the PF (or PIC) curtly orders, "Complete the list to yourself," does this satisfy the purpose of the checklist?

Short Circuits in the Process   Perhaps the biggest problem with checklists comes from being distracted before you complete them. Either the list is never completed or, more insidiously, it is resumed after one or more items is skipped over. The culprits? Malfunctions, traffic watch, ATC communication, approach charts, and radar monitoring. Each of these items somehow demands greater or more immediate attention than does the completion of the checklist. These events should raise alarms in the minds of aircrews because of their distraction potential.

Next in line for checklist subversion is complacency. It's the seventh leg of the day and the 22-item taxi checklist begins to blur into the background—"All of the items were set properly at shut down, why would they change?" Or, forgetting to initiate the descent or approach checklist may catch the cockpit crew napping or unaware. Reciting the checklist from memory is another favorite method of accelerating the checklist, especially for short segments like takeoff and landing. But doing so undermines the entire rationale of the checklist—so, why bother?

Checklist "holds" are another way to lose your way through a critical sequence of events. ATC demands, aircraft configuration changes, malfunctions, and operational delays create natural stopping points in the checklist litany. But how do you know where to restart the list?

The most insidious of all checklist faults may lie in not hearing the words, "[segment] checklist complete." This statement should spell the end of one chapter of aircraft operation and signal that there may be another phase of flight requiring another checklist segment close behind it. Inquiring whether a particular checklist segment was completed should trigger real concern in the mind of the cockpit crew—doing it all over again may be the only way to resolve the question.

Designing a Good One   Checklist design figures heavily into efficient and effective checklist usage. Are there too many items on the list? If certain items were left off the list, what would be the effect on safety and security of the aircraft and its passengers? Should there be a turn-around checklist for quick turns? Is the checklist unambiguous? For example, does it call for a *specific* setting, not just "Checked and set"? Are all items in the proper sequence, that is, one that follows a logical flow? If your answers to these questions leave you in doubt, consider changing your checklists to the point where they make sense for your operation and fit your operational culture and philosophy as stated in your Flight Operations Manual.

Many regulatory authorities permit operators who are not required to hold an aircraft operators certificate to modify the normal portions of the manufacturer's checklist without notification. Most authorities will not permit modification of the manufacturer's emergency checklists without approval. If in doubt, check with your regulatory authority.

The most important issue regarding checklists is having the discipline to use them consistently during all phases of flight. The aircraft checklist is an integral part of a *system* of references, procedures, and actions that is designed to operate a specific aircraft safely and efficiently. Removal of this essential part of that system invites misunderstanding, errors, and potential disaster.

The Tyranny of Automation

Are cockpits becoming over-automated or overly sophisticated? Are cockpit crews becoming overly enamored with the new bells and whistles appearing in cockpits? Are flight crews drawn to new automation features because of their Nintendo™ appeal?

These questions have been and will continue to be debated hotly in flight departments and airlines around the world because of an increasing number of accidents and incidents associated with automation. But manufacturers continue to automate and refine automated devices within the cockpit. Obviously, the people who build, certify, and buy these devices believe that automated systems are beneficial to the operation of modern aircraft. Is there a solution to this supposed tyranny of automation?

Problems   Automation-related incidents are common knowledge because of the increasing frequency of their occurrence. They fall into several broad categories, including:

- *Data Entry Errors.*   The KAL 007 shoot-down is the most notable of this type of error that pits the crewmember's accuracy and spell-check skills against a device, usually a flight management system (FMS) entry key pad, that is unswervingly faithful to whatever it receives. A more common occurrence is entering the wrong altitude in an altitude alerter.

- *Monitoring Failures.*   Missing an FMS message, annunciator signal, or FMS page or mode change are common faults under this scenario. Not knowing what to look for or how to find it are the principal causes.

- *System Workarounds.*   You can't get the darned thing to do what you want, so you find a clever, albeit dangerous, way to "fool" the machine into getting it to do what you want. For instance, entering manual top of descent locations into the FMS to alter the computed points may yield other, unwanted, effects.

- *Mode Misapplication.*   Autopilot vertical speed being selected instead of indicated airspeed or Mach number is a common fault. More insidious is working with a very complicated system that yields unsuspected or unanticipated effects such as the recent Airbus crashes involving engine control and instrument approach effects.

These faults rarely occur in isolation or without contributing factors. Haste, inattention, fatigue, distraction, and other system factors may begin the sequence of events that lead to the incident. As with virtually every type of accident, automated or not, it is a *combination of failures* that leads to the ultimate end result. In the final analysis, this means that there are inadequate checks and balances available to the crew to detect and interrupt the fault chain. Awareness, judgment, and good standard operating procedures will most likely serve as chain breakers.

Some of the conditions that may lead to unsafe conditions in advanced technology aircraft (or just about any machine) are:

- *Mode Awareness.*   Operating in one mode when you think you are in another: heading hold instead of nav track, vertical speed instead of indicated airspeed hold, or fuel computed to an intermediate point instead of destination. Mixing a desired effect with an improper mode (or unknown/unanticipated mode) is the root cause.

- *Situational Awareness.*   Where am I, where am I headed, and where do I want to be? Navigational errors are popular consequences of lack of situational awareness. This may stem from too much attention to an aircraft problem or automation device. More than just a few seconds attention to a nonaircraft attitude or nonnavigational problem will rapidly put the crew out of the picture.

- *Systems Awareness.*   Electrical, hydraulic, and fuel systems may be complicated but, under normal circumstances, the automated systems manage them quite well. However, poor monitoring or poor knowledge of systems and their automated masters may lead the unwary crew astray. Failure of the electrical or the hydraulic systems or fuel exhaustion are common consequences.

- *Increased Heads Down Time.*   Automation may demand so much care and feeding that both crewmembers spend uncommon amounts of time ministering to the devices that are supposed to relieve them of routine chores. Result? No one is minding the store. Aviating, navigating, and communicating are sacrificed to undivided attention to the little black boxes, and nobody is looking outside the cockpit for other traffic. (TCAS hasn't solved all mid-air collision problems just yet.)

- *Overdependence on Automation.*   No system is infallible, even one that is designed to detect failures in other systems. Certain subtle or insidious failure modes may allow unexpected events to occur. Fuel balancing computers and navigation system comparators are examples of automated systems that can lead us astray when they malfunction. But when the complexities of a very sophisticated aircraft conspire to hide or confuse complex situations, the pilot becomes more a passenger than a human commanding the situation.

- *Interrupted Crew Coordination.*   When an automated system requires or captures the full attention of both pilots, the stage is set for disaster. The object of crew coordination and crew resource management is a system of checks and balances designed to help the crew to complement each other. When this link is compromised, the team is dissolved, and the two crewmembers become a single entity; it's as if one crewmember left the cockpit.

Solutions The antidotes to these faults all address human frailties and shortcomings in much the same manner as aviation safety practices have evolved over the past 100 years. Perhaps they have a new spin to them because of the automation devices involved, but they still address awareness, training, standard procedures, and crew coordination.

- *Awareness.* Sure, we all know of the hazards associated with automated cockpits, but how often are we lulled into a false sense of security by these marvelous and clever devices? Although the analogy may overstate the case, it's almost like a human's relationship to a domesticated wild animal: the animal is usually friendly but it has the capacity to bite when least expected.

- *Training.* Do all crewmembers receive initial and recurrent training in all installed automated systems? Are *all* modes of operation and potential pitfalls practiced and explored? If the training is received on part-task or computer-assisted training devices, do these devices adequately replicate the system's interface modes and faults as they appear in the real aircraft?

- *Standard Operating Procedures.* Good SOPs can solve many automation problems if it is devised (and revised) to reflect actual operating conditions and practiced by all crewmembers. Typically there will be one or two crewmembers who are more "creative" than others and who devise their own "SOP." The result may not be chaos, but the lack of predictability for the remaining crewmember greatly diminishes the effect of the checks and balances afforded by having another set of eyes (and a brain) monitor the principal crewmember's actions. Instituting good SOPs and ensuring their use is one of the principal challenges of modern flight department managers.

- *Crew Coordination.* When the regulatory authority issues an aircraft certificate, considerable time has been spent in determining whether a second crewmember is required according to the anticipated cockpit workload. If it's too much for one crewmember, the mandatory minimum is raised because the aircraft's systems are so complex and the dynamics so rapid that one person can't handle it all. The introduction of increased amounts of automation has done little to change the certifying authority's opinion about the necessity for a second crewmember on modern turbojet aircraft. The second person is there not only to divide up and share a large number of tasks, but also to serve as a backup to the pilot and as a safety and systems monitor. If a second person is present, ensure that he or she is being used for the intended purpose.

The concept of pilot as monitor has been popularized in the joke stating that aircraft will eventually be crewed by a pilot and a dog: the pilot to monitor the automated systems and the dog to bite the pilot if he touches anything. The second pilot doesn't have to bite; a word will do.

New Problems The emergence of totally electronic and fly-by-wire systems has greatly increased the complexity of today's aircraft. The behind-the-scenes automated responses of the aircraft to abnormal and emergency situations may remain largely unknown or unappreciated by the flight crew, leaving them figuratively, and sometimes literally, in the dark regarding the reconfiguration of systems within the aircraft. Since dozens or scores of actions may be taken by the aircraft automatically in response to a single failure, such as a generator going offline or a hydraulic system malfunction, the consequences of these "hidden" actions may elude the flight crew and create other potential problems. Therefore, readily accessible information regarding system degradation, load shedding, system sharing, intra-system activity, and other consequent effects must be readily available to flight crews to enable a full understanding of the shape and size of aircraft system responses.

Unfortunately, very complex aircraft reactions to failures may test a flight crew's knowledge or ability to adequately assess the immensity and implications of the aircraft's largely hidden actions. Unfortunately, the appearance of multiple component and system failure indicator lights may tend to confuse and conceal rather than inform. This makes initial, recurrent, and interim or ongoing training essential for dealing with modern, sophisticated aircraft systems.

Aircraft Airworthiness

ICAO and most country's regulations require the pilot in command to determine that the aircraft he or she is about to fly is airworthy. This means that the aircraft inspection, life-limited items, and airworthiness directives status are current in accordance with the manufacturer's inspection program and applicable regulations. Additionally, an aircraft maintenance technician should have cleared any aircraft discrepancies. Unfortunately, all of these items are often delegated to the AMT caring for the aircraft, leaving the pilot clueless about these critical airworthiness elements. A system should be devised to formally communicate these items, preferably in writing, to the pilot (see "Airworthiness Determination" on page 269 in Chap. 7).

Aircraft Discrepancies   It's often said with tongue in cheek that aircraft discrepancies only occur on the final leg home, never on an outbound leg, where the discrepancy might ground the aircraft. This miraculous occurrence certainly keeps the department's reliability rate high and provides the pilots with more flight time. Yet, in the process, the FAR Part 91 inoperative equipment rule may be stretched to the breaking point.

Gridlock Solved

A large, multinational company based in the Chicago area was plagued by the fact that company executives could not readily get from one location to another in the area due to heavy road traffic. A 15-mile trip undertaken during the wrong time of the day could take more than an hour.

An experiment using a chartered helicopter led to the acquisition of a five-passenger helicopter capable of flying in most weather conditions. The 15-mile trip rarely takes more than 10 minutes, except when the ceiling and visibility necessitate an air traffic control clearance. Even then, 25 minutes is tops for the trip. One of the frequent flyers estimates that the helicopter saves him 5–8 hours a week. The way he puts it: "The helo gives me an extra day per week."

Handling aircraft discrepancies is a highly individualized process in flight departments. Some flight departments never write down a discrepancy for fear that an inquiring regulatory inspector will discover an "open" discrepancy and issue a violation. Still others insist that all squawks be committed to writing, no matter how minor. These departments usually have a well-developed, deferred-discrepancy procedure to accommodate the minimum equipment list (MEL), too. Most flight departments come down somewhere between these two extremes, often recording only "downing" gripes.

Those departments that don't get much practice writing up discrepancies depend upon verbal communications between pilots and technicians to convey the news about aircraft airworthiness. While this may work well for operations that have just one flight crew and one technician, when these numbers grow, communication regarding aircraft fitness begins to suffer. There is much to be said for maintaining a permanent record of aircraft discrepancies and, perhaps the most important element, their write-off by the technician. Without this final detail, the pilot may not be certain of the aircraft's airworthiness status.

Recording discrepancies provides a complete communication between flight crew and technician: The technician understands the discrepancy and the flight crewmembers are apprised of the status of the discrepancy, which is either repaired or deferred in accordance with the FARs.

Ideally, a maintenance technician will meet each returning aircraft to discuss discrepancies with the flight crew. Whether this happens or not, the need for complete and legible discrepancy write-ups is essential to a safe and efficient flight department operation.

Helicopter Operations

The operational flexibility of rotary-wing aircraft provides a great advantage to some corporate operations. This same flexibility, however, gives rise to more *ad hoc* operational characteristics and less control over their operations. While the corporate helicopter safety record is good, the number of incidents among this community remains high for the number of flights.

The independent and largely unmonitored nature of these operations makes the use of experienced pilots a necessity. The lack of a flight plan and the short leg lengths for most flights contribute to the *ad hoc* image of helicopter operations. This, coupled with the ability to operate into unprepared landing sites, make helicopter operations a candidate for closer monitoring and controls. Therefore, additional standards and limitations unique to helicopter operations must be devised to ensure that sufficient guidance is available to pilots.

Passenger control in and around helicopters presents special problems due to the nature of operations, especially at remote sites. Boarding and discharging passengers with the rotor turning should only be allowed

Suggested Helicopter Operating Standards

- Minimum experience levels for pilots
- Landing site restrictions
- Minimum altitude/speed restrictions
- Flight following
- Passenger manifest control
- Weather operating minimums
- Required fuel reserves
- Night flying restrictions
- Passenger handling procedures
- Special/additional maintenance requirements
- Special safety procedures/precautions

if the passengers are accompanied by an experienced person who can guide them away from danger areas. Once passengers are aboard, ensuring that seat belts and doors are secure is often difficult, given the configuration of the interior and the other duties of a single pilot.

The unique capabilities and short flight duration of helicopter operations may require different management techniques from those used in fixed-wing operations. The Helicopter Association International and the International Oil and Gas Producers offer operational and management guidance for rotary-wing operations.

IS-BAO now contains detailed standards for helicopters and provides an additional qualification, if desired. This qualification incorporates mission specific standards (MSS) for individual types of operations, such as emergency medical transport, law enforcement, tour operators, offshore oil operations, and eight more specific types of operations. These additional qualifications are added for those who are IS-BAO registered when audited by specially qualified auditors.

Chartering Aircraft

Flight departments go to great lengths to give their companies the best possible service in air transportation. Scheduling, management, maintenance, crew training, and equipment upgrades are all given a high priority to ensure safe, efficient, and reliable service for the customer. This level of service usually builds a level of trust and confidence between members of the flight department and their passengers, one that serves them both well. But what happens when the boss has to get to an important appointment and a department aircraft is not available?

Of the two possibilities, airlines and charter, the charter option is often considered preferable. The service is in theory similar to that provided by the flight department, and there are certainly plenty of on-demand air taxi companies available, with more than 2500 operating certificates current in North America alone. But is the service similar to the flight department's, and if it isn't, what are the implications for the flight department?

If one of your corporate executives is exposed to a charter company that does not stick to the prearranged schedule, shows up with a old, dirty aircraft, gives an unnecessarily rough ride, and does not show the proper courtesy, the executive will undoubtedly be unhappy. While this should make them appreciate your operation, it may turn them off to all on-demand air transportation. High-class passengers expect the same level of service, whether you provide it or some outside agency does; the effect on the passenger should be the same. If it isn't, the passenger is unhappy.

Many flight departments consider it their responsibility to provide all on-demand air transportation, not just that provided with their own assets. Some go to great lengths to make the operation appear the same to the passengers, arranging for an aircraft similar to theirs and putting one of their pilots in the right seat of the chartered aircraft (that pilot must meet the air taxi operator's requirements and be named on his certificate). As a minimum effort to ensure high quality charter transportation, the flight department conducts an evaluation of the charter operator.

This does not need to be a full audit, but a check to see that you are receiving something above the minimum mandated by the FAA. The effort will require some background research and a brief visit to the operator. Some items to check for in a prospective charter operator:

- FAA FSDO for certificate actions against the operator
- Experience levels of the director of operations and maintenance
- Type and physical condition of the aircraft to be used

- Flight hours in type and within the last 90 days for the crews to be assigned
- Type and frequency of training for flight *and* maintenance crews
- Type of maintenance and inspection program used
- Inspection, airworthiness directive, and life-limited components status records for the aircraft to be assigned
- Reputation of the operator among other flight departments and his peers
- Competitiveness of the operator's charter rates
- Would you allow your family to use the operator?

The fact that you are asking these questions should tell the prospective operator that you expect superior service; he has lots of competition. (See also "Aircraft Charter" in Chap. 3.)

Knowing the Regulators

The average flight department chooses to avoid any voluntary interaction with their national aviation regulatory authority responsible for their activities. They are generally viewed as the "police" and, as such, are not subject to normal interaction, social or professional. In fact, some flight departments go for years, even decades, without seeing or hearing from the regulators in any capacity. When the interaction comes, it may be something as dreaded as a ramp check, wherein the basic aircraft and pilot certificates are checked for their existence and validity. (Most flight crews survive ramp checks without hassle or fines.)

What happens if there is a need to contact the regulator in some official capacity such as to obtain an MEL, ferry permit, or major repair authorization? Not knowing the inspectors and administrative personnel at the local office (FAA Flight Standards District Office [FSDO] in the U.S.) when you need something from them can be a disadvantage. Being acquainted with an experienced and accommodating operations or airworthiness inspector who knows your operation can be a big help when time is short and the stakes are high. Something as mundane as getting the correct interpretation of a regulation or provision in an inspector's handbook will be greatly eased if you can connect a voice with a face at the regulator's office.

There may be some not-so-accommodating inspectors at the local regulator's office, but their reputations usually precede them and they can be avoided. While you don't have to become best friends with your local government representative, you should take time to introduce yourself and other key personnel within your department. The investment of your time may later prove beneficial.

Training

Initial and recurrent training in specific aircraft types, employing vendors approved by the aviation regulatory authority, has become a routine feature of corporate flying. Often this is mandated by insurance requirements, although the frequency and type of training has in large part been borrowed from airline practices. The regulator's requirements for currency and recurrent training are minimum requirements and are normally exceeded by employing high-quality training vendors.

Flight   Virtually all initial and recurrent flight training is performed in sophisticated flight simulators, which are certified to replace the actual aircraft in most training operations. The frequency of recurrent training is an often-debated topic within private and business aviation. Most insurance requirements (and NBAA corporate membership) specify at least annual recurrent training. Many companies have adopted the airline practice of requiring recurrent training at six-month intervals and employing both ground and simulator training at all sessions. For pilots qualified in more than one type of aircraft, an attempt is made to get them to take recurrent training in each type, at least annually. Before any decisions are made concerning pilot recurrent training, the requirements of the company's insurer should be noted.

Line Checks The real question to ponder when considering recurrent flight training is, "What are the proficiency levels of the pilots?" Perhaps a low flight-time pilot may need more training than the more seasoned senior just to make up for lack of experience. Perhaps a high flight-time captain flying multiple aircraft types needs more training just to stay abreast of the differing characteristics of the various aircraft. The concept of training to proficiency has been around for many years and is employed by most training vendors, but the real issue may be, "How long does the individual remain proficient?" The only way to determine this is through objective observation and the testing of individuals. That is why the airlines use line checks at intervals after recurrent training has been completed.

Ideally, line checks should be conducted at least annually by the chief pilot or individuals responsible for standardization or training. The object of the line check is to determine if the pilots follow the regulations and company policy, are proficient in normal aircraft operations, and exhibit good crew resource management techniques. Because people performing the line check need to concentrate on observing the flight crew, they should not act as a crewmember during the check, but should function as an observer and be seated in the jumpseat.

If the line check is satisfactory, the flight crew should be given a comprehensive debriefing on the details of the flight and their performance as it is recorded in their training records. If either crewmember's performance is unsatisfactory, he or she should be provided training to remedy the unsatisfactory portions of the flight and then rechecked.

Flight Department Training

General:

- Management
- Company administrative courses
- Finance/budgets
- Environmental controls
- First aid/CPR
- Fire fighting
- Safe management systems
- Security
- Personal development
- Customer relations

Operations:

- Aircraft initial/recurrent
- Crew resource management
- International operations
- Emergency evacuation
- High-altitude physiology
- Survival
- Weather radar
- Cabin emergencies
- Scheduling/trip planning

Maintenance:

- Aircraft initial/recurrent
- Power plant initial/recurrent
- Specific systems
- Maintenance resource management
- Troubleshooting
- Avionics
- Manufacturer's seminars/workshops

Ground   Unfortunately, many flight departments consider recurrent training in specific aircraft to be the only training necessary for their flight crews. Training should be viewed not just as a means of satisfying an insurance requirement or preparation for specific events but also as a concept to be used in the growth of both the department and the individual. Training should be considered an investment in the future of the company and the people who will be used to gain that future. When training is provided for an individual, it sends the message that you place a high enough value on that person's services to invest in him or her.

If you need further convincing or justification for training, note the training accomplishments required of air taxi and air carrier operators in addition to their aircraft-specific pilot training. Since you consider the level of your services at least as safe and efficient as those of a charter operator, the training required of that operator should be the minimum level you specify for your operation.

To be effective, the department's total training requirements should be set forth in a two- or three-year plan that both schedules and budgets these activities. Without a plan, there is little opportunity or incentive to get people to the proper training. And if training is not in the budget, the chances of it happening are even slimmer.

Preparing to Train   The need to be constantly refreshed about routine and not-so-routine systems and procedures has become an article of faith for personal and business flight department personnel. Executives, insurance companies, and even associations have recognized the value of such training to the point where it is expected and made mandatory through a variety of mechanisms.

But once crewmembers or technicians leave the hangar for recurrent training, what are the expectations levied upon both the individual and the training organization regarding that training? Most managers don't give the subject much thought—they have executed a contract with a reputable training organization and are sending a trusted employee to reap the benefits of the contracted services. What more could one ask?

Attitudes toward Training.   Those of us who have been to recurrent training more than once know that individuals can approach the experience in many ways:

- "Something I gotta do."
- "Ho-hum—more V_1 cuts at Denver…"
- "I have some questions about engine and FMS operation."
- "I'm going to ace every simulator session and do all the enhancement exercises."

Somewhere within this spectrum of expectations lies a preconceived notion of each recurrent training participant. Where the participant lies within this continuum is largely determined by the self-motivation and curiosity of the individual. But should it be left up to the individual?

The safety and professionalism of the department are at stake here. Training costs the company in terms of productive time lost and fees paid for the service. We make sure we get our money's worth for maintenance, upgrades, and FBO services—why not for training. We create administrative, financial, and operational expectations for our employees to follow in the form of standards and limitations, why not create expectations for the training experience, too.

Expectations.   For people going for their first or second recurrent training session, most will approach the experience with the anticipation that they will learn something they never knew before. In general, they go ready to absorb whatever they can about the aircraft or subject at hand. It is when they have been a few times that the anticipation level decreases and a sense of duty regarding the experience emerges. However, this need not happen if the individual is properly prepared to accept what is offered.

Prepare the individual for the experience by asking him or her the following questions:

- What would you like to know more about: the aircraft or the procedures associated with it? Do you fully understand the electrical bus transfer and load-shedding system? Have you done a circling approach in winds above 20 knots lately? Are you up-to-speed on the new FMS software changes?

- What training enhancement sessions would you like to receive? Training providers offer participants a variety of subjects that go beyond the normal recurrent subjects. These topics may occasionally be substituted for normal syllabus subjects, too.

- At what skill level do you expect to perform during the ground school and simulator sessions? Could you pass the FAR 61.58 proficiency check on the first simulator ride? Could you pass your ATP check ride again on the first simulator session?

- Are you prepared for the experience? Have you reviewed your regulations, AFM, technical manuals, and SOPs so that you can hit the ground running on the first day of training?

- What do you expect to come home with as a result of the recurrent training? What will you have gained that will make you a better pilot/flight attendant/technician?

Training is an investment that benefits the department, the individual receiving the training, and everyone else within the department. Make sure you invest wisely and prepare individuals to profit from the investment. Moreover, know what you are receiving for your costly training. Understand what *value* you receive rather than just checking the training boxes for each department member. More importantly, remember that training is not a static, same-thing-every-year proposition; new vendors, delivery systems, and operational modes for the department should trigger an annual review of the department's training program. In this review, consider continuing applicability, value received, and innovation. Training is expensive—make it work for you.

Where to Train   Most flight departments obtain their initial and recurrent flight training from large training organizations that feature professional instructors, well-developed curriculums approved by the state regulatory authority, and high-fidelity aircraft simulators to provide realistic normal and emergency procedure training. Some continue to hire aircraft type experts and use their own aircraft for in-flight training. While this may be an adequate method for initial exposure to an aircraft, it may not prove effective or safe for abnormal and emergency training, which should not be attempted in an aircraft.

Annual or more frequent aircraft training by providers who employ professional instructors and certified simulators does get expensive, but most people think that this is a justifiable expense, given the professional instruction and realism provided. Discounts are available on frequent training or multi-pilot contracts.

Real savings for all types of training can be realized by using virtual training methods that draw on computer technology, readily available subject matter experts, and in-house personnel. Online training courses for aircraft familiarization and systems and procedures training make ideal initial, refresher, and enhancement training methods. Crew resource management, high altitude physiology, risk management, and a variety of other subjects are also available. These courses are self-paced. They contain tests for comprehension and scoring methods that make them particularly useful. Best of all, they have been tailored to aviation personnel and are, for the most part, interesting and satisfying. A number of training providers may be found through an Internet search for providers or by relevant topics.

Inexpensive or even free instructors may be available within your community, company, and especially, at the airport. Air traffic control, fire fighting, environmental, security, emergency services, aircraft maintenance, safety management, and other subject matter experts are there simply for the asking. The added advantage of using these local resources is that they will learn about your organization and what your company has to offer as well.

Last but not least, use your own flight department personnel to provide in-house training. While they may be reluctant to share their knowledge or research a topic, they will benefit from the experience as will the rest of the flight department. Ideally, subject matter experts can be created in-house by sending individuals to schools to learn the newest safety management systems, international operations, aircraft systems maintenance, dispatch, and so on. When they have completed their training, have these individuals train the other flight department personnel.

International Operations

The performance characteristics of modern corporate aircraft have made most of the world accessible to today's flight departments. This, coupled with the rapid expansion of the global economy, has caused a significant increase in international operations for corporate aircraft in recent years. Intercontinental flights have become commonplace.

The International Civil Aviation Organization (ICAO), a specialized agency of the United Nations, was created in 1944 to promote the safe and orderly development of international civil aviation throughout the world. It sets standards and regulations necessary for aviation safety, security, efficiency, and regularity, as well as for aviation environmental protection. The organization serves as the forum for cooperation in all fields of civil aviation among its 191 member States.

The ICAO Air Navigation Commission (ANC) considers and recommends, for approval by the ICAO Council, the Standards and Recommended Practices (SARPs) and Procedures for Air Navigation Services (PANS) for the safety and efficiency of international civil aviation. The Commission is composed of nineteen persons who, as outlined in the Convention on International Civil Aviation, have "suitable qualifications and experience in the science and practice of aeronautics." Commission members, who act in their personal expert capacity, are nominated by contracting States and are appointed by the Council of ICAO.

The SARPS developed by ICAO apply when aircraft operate over the high seas and, by treaty, are to be included in each State's air law; however, exceptions to this last feature are permitted if the State notifies ICAO of the differences involved. When operating in foreign airspace, each civil aircraft is required to comply with that State's aviation operating regulations.

Being Prepared

Since the aircraft flies the same, regardless of the nationality of the airspace in which it operates, the only differences between domestic and international trips are found in the in-flight procedures and ground handling. Therefore, preparation for the flight is key to the success of the mission. Experience has shown that the amount of flight planning and *backup/ contingency* planning that goes into the trip is directly proportional to the success of the trip. Backups are important because of the many surprises one is likely to encounter when operating outside of a familiar environment and culture. The *NBAA Management Guide* contains

International Operations Essential Information

- ICAO
  - Standards and Recommended Practices (SARPS)
  - Procedures for Air Navigation Services (PANS)
  - International Flight Plan Procedures
- International flight information publications
- Oceanic operations
- International flight planning techniques/training
- Customs and cabotage information
- Security information
- Emergency procedures
- Extended range considerations

a comprehensive international operations checklist that is a useful guide for beginning the planning process.

If a foreign destination is not regularly visited or has insufficient facilities, a flight planning/aircraft handling service is normally used to reduce the number of variables and unknowns associated with that destination and the en route structure. International flight plans, weather briefings, preferred routings, overflight permits, regulations, fuel service and payment provisions, visas, and ground transportation requirements are best handled by those who provide those services on a daily basis. The vagaries of international flight are too unpredictable and consequential to be performed by the uninformed.

Those who have flown in foreign airspace or oceanic routes know that the differences are sufficiently large to be bewildering, or at least confusing. Several corporate aviation-training vendors provide excellent introductory courses for international operations that are well worth the time and cost. NBAA provides a wealth of essential information in the international operations section of their website. Additionally, the NBAA International Operators Conference, held in the spring of each year is a must-attend event for those who are serious about operating internationally.

Companies that conduct a large number of international operations often develop a separate international operations manual, which may provide additional and unique information about administrative, operational, and maintenance procedures.

Airports

Airports always seem to be there, in the background, with few issues to concern users. Since airports are normally operated by municipalities and governed by federal, state, and local laws, issues regarding the airport rarely arise. Certainly issues such as access badges, perimeter security, and disposal of toxic substances are always present, but they are mostly minor and don't materially affect one's flight department.

Inevitably, issues that will have a significant impact on business aviation do arise. Operational and capital funding, user fees, noise impact, wetlands and wildlife concerns, operating curfews, and land use compatibility are among the more popular concerns confronting both user and airport operator alike in recent years. Each of these potential problems is capable of severely restricting operations or closing the airport permanently; these are serious issues that cannot be ignored.

Unfortunately, airport users hear little of these problems until they are well advanced. Users customarily become aware of forces affecting their access only when they have become a real threat: legislation or ordinances have been introduced, funding measures have been approved or disapproved, or studies have been initiated that will lead to unwanted conclusions. Traditionally the users become aware in the eleventh hour of the debate and scramble to catch up and become advocates for their position. Often, these last minute stands are ineffective because of the force of hardened public opinion, credible evidence presented, and significant political pressure. These situations can be avoided with a little foresight.

Many threats to airport use come from outside the aviation community. These threats may often be defused by early action with local government or community groups. The investment of one's time in airport issues is time well spent.

If an airport user and community partnership can be formed that will address the needs of all members, the future of the airport will be bright, indeed. Each airport user should become an advocate for the airport and its operations.

Airport User Group Goals

- Receive and provide timely and accurate communications about airport issues.

- Improve relationships with airport management and community leaders.

- Understand community needs and concerns regarding the airport.

- Educate politicians and community leaders about the benefits of aviation and the airport.

- Promote and preserve desirable features of the airport.

- Act to improve airport operating conditions.

Involvement

Sooner or later, each airport user will become involved in airport affairs, willingly or unwillingly. Why not preempt the process and become involved in a structured and timely manner? Doing so will reduce the number of surprises associated with the airport and provide knowledge of the players who will become the advocates and, perhaps, adversaries on specific issues.

Forming or joining an existing airport support group is highly desirable for the individual user, the airport, and the surrounding community. This support group will be able to deal with emerging issues and educate the community regarding airport benefits and to set the record straight about aviation issues.

Organization

User groups normally form the core of airport associations. Some enlightened associations include airport management as an integral part of the association.

While business aviation needs may be unique among all airport users, a more powerful and diverse group will be formed if *all* airport users are included in the group; this includes nonaviation users. If an issue specific to business aviation interests arises, subsets of the total organization may readily be identified from among the full membership to provide a more focused approach.

To realize the benefits that aviation provides for our nation, airport management and aircraft operators must work together. —J.W. OLCOTT, FORMER NBAA PRESIDENT

Security

Security has always been important to business aviation operations due to the nature of the missions and the types of passengers involved. The need to ensure both the confidentiality and security of on-demand air transportation operations has figured prominently into the rationale for engaging in this type of transportation. Not only do the reasons for a trip need to remain confidential and secure for the future of corporate operations, but so too do the embarked company executives, who are valued for their role in companies' ability to continue operations and grow to meet market forces and opportunities.

Terrorist activities have become a pressing issue for all types of transportation and must be accounted for in the daily planning and execution of the flight schedule and facility activities. Like safety, security is a mind-set, a culture that must be cultivated within the organization.

Security is a complex and highly specialized activity requiring background and expertise not resident in most flight departments. For this reason, a security audit is recommended for all flight departments, regardless of activity level or aircraft type. Parent companies often have security departments that may be able to perform the audit.

Assessing and Preventing Security Threats

The first step in the development of an effective security program is to assess the threat against the operator, and its personnel, aircraft, and facilities, and to assess the operator's vulnerabilities. Threats may relate to the nature of the business the organization conducts, where that business is conducted, the nationality of the organization, the nationality of the operator's aircraft, the profile of passengers carried, and the value of goods carried. Information on the various kinds of threats the operator is subject to will come from a variety of sources. In developing and maintaining a current threat assessment for areas of operations, the manager should use the following resources as appropriate:

a. National and local security officials

b. National and local law enforcement officials

c. The organization's security officer, if applicable

d. National and international trade associations

e. Air security assessment and intelligence service providers

f. Local and foreign media reports

g. Organization officials posted in foreign locations, if applicable

Security professionals can provide assistance in determining and assessing the operator's vulnerabilities.

The focus of preventive security measures will be to

a. prevent unauthorized access to operator aircraft and facilities;

b. prevent the unauthorized introduction of weapons or explosives onto company aircraft and into the operator's facilities; and

c. prevent the use of operator aircraft to commit unlawful acts, such as the transport of illicit drugs.

The security measures implemented by the operator should be proportional to the threat. Procedures and training should be in place to implement enhanced measures when the threat is increased and to implement reduced measures when the threat is reduced. (See Table 6.3 for a list of best practices for aviation security.)

People

- Establish a security champion role (much like the safety champion's role).
- Establish and maintain a communications link with the company security department or the equivalent.
- Have flight department personnel complete annual security training.
- Remain diligent to changes in emotional well-being and health of all crewmembers, ground personnel, and passengers.

Facilities

- Ensure the home facility perimeter security with effective fencing, lighting, security patrols (as appropriate), gates, and limited access areas.
- Ensure that the street-side gates and doors are closed and locked at all times.
- Require positive access control for all external gates and doors.
- Close and lock hangar doors when that area is unattended.
- Secure all key storage areas (food and liquor, parts and tools, etc.).
- Have an access control management system for keys and passes.
- Confirm the identity and authority of each passenger, vendor, and visitor prior to allowing access to facilities and aircraft.
- Escort all visitors on the ramp and in the hangar area.
- Use a government issued photo ID to verify identity of any visitor or vendor.
- Post the emergency numbers prominently around the facility.
- Ensure easy access to phones or panic buttons in various facility locations (break room, hangar bay, etc.).
- Confirm the security of destination facilities.
- Be aware of your surroundings and do not be complacent—challenge strangers.

Aircraft

- A flight crewmember must be present at all times when the aircraft is being serviced (fueling, catering, etc.).
- Check lavatories, baggage compartments, and all cavities for unauthorized people or objects prior to every departure.
- Use the aircraft's security system (locks and alarms) whenever it is unattended to prevent unauthorized entry.

Procedures

- Require that the aviation department members participate in security training.
- Maintain a security information program.
- Require an accurate and accessible passenger manifest for all trip legs.
- Ensure that only company personnel and authorized guests, identified in advance, are allowed to board a company aircraft.
- Require passengers or flight department members to maintain positive control of luggage.
- Positively identify all luggage and match luggage to specific passengers (color-coded bag tags can be helpful).
- Require crewmembers to display photo IDs.
- Have a security plan specific to your location and operation.
- Develop, maintain, and exercise an emergency response plan and its associated resources.

TABLE 6.3 NBAA Best Practices for Business Aviation Security

CHAPTER 7
Maintenance

One of the axioms of golf states, "You drive for show and putt for dough," implying that the glory goes with the first activity and the critical substance of the game goes with the second. The same is often said of the difference between flying and maintaining in the world of business and personal aviation.

The real essence of this type of aviation is having an aircraft ready when somebody wants to go somewhere. The pilots are supposedly always available via their pagers, but what about the aircraft? If it is not ready to go virtually all the time, the basic reason for having the aircraft is compromised. The aircraft may be officially out of service for certain periods of scheduled maintenance, but even then, it is usually understood that the aircraft can be returned to service within an hour or two and put on the line for a trip.

The point is that the maintenance function is quite critical. The boss does not understand aircraft dispatch rates of 99.5 percent; anything less than 100 percent represents unsatisfactory performance in his or her eyes. Passengers kept waiting for an hour while last-minute work is performed on the aircraft are not particularly understanding of the vagaries of mechanical performance or the stringent airworthiness requirements placed on their aircraft; getting to the destination in a timely manner is all they understand.

However, the most important aspect of good maintenance is safe operations. The complexities of modern turbine-powered aircraft are bewildering to the layperson and challenging to the aircraft maintenance technician (AMT). Yet all the critical aircraft components have to be working all the time for the department's safety record to have even a chance at 100 percent. The trust and responsibility given to the maintenance crew are great, and they have proven to be well-placed. The maintenance safety record for professionally flown on-demand air transportation is near perfect.

Mission

The mission of the maintenance department is to provide airworthy aircraft, available when needed, at a reasonable cost. We are committed to inspect, maintain, and repair our corporate aircraft to ensure their safe operation and maximum dispatch reliability.

All this serves as preface to the fact that the maintenance function is very important and should be treated as such. The predominantly superior performance of aviation maintenance departments tends to make continued aircraft airworthiness look easy, something to be expected. However, it all comes at the cost of the hard work of dedicated professionals. Constant attention to and recognition of these efforts should be expected as well.

Contract or In-House Maintenance

Most airports have some form of aircraft maintenance available for personal and business flight operations. Be it a technician working independently, a fixed-base operator (FBO), or an approved maintenance organization/repair station, there is usually some form of maintenance available. Therefore, the question arises, should the department perform its own maintenance using company employees or an outside (contract) vendor?

The decision is not as clear-cut or simple as it seems. Every maintenance operation has its limits—work beyond its technicians' experience levels, tasks that require special tools, or processes or jobs that would keep the aircraft in an unserviceable condition for too long due to the nature of the work to be performed. Therefore, virtually all in-house maintenance departments must seek outside assistance at some time during the life cycle of the aircraft. However, the same is true for even the most capable and experienced FBO or repair station. The point is that regardless of the source of most maintenance, no one group can do it all.

> To be airworthy, an aircraft must conform to its type certificate as well as be in a condition for safe operation.   —FAA ORDER 8900.1

However, an experienced maintenance crew can perform most of the work required for the continued airworthiness of an aircraft. Given the right set of circumstances, the reliability, quality, and expenses involved may be superior to using an outside agency. Conversely, some operators find that contracting for all maintenance makes sense. An informed decision between the two requires a sharp pencil, knowledge of maintenance practices, and some hefty intuition.

Contract Maintenance

Contractors may perform maintenance under two different sanctioned methods: an individual AMT's certificate (license) or under an air agency certificate known as an approved maintenance organization (AMO) or aeronautical repair station. While there are many capable individual AMTs working on corporate aircraft, it is difficult to beat the advantages of a repair station. A properly certificated repair station has the advantage of written minimum personnel qualifications, supervisory methods, quality control programs, and a higher level of surveillance from the national regulatory authority. Further, the depth of personnel and experience for a given task are usually greater for an AMO. And if this organization is chosen for its expertise in specific types of aircraft, it should provide a higher level of service, if not confidence in its labor.

All this is not to say that non-AMO/repair station sources should not be used. However, the relative levels of experience and expertise possessed by the repair station should produce an advantage for that method

Vendor Monitoring

When an aircraft is undergoing maintenance at a contractor facility, close contact will be maintained with that facility to ensure that the contracted work is performed completely, to high quality standards, and in a timely manner. Whenever possible, a maintenance technician will accompany the aircraft and monitor the work performed.

Regularly used vendors will be monitored for compliance with national regulations, manufacturers' standards, and good operating practices on an ongoing basis.

over that of the individual technician. However great the advantage of the repair station, it should still be investigated prior to use. Table 7.1 presents some vendor selection guidelines.

Potential contractors should be visited prior to work being performed. During the visit, interviews with key personnel (not just the owner) will yield useful information regarding

Selection process
- The right facility for the job. Are they familiar with your
  - Aircraft type
  - *Specific* job requirements
- Cost
- Availability
- Reputation
- References for specific tasks/aircraft types
- Visit/audit the facility
- Corporate purchasing/legal clearance

Request for proposal
- Be specific about work to be performed
- Specify work/completion standards (usually those of the OEM)
- Specify worker qualifications/experience levels
- Require milestones/completion date(s)
- Include recourse provisions

Quality quotes from vendor
- Date of completion (and milestones, if applicable)
- Separate quotes for each requested task
- Warranty information
- Tax information
- Consumables/parts cost breakdown
- Administrative/certification/engineering costs
- Labor rates by employee/skill type
- Handling/storage/shipping/hazmat charges

Working relationships
- *Always* send a technician to oversee major work
- Review induction work order
- Participate in induction inspection
- Identify a *single* vendor contact for all communications
- Define/discuss when you will need to be called
- Define quality expectations—parts, hardware, cleanliness, etc.
- Delivery/paperwork inspection (must have provisional invoice)

Follow-up
- What's good, what's not—provide feedback in writing—always!
- Final invoice inspection
- Warranty claims

Seek to establish a lasting relationship with the vendor.

TABLE **1.1** Vendor Selection/Relations

the station's level of expertise and experience on the type of work to be performed. Ask to see personnel qualification and training records, observe typical work in progress, and inspect parts ordering, receipt, and issue procedures. However, the true test of a vendor's capability is customer satisfaction; call several recent customers chosen at random from the organization's files to see if they were pleased with the service they received.

Questions for Prospective Vendors

- How long have you been in business?

- How much experience do you have with my type of aircraft?

- What type and amount of initial and recurrent training do your technicians receive relative to my class and type of aircraft?

- Describe the process you use to obtain spare parts for aircraft? (Quality and price are important.)

- Do you use written work orders and provide a detailed breakdown of work completed on your invoice?

- Describe the warranty you provide for both labor and parts.

- Provide the names of three recent customers flying my type of aircraft to use a reference.

Like so many other service provider decisions in life, the provider's reputation and word-of-mouth recommendations associated with it probably will provide you with the best sources of information. However, this choice must be tempered with your individual values and expectations. While cost should be a major consideration in the selection process, quality and reliability of the work preformed should rank at least as high.

Before any work begins with an outside contractor, ensure that a clear understanding exists (in writing) regarding the type and scope of work to be performed, including completion dates for individual tasks. Additionally, there should be a written agreement regarding the responsibilities of the vendor in delineating actions to be taken when unusual items are found during the course of work and when items are found that will extend the period of work performance. Specify that statements of work completed and invoices must contain sufficient detail to determine the type and amount of work performed and the parts used in performing it. Finally, the work performed and parts used should be subject to a written warranty.

All this applies whether the work is performed by an individual technician or a repair station. Setting up these procedures with a prospective vendor may require some effort the first time but will pay dividends later if the relationship continues.

A considerable amount of time will have to be allocated to scheduling, coordinating, and communicating with the vendor. Therefore, the director of maintenance, maintenance coordinator, or a designee will have to carefully integrate his or her maintenance oversight duties with his or her regular tasks if the relationship is to work correctly. If this is done properly, the flight department maintenance contact may spend many hours marrying department requirements with the capability of the vendor. If the coordinator is inexperienced in the field of maintenance, the learning curve in this new role will be quite steep.

Owner's Maintenance Responsibilities

2.6.1.1 The owner of an aeroplane, or in the case where it is leased, the lessee, shall ensure that, in accordance with procedures acceptable to the State of Registry:

a. *the aeroplane is maintained in an airworthy condition;*

b. *the operational and emergency equipment necessary for an intended flight is service-able; and*

c. *the certificate of airworthiness of the aeroplane remains valid.*

—ICAO Annex 6, Part II, Operation of Aircraft—International General Aviation—
Aeroplanes

In-House Maintenance

As stated earlier, an in-house capability will never permit all maintenance to be performed within the department. Nondestructive testing, component repair, many powerplant procedures, and avionics are examples of maintenance functions beyond the capability of most departments. However, most scheduled and nonscheduled maintenance should be within the capability of inhouse personnel.

Proximity to an aircraft service center or major repair station may be a good reason to have in-house maintenance personnel, too. If the maintenance contractor is even a half-hour flight away, the inconvenience and wear on the aircraft just to get it to where it can be worked on may be more trouble than it is worth. And what happens if the aircraft cannot be flown to the vendor's site?

Having reasonably high experience levels of department technicians is critical to efficient and effective maintenance practice. AMTs working on sophisticated turbine-powered aircraft need both general and aircraft-specific training and experience to be truly effective. Without this background, performance of required tasks may require excessive amounts of time.

If an in-house maintenance capability is chosen, the type and level of work to be performed should be decided on. If just routine, low-level scheduled and nonscheduled tasks are to be performed, minimum staffing and experience levels will be required. This will be accompanied by the need to select appropriate vendors for higher-level, nonroutine tasks. If a comprehensive self-contained capability is desired, staffing and equipment requirements grow accordingly, with a corresponding reduced need for contractor support.

Contract Maintenance

All maintenance and inspection functions performed on company aircraft will use in-house resources whenever possible. When time restrictions, resident expertise, or tooling are a limiting factor, only vendors approved by the chief of maintenance will be used.

Vendors will be selected based on their Federal Aviation Administration (FAA) certification status, reputation, work history, and pricing policies. Although the cost of the work to be performed is an important factor, the most significant measures of suitability are the value and quality that corporate aviation receives.

In-house capability may be supplemented through the judicious use of outside sources of information and advice there for the asking. Manufacturers' representatives and customer service personnel are often available to assist their customers with a wide range of services. Such services are normally free of charge and only a telephone call away. A strong local network of other technicians and repair station personnel, particularly those who operate similar aircraft, often will provide answers to difficult questions and yield some

Reasons to Send Your Technician with Your Aircraft to a Maintenance Vendor

- Sends a strong message to vendor regarding a desire for quality
- Ready point of contact for vendor
- Keeps downtime to a minimum
- Excellent training resource for the technician
- Ensures quality control
- Saves money

on-site advice. These methods should be exploited by the operators choosing in-house maintenance to enhance their capabilities.

I believe that one of the greatest advantages of having your own technician is for the times when someone else is working on the aircraft. Having your team member monitor the work being done on the aircraft is a service you should always have. This is a management and oversight job that will guarantee you getting the best parts, skill levels, and service when someone else is doing the work.

Good technicians have the ability to save their own wages every year by ensuring that outside work is done economically and correctly and that best prices are obtained on parts and services. This is especially true for more "mature" aircraft. If your maintenance costs run, say, $500 per hour, there are a lot of savings to be had in a 500-hour year.

In the final analysis, however, it is all about quality and control. How much does your company want?

Organization

While organizations may seem to be the same, regardless of the size of the flight department maintenance operation, significant differences arise.

Small Flight Departments

Small maintenance sections (generally two or fewer technicians) have little difficulty devising an organization because of the relative simplicity of their operation. A single AMT may seem to have the easiest task of getting organized, but because of lack of backup and depth, there are organizational factors to consider.

Just because a maintenance section has few people does not mean that there are not valuable resources available to them that will make their job easier, more productive, and more accurate. One or more of the pilots within the department may be assigned to assist the maintenance section during heavy work periods, especially if those pilots have either an Airframe and Powerplant (A&P) certificate or some maintenance management experience. Part-time or vendor-supplied technicians and semiskilled and apprentice personnel should be considered when developing an organization. All these additional resources may be used to good advantage during heavy work, vacation, and training periods. They may be considered a part of the maintenance organization if written controls and limitations are specified for their work and adequate supervision is maintained. Aircraft insurance policies and the insurance broker should be consulted to determine the legal limits placed on non-A&P and nondepartment personnel. Further, regulatory authorities strictly limit the privileges of nonlicensed maintenance personnel.

Larger Flight Departments

Larger departments should be organized to make the most of their collective and individual talents. The object of a well-organized department is to form a team that exploits the best features of all members of the section. While there should be definite divisions of work designated by job descriptions, there should not be so much compartmentalization of tasks that there is insufficient task overlap to allow individuals to cover for those who may be not present. Similarly, there should be as few as possible supervision and management levels within the section to allow for free communication and coordination of tasks without having to deal with an excessive bureaucracy.

Larger departments, i.e., those operating more than two aircraft, will require additional personnel, including some or all of the following.

- Maintenance department manager
- Chief inspector/quality control inspector
- Lead technician
- Crew chief
- Avionics technician
- Parts clerk
- Records clerk
- Aircraft line service technician
- Aircraft cleaner

All these titles may not be necessary for every department, but the functions will be required. Again, some of these functions may be performed by vendors, or they may be assigned as collateral duties for section personnel.

Quality Control Teams Make Time

A large mail-order clothing retailer located in the Midwest used its two Learjets to good advantage by transporting company executives on a variety of high-profile missions. The executives' busy schedules effectively precluded other employees, including middle managers, from using the aircraft.

As the company grew, its requirement to conduct quality control inspections on manufacturers placed an increasingly heavy burden on its inspection teams to maintain the same level and frequency of inspections. Further, manufacturing facilities were moving offshore in increasing numbers, making the inspection schedule more difficult to maintain.

The head of quality control finally determined that the inspection schedule could not be maintained without hiring an additional 15 inspectors. The vice president of manufacturing was dedicated to maintaining the quality of the products but did not want to put an additional burden on his already stressed budget.

As an experiment, one Learjet was assigned to transport a quality control team on its scheduled rounds to determine the time savings involved, if any. To the surprise of all concerned, the team's productivity was doubled for domestic visits and tripled for Central American and Caribbean visits.

The two aircraft now are used more than 75 percent of the time by quality control, marketing, and audit teams—top-level executives often use commercial airlines or charter.

Personnel

As with other sections of the flight department, attracting and retaining properly experienced personnel that can fit into the organization's culture are essential parts of ensuring a high-quality operation. Chapter 3 listed basic guidelines for finding the right people for the maintenance section of the department. They are repeated here for convenience. In deciding the number of technicians, it is important to know how much time the aircraft will be at home base; if it is always on the road, it will be difficult for in-house personnel to adequately maintain the aircraft. In a related matter, extensive foreign operations may dictate the need for a technician to travel with the aircraft, further taxing personnel left behind to maintain aircraft remaining closer to home.

The first AMT hired normally will be designated the director of maintenance, regardless of the size of the flight department. Because this person may head the maintenance section as it grows and gathers more technicians, he or she should exhibit some leadership and managerial abilities in addition to being a good technician.

The director of maintenance reports to the director of aviation (aviation department manager) and is responsible for maintaining department aircraft in accordance with the manufacturers' standards and regulatory norms [Federal Aviation Regulations (FARs) in the United States]. Duties include

- Ensuring that all maintenance operations are conducted safely.
- Providing guidance and direction for those assigned to the maintenance section.
- Maintaining the company aircraft in accordance with the manufacturers' continuing airworthiness instructions, FARs, and the highest standards of airworthiness and appearance.
- Assisting with flight department planning issues, including budgets.
- Ensuring that maintenance personnel are properly qualified and trained for their assigned duties.
- Maintaining adequate spare parts and consumable inventory.
- Responsibility for contracting with vendors to provide maintenance services, as needed.
- Ensuring that the hangar and associated facilities are properly maintained.

AMTs are responsible for ensuring that the work performed on department aircraft is done in accordance with regulatory standards and good operating practices. Their duties include

- Performing maintenance and inspections in accordance with manufacturer's and FAA-approved methods and techniques.
- Keeping abreast of and maintaining a current working knowledge of the procedures, parts, and tools needed to perform maintenance and repair work on the company aircraft.
- Complying with the manufacturers' technical data, FARs, and other applicable safety rules.
- Being responsible for strict adherence to the inspection system employed for assigned aircraft.
- Shipping, receiving, and inspecting aircraft parts.

Guidelines for finding a qualified AMT include

- Aviation maintenance technician/engineer licensed by the national regulatory authority
- At least one year (ideally two) working on the same class of aircraft
- Specialized training for the type of aircraft operated (may be provided after hire)

As with all other department personnel, technicians should be hired for their ability to become a member of the flight department *team*. Interpersonal skills, a service orientation, communications ability, and overall ability to get along with their fellow workers must be considered as important, if not more important than their technical skills. See Appendix E for a list of performance criteria considered important by the National Business Aviation Association (NBAA).[1]

Most small flight departments exist with one or two AMTs and no other personnel to assist with the maintenance work. As the department grows, other specialties are added to the maintenance section (see "Larger Flight Departments," on page 266).

Determining the proper number of technicians to employ for a flight operation is difficult due to the variety of operational modes and tempos of operations for various departments. Basic factors in determining this include

- Type and number of aircraft
- Flight operations tempo
- Aircraft hourly utilization
- Dispatch reliability expectations
- Amount of contract (vendor) maintenance used

This last factor is probably the most important feature in determining the number of technicians required. If all scheduled maintenance, including minor inspections, is performed by outside maintenance vendors, then few technicians will be required to perform routine dispatch and unscheduled maintenance. If, however, a flight department attempts to perform all but the most major inspections in-house and expects technicians to be present for each launch and recovery, then additional technicians will be needed. As a starting point, see Table 7.2 for the method recommended by the NBAA.

Maintenance Operations

A number of maintenance functions are required of a flight department, common to all types of operation, regardless of size or type of aircraft flown. Understanding and preparing procedures for these processes will make life easier for the flight department and will make the maintenance section more efficient and effective.

Airworthiness Determination

In the United States and most other countries, the owner/operator of an aircraft is responsible for maintaining the aircraft as prescribed by regulatory authority. Moreover, each pilot in command (PIC) must determine that the aircraft that he or she is about to fly is airworthy. A formal maintenance release is required for airline operations to communicate the state of airworthiness between maintenance personnel and the pilots. Yet, in most corporate

The following formula is provided for determining the number of maintenance personnel required. A constant of 4 work hours of maintenance per 1 hour of aircraft flight is used. Each company can revise this number as necessary. The other constant is 1577 maintenance hours a year per person.

- Work hours per year per maintenance technician: 2080
  (52 weeks per year × 40 hours per week)
- Subtract hours not available:
      Vacation: 120 (15 days × 8 hours)
      Holidays: 88 (11 days × 8 hours)
      Sick leave: 40 (5 days × 8 hours)
      Training: 80 (10 days × 8 hours)
- Subtotal of hours not available: 328
- Hours available for duty: 1752 (2080 hours − 328 hours)
- Nonproductive time: 175 (10 percent of 1752 hours) (Nonproductive time includes training, breaks, and cleanup time.)
- Total maintenance hours available: 1577 (1752 hours − 175 hours)
- Maintenance work hours per one flight hour: 4
- Estimated flight hours per year: 850
- Estimated maintenance personnel hours needed: 3400 (4 maintenance work hours × 850 flight hours)
- Number of maintenance personnel needed (rounded):   2 (3400 maintenance work hours/1577 work hours per year)

TABLE 7.2   Sample Calculation for Maintenance Personnel Required (*NBAA*)

aviation operations, the actual determination of airworthiness is delegated informally to maintenance personnel, with no formal list of the items constituting an airworthy aircraft provided to the flight crew prior to flight. Therefore, the PIC normally reviews the disposition of a few recent aircraft discrepancies prior to conducting a preflight inspection and flying away.

Airworthiness Release

The aircraft inspection, life-limited items, and airworthiness directives status are current in accordance with the manufacturer's inspection program and applicable FAA directives.
      /signed/AMT

Ideally, a flight crew briefing is provided by the director of maintenance or other technician regarding the airworthiness of the aircraft. However, there also should be an independent means for the flight crew to ascertain the inspection, life-limited items, airworthiness directive, deferred maintenance, and discrepancy status of the aircraft. This is usually accomplished by placing a discrepancy, inspection, and life-limited items status sheet in the aircraft maintenance (discrepancy) log for the benefit of flight crews. Alternatively, a formal aircraft postflight form completed by an AMT could contain a statement attesting to the aircraft status (see the sidebar). A portrayal of this and the maintenance control process is shown in Fig. 7.1.

Maintenance Planning

All aircraft, particularly turbine-powered aircraft, are required to be inspected at regular intervals, and certain components must be inspected or changed when time limits are reached. Inspections may range from simple visual inspections of external components to complex inspections involving high-technology nondestructive testing methods. They may

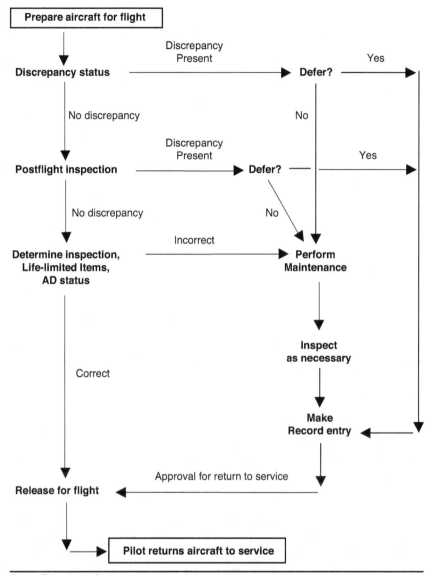

FIGURE 7.1   Aircraft discrepancy control process.

require only minutes to perform or weeks for major inspections. All these items must be scheduled to fit into the aircraft's operational schedule so as to disrupt its utility as little as possible. Maintenance planning is a critical part of the entire flight department's operations and should receive sufficient attention outside the maintenance group as well as within.

The task of maintenance planning or scheduling usually falls to the director of maintenance, who must coordinate the desired work with the chief pilot, aircraft schedulers (and indirectly, with potential passengers), vendors, and in-house work force. This work is complicated by the fact that some required inspection and removal items are predicated on calendar and some on flight-hour requirements. Simple 36-month inspection requirements may be difficult to coordinate with a 600-flight-hour inspection interval because it is not easy to determine on what day the 600 hours will have been reached. Computerized maintenance record tracking systems can assist with the planning process by projecting estimated flight hours onto calendar requirements to reach a comprehensive view of an inspection and replacement schedule.

If a vendor is to perform a scheduled maintenance action, an additional scheduling variable is introduced, that of the availability of the vendor to perform the action. Care should be taken to ensure that the vendor has the proper type and amount of labor available to accomplish the desired work in the minimum time possible.

Scheduling maintenance must be done in coordination with the department manager/chief pilot and the scheduler if proper customer service is to be realized. A scheduled maintenance plan should be devised and published for flight department personnel to alert them to expected dates of inspection and maintenance items based on anticipated utilization rates. If this schedule interferes with required aircraft use, adjustments may be made to accommodate them. Importantly, these dates must be updated continually to reflect both utilization rate changes and varying demand for the aircraft.

Maintenance Control

The functions of anticipating airworthiness requirements and dealing with discrepancies that render an aircraft unairworthy are the principal tasks of the maintenance control process. This function ensures that required maintenance is accomplished on aircraft and that the aircraft is only released for service in an airworthy condition. The function is usually performed by a single individual who acts as a work and paperwork controller to designate what work is performed at what time and by whom, and then finally ensures that the work is accomplished by monitoring the flow of paperwork. The function is considered important enough to often be performed by the director of maintenance.

The maintenance schedule is used to trigger inspections and time-controlled item actions, whereas aircraft discrepancy reports are used to initiate repairs. In large maintenance departments, work orders are commonly used to both initiate actions and determine their progress toward completion. Only after the work has been signed off and inspected should the aircraft be released to other work or approved for return to service. Ideally, when required maintenance has been completed, a maintenance release process will certify that the aircraft is airworthy and ready for service (flight).

Discrepancies

The principal source of maintenance actions outside of scheduled maintenance comes from aircraft discrepancy reports. These written reports may be received from either AMTs who have discovered the fault while working on or inspecting the aircraft or from flight crew members who have discovered a fault during flight. These discrepancy reports should be recorded on a suitable form that provides not only a section for recording the fault but also a companion section to indicate maintenance actions taken to correct it. Dates and signatures should be required for both sections (see Fig. 7.2).

Aircraft discrepancy recording and control are important functions within the flight department. A complete discrepancy and corrective action log provides both flight and maintenance crews with a complete record of a recent airworthiness history of the aircraft. Further, the corrective-action-taken section of the log may be used as a portion of the official aircraft log.

Many flight departments have instituted a procedure whereby an AMT is designated to meet every flight on its return to base. This procedure allows the flight crew to explain the written discrepancies and discuss them with maintenance personnel to ensure that a complete communication has taken place between the writer and the person who will correct the fault. The procedure provides another opportunity for maintenance and operations to interact, furthering good working relations between the two.

Aircraft _____ **Time in Service**_____ **Cycles**_____

Discrepancy		Corrective Action	
Pilot	Date	Technician	Date
Part # off	# on	Part S/N off	S/N on

Discrepancy		Corrective Action	
Pilot	Date	Technician	Date
Part # off	# on	Part S/N off	S/N on

FIGURE 7.2 Sample aircraft discrepancy record.

It is important for pilots and technicians to understand that the act of recording and clearing discrepancies is a critical communications process essential to the safety and performance of the flight department. Complete, comprehensive, and legible entries are important to the process. Most important, *all* discrepancies must be recorded and properly cleared by authorized personnel—verbal or Post It note discrepancies are not adequate.

If a discrepancy affecting the airworthiness of the aircraft cannot be corrected prior to the next flight, the aircraft minimum equipment list (MEL) should be consulted to determine whether the aircraft may be flown with the discrepancy outstanding. If possible, the discrepancy should be deferred using approved procedures and recorded as such. Although general aviation operators may defer discrepancies until the next scheduled inspection, an anticipated correction date for the discrepancy should be specified for planning and control purposes.

The MEL is used infrequently in most flight departments and, therefore, should be subject to training and discussion by both flight crews and technicians. Additionally, the technician responsible for ensuring that the maintenance technical publications are up to date also should check the currency of the master MEL (MMEL) at least quarterly.

Minimum Equipment List

If you operate turbine-powered aircraft, with certain exceptions, it should be covered by an MEL. Granted, if *everything* of consequence on the aircraft is operable for *every* flight, an MEL is unnecessary. However, credibility regarding perfect aircraft reliability falls into the same category as the Tooth Fairy and Santa Claus.

For the United States, FARs covering inoperative instruments and equipment essentially state that everything has to work on the aircraft prior to flight unless specific procedures are taken. In the case of turbine-powered aircraft, this almost always means using an MEL issued to a specific operator and aircraft.

A U.S. operator who wishes to conduct operations with an MEL must contact the Flight Standards District Office (FSDO) that has jurisdiction over the geographic area where the aircraft is based and make an appointment. For part 91 operators, the FSDO will assign a Flight Standards inspector to advise the applicant operator about regulatory requirements pertinent to using an MEL. During the initial appointment, the applicant will likely be dealing with a team of inspectors from the operations, airworthiness, and avionics units.

The operator must develop the Operations (O) and Maintenance (M) procedures using guidance contained in the manufacturer's aircraft flight and/or maintenance manuals, the manufacturer's recommendations, engineering specifications, and other appropriate sources. An operator may consult FSDO inspectors for advice or clarification, but the operator is responsible for preparing the document. The FAA does not approve procedures documents for part 91 operators with an MEL authorization.

Notification of revisions to the MMEL will be sent by the FAA. If installed equipment is not covered in the MMEL, you must apply through the FSDO for modification to your MEL or for an amendment from the FAA Flight Operations Evaluation Board (FOEB).

The MEL provides operators with greater flexibility by permitting an aircraft to be operated with certain types of installed equipment inoperative. Once pilots and maintenance technicians have studied and understand its provisions, the MEL allows for legal operation despite minor malfunctions.

Nonessential Equipment and Furnishings.   Currently, master minimum equipment lists (MMEL) provide relief only for passenger convenience items located in the cabin, galley, and lavatory areas. Other areas of the aircraft have items installed that are not captured by the MMEL and which are non-safety-of-flight items that must be repaired before further flight if found inoperative. The MMEL Industry Group was tasked to develop a procedure that gave the operator a path to incorporate these relief items into what is now called the Nonessential Equipment and Furnishings (NEF) Program. The NEF relief item will replace the current passenger convenience item and will be incorporated into ATA chapter 25 of the MMEL. However, the NEF Program is generated by the operator and separate from the aircraft Minimum Equipment List (MEL). For private operations, the NEF Program is not approved by FAA.

Maintenance Away from Home Base

When an aircraft becomes unairworthy while away from home base, flight crews should consult with the maintenance manager or chief inspector via telephone concerning possible courses of action.

Minor discrepancies not requiring extensive aircraft-specific knowledge may be corrected by an aircraft maintenance facility where the aircraft is located. The company maintenance representative responsible for selecting the maintenance facility should be familiar with the facility, but if this is not possible, at a minimum, the facility should possess an FAA repair station certificate.

Resolution of major discrepancies or scheduled heavy maintenance away from home base may require the presence of one or more company AMTs to oversee the work performed. However, an aircraft maintenance facility having experience with the aircraft type should be considered adequate to perform the required work if the previously stated criteria are met.

Aircraft Maintenance Reference Materials

Reference materials required by the manufacturer's maintenance program and FAA regulations are required to be kept complete and current. Each manufacturer publishes a document that provides the current editions of and changes to its technical publications. This should be used to determine the currency and completeness of the airframe, powerplant, accessory, and appliance manuals required for department operations. Additionally, all FAA-required publications, including applicable airworthiness directives, advisory circulars, and regulations, should be kept current via FAA document checklists.

A system should be devised to ensure that these publications are kept up to date on a continuing basis. Ideally, an individual will be assigned to the task and review the controlled publications on a monthly or quarterly basis. This process may be handled most easily by subscribing to a service that will keep the documents up to date via the Internet. Yet

> **Reliability**
>
> Reliability is an important aspect of the corporate aviation mission. Therefore, this department shall provide mission-ready aircraft to meet flight schedules so that every flight is launched within 15 minutes of its scheduled time or upon passenger availability. This is to be accomplished for a minimum of 99 percent of all scheduled aircraft launches.

these services may not fully cover all appliances, accessories, or STC items; therefore, a process must be devised to track all required reference materials. Even with subscription services, someone must ensure that the periodic subscription is received and not lost in the mail or subject to a subscription expiration.

Each person performing maintenance...shall use the methods, techniques and practices prescribed in the current manufacturer's maintenance manual or Instructions for Continued Airworthiness prepared by the manufacturer.   —FAR PART 43.13

Parts Inventory and Control

The number of spare parts, both consumable and rotable, should be kept to a minimum. Parts are expensive to keep on hand and account for, and they may become obsolete. Most parts, even major or rotable items, can be obtained within 24 hours or less through a number of vendors. Aircraft-on-ground (AOG) parts are usually available even sooner. These "on-demand" parts services have allowed flight departments to reduce their inventories to a minimum, freeing up both storage space and parts personnel and reducing inventory costs.

Parts should be ordered only from vendors who have been investigated properly and found to have a good reputation and to charge competitive rates. New and additional vendors should be constantly sought to ensure that the best possible product is available at the best possible price.

Parts ordering and control should be accomplished by an individual who will be responsible for all parts-related functions. The advantage to having a single individual perform these tasks is one of control and quality control. If more that one individual attempts to work with parts, there inevitably will be gaps in communication and miscommunications that inject errors into the system.

A parts usage and inventory system is essential to maintain adequate stocks of parts. Available software programs for ordering, tracking, and billing may simplify the process of parts control and certainly will provide better records of parts transactions. However, small inventories may be maintained using paper records with only a minimal effort.

All parts must be identified properly to ensure that they are approved by the regulatory authority. A receiving inspection and verification system should be instituted to verify the authenticity and proper documentation of all aircraft parts coming into the department. Similarly, an outgoing identification and shipping and inspection function should be performed to ensure that the receiving vendor is properly supplied with authentic documentation (see Table 7.3).

Duty Time

Flight time and duty time are always hot topics for flight crews but rarely heard of when speaking of AMTs. Traditionally, technicians work until the job is complete, be it a critical repair that affects the airworthiness of the aircraft or a single inspection work card. Many late nights are spent in lonely hangars by single technicians working to make the morning launch. However, if a pilot cannot be expected to be productive, alert, and safe after 12 or 14 hours (during which some rest time may have been available), is it possible for a technician to be safe after the same interval?

Technicians work with complex machines that require high levels of understanding; theirs is a world of complicated detail. It is reasonable to assume that their comprehension of the task at hand and the detailed procedures to complete it require great insight and concentration. Fatigue, poor lighting, heat, cold, humidity, noise, and frequent distractions are the

- Procedures should include a means of identifying suspected unapproved parts during the receiving inspection and prevent their acceptance. Suggested steps include the following:
    - Confirm that the packaging of the part identifies the supplier or distributor and is free from alteration or damage.
    - Verify that the actual part and delivery receipt reflect the same information as the purchase order regarding part number, serial number, and historical information (if applicable).
    - Verify that the identification on the part has not been tampered with (e.g., serial number stamped over, label or part/serial numbers improper or missing, vibro-etch or serial numbers located at other than the normal location).
    - Ensure that the shelf life and/or life limit has not expired, if applicable.
- Conduct a visual inspection of the part and supporting documents to the extent necessary to determine if the part is traceable to an FAA-approved source. For detailed guidelines on the identification of replacement parts, refer to AC 20-62(E). The following are examples of positive forms of identification:
    - FAA Form 8130-3, Airworthiness Approval Tag
    - European Aviation Safety Agency (EASA) or Transport Canada Civil Aviation (TCCA) Authorized Release Certificate (equivalent to FAA Form 8130-3)
    - Maintenance records or release document with approval for return to service
    - FAA TSO markings
    - FAA PMA markings
    - Shipping ticket/invoice from Production Approval Holder (PAH)
    - Direct ship authority letter from PAH
- Evaluate any visible irregularities (e.g., altered or unusual surface, absence of required plating, evidence of prior usage, scratches, new paint over old, attempted exterior repair, pitting, or corrosion).
- Conduct random sampling of standard hardware packaged in large quantities in a manner that corresponds to the type and quantity of the parts.
- Segregate parts of questionable nature and attempt to resolve issues regarding the questionable status of part (e.g., obtain necessary documentation, if inadvertently not provided, or determine if irregularities are a result of shipping damage and handle accordingly).

TABLE 7.3 Aircraft Parts Acceptance and Identification (*FAA AC 21-29*)

enemy of this type of work, all conditions common to a technician's work life. Controlling these factors should make the technician a more productive, safer worker. Fatigue is the perhaps the easiest to control.

A number of flight departments have placed duty time limits on their technicians' work in much the same manner as they have for pilots. The most common limit seen is 12 hours, with provision for finite extensions at the discretion of a lead technician or the director of maintenance. If an early launch and late recovery are planned, wherein the technician is required to be present, credit can be given for time off during the day to extend the duty day. However, this should not be done for more than 2 or 3 days in a row. Further, consideration must be given to when work

Duty Time

AMTs will work not more than 12 continuous hours in any 24-hour period. The minimum acceptable rest period following a duty period is 10 continuous hours. Each AMT should be relieved from all duty for at least one period of 24 continuous hours during any 7 consecutive days.

begins and ends. If the technician arrives at 5 A.M. to launch an aircraft, a normal work day should end at 1 to 2 P.M. Similarly, this technician will have difficulty performing/resting if the next day begins at 4 P.M. and goes to midnight.

Fatigue Risk Management Principles.   In recent years, comprehensive fatigue risk management approaches have been adopted in aviation and road transport, supplementing, or in some cases replacing, older hours of service approaches. Fatigue Risk Management Systems (FRMSs) have been promoted by a number of international and national aviation organizations (Internet available), especially for aviation operating certificate (AOC) holders. The FAA has defined FRMS as: "...a data driven and scientifically based process that allows for continuous monitoring and management of safety risks associated with fatigue-related error. It is part of a repeating performance improvement process. This process leads to continuous safety enhancements by identifying and addressing fatigue factors..."

This parallels modern approaches to flight crew FRMS.

Quality Control

This is a safety issue that is difficult to resolve in smaller departments. Airlines and repair stations require that certain critical maintenance actions be inspected by a designated inspector prior to the aircraft, powerplant, or appliance being approved for return to service. These required inspection items (RIIs) are listed in the operations specification or general maintenance manual. While there is no requirement to do so in most corporate operations (U.S. FAR Part 91), some departments have instituted their own RIIs that call for a designated inspector, usually a lead technician or director of maintenance, to inspect certain items of work before the aircraft is released; some departments go so far as to double inspect all work performed. This is especially difficult for a single technician working on a weekend or late at night. This situation should be considered when setting work rules for technicians.

Continuing quality requires that the quality be built into each procedure used by the technician. That is, instead of relying on inspection procedures to ensure quality work, the process that created the work must have procedures and checks built into each action to ensure accuracy and safety. While many manufacturers' maintenance instructions strive to accomplish this in each work card and procedure, some fall short and must have procedures devised to preclude error.

Did It Get Done?   This question normally comes to mind when you are driving home from the hangar at 2 A.M., wondering whether you safety wired the final bolt on the hydraulic pump you just replaced. This is an occupational hazard of those who perform solo maintenance work.

To an even greater degree, this question comes to mind when a high-priority nonscheduled maintenance task has been completed, and it is time to return the aircraft to service. The question refers not just to the many actions required to accomplish the actual maintenance task but also to the paper trail that will prove that the action has occurred.

If you operate small aircraft with no maintenance record tracking system, work exclusively by yourself, or have a maintenance paperwork checklist you use regularly, this section may not apply to you. However, if you are like the great majority of technicians who work with others, perform complex tasks, and have a maintenance recordkeeping system in addition to the actual aircraft, powerplant, and accessory logs, this section is probably for you.

Traditionally, the paperwork is saved until the maintenance task has been accomplished and tested and the aircraft is given to the pilots to fly. (Do not forget, however, that the aircraft may not be approved for return to service until an official logbook entry has been made—paperwork first.) In the heat of battle to get the aircraft fit to fly, certain items may get overlooked. For instance:

- Was the received part logged in correctly and its inspection recorded properly?
- Was the serviceable part tag attached to the work order/work card prior to filing?
- Did the unserviceable part tag get filled out completely and attached to the part prior to return/disposal?
- If consumable or standard parts were used, was an inventory debit made?
- Was the proper maintenance record tracking system work card completed/entered into the system?
- Was the aircraft/powerplant/accessory logbook entry made (strange as it may seem, this important step is sometimes neglected, yet all the ancillary paperwork has been accomplished).
- The airworthiness directive (AD) was accomplished and recorded, but was the AD status record updated?

This is a representative sample of some of the many records that should be kept to ensure a comprehensive view of accomplished maintenance actions—do you have a means for making sure that all necessary recordkeeping is accomplished?

Airlines and repair stations are required to have systems in place to ensure that all their record entries are accomplished. These systems are set forth in their general maintenance/inspection manuals and may provide checklists for technicians to use. In the fully integrated recordkeeping systems, internal edits prevent certain entries from being made if other, more fundamental record actions were not accomplished first.

Some flight departments use recordkeeping checklists temporarily attached to maintenance action packages or simple checklists posted at all normal recordkeeping locations around the shop to ensure all the i's are dotted and t's crossed. Others impose a strict sequence of recordkeeping actions that forces technicians to complete all required actions. Still others use either a single individual to make all record entries or a quality control person to ensure that the recordkeeping actions of others are complete and correct.

Maintenance Inspection Accuracy. The airworthiness of aircraft depends heavily on the scheduled maintenance that they are required to undergo. The items to be inspected and the frequency with which they must be inspected are specified in the manufacturer's continuing airworthiness instructions and the applicable regulations. However, a critical part of the inspection process is the ability of the inspector to reliably and correctly perform the required inspection task.

If the aviation maintenance technician is inadequately trained in inspection techniques, does not know the possible fault conditions specified in the task, is not psychologically or physically prepared, or is performing the inspection in less than favorable environmental conditions, the outcome of the procedure may be in doubt. Inherent in inspection task design is the notion that human beings are fallible and that some are better prepared to detect fault conditions than others; defining the specific abilities and inabilities to perform is the subject of ongoing maintenance human factors systems design studies.

A recent study[2] examined the ability of a number of technicians to detect surface fatigue in aircraft fuselage skins using high-frequency eddy current inspections. The experiment was taken to nine maintenance facilities, where inspectors examined a large number of rivet sites for skin cracks known to the test conductor. Various controlled factors were incorporated into the experiment, including painted versus bare surfaces, accessibility to the inspection site, and type of specimen. Other, uncontrolled areas of environmental and personnel factors also were considered for their impact on the task. Environmental factors included noise level, instrument calibration, temperature, lighting, procedures used, and scanning techniques. Personnel factors included both physical and psychological factors, including age, amount of sleep, time on duty, work patterns, type of training, attitude, education level, and recency of inspection experience.

Results were predictable in some cases and surprising in others. For example, the length of the crack to be detected determined the probability of detection (small cracks were more difficult to detect), and painted surfaces made reliable crack detection significantly more difficult. In this latter case, the number of false detections increased dramatically when painted surfaces were inspected.

The most interesting results came from the uncontrolled personnel and environmental factors. A number of these included the following:

- Inspectors overcame poor lighting in the inspection area by using flashlights, but when portable lighting was used, the process speeded up significantly. Quality and amount of lighting did not, however, appear to affect the accuracy of the inspection tasks.

- A wide range of eddy current calibration techniques and inspection procedures were observed. Some nonstandard and substandard calibration techniques measurably affected the outcome of the tests.

- Fatigue caused performance to suffer, especially for those who had already worked a full shift prior to engaging in the test. However, a direct correlation between amount of recent sleep and inspection performance was not possible.

- Younger inspectors tended to perform better than older inspectors. While not fully verified, the training received by the younger inspectors may have caused the difference.

- A less desirable behavioral climate in an individual facility or the perceived poor attitudes of management did not appreciably affect inspector performance because of the apparent personal and professional pride they exhibited in their work.

- Even formalized on-the-job training did not seem to produce the quality of inspectors that formal classroom training produced.

- Inspectors who took more work breaks tended to be more accurate than those who worked straight through.

- The quality of inspection teams (two-persons) was dragged down by the poorest member of the team rather than being pulled up by an excellent team member. Preliminary findings indicate that single inspectors were better and more effective.

Be aware of these factors when assigning and engaging in inspection activities. There is more to it than meets the eye.

Maintenance Error.   Maintenance errors occur through lapses in attention, distractions, racing to finish a job, and just plain carelessness. Traditionally, quality lapses, unsafe acts, and less than adequate maintenance performance are sorted according to their adverse impact on the aircraft. These error types fall into two general classes:

- The introduction of an aircraft discrepancy that was not there before the maintenance activity began.
- The failure to detect damage or incorrect components during maintenance inspections.

Errors in Maintenance Caused by Human Factors. Due to the occurrence of a large number of maintenance-related aviation accidents and incidents, Transport Canada identified twelve human factors that degrade people's ability to perform effectively and safely, which could lead to maintenance errors. These twelve factors, known as the "dirty dozen," were eventually adopted by the aviation industry as a means to discuss human error in maintenance:

1. Lack of Communication
2. Lack of Teamwork
3. Lack of Assertiveness
4. Complacency
5. Fatigue
6. Stress
7. Lack of Resources
8. Lack of Awareness
9. Lack of Knowledge
10. Pressure
11. Distraction
12. Norms

Another study conducted by the British Civil Aviation Authority in 1992 found that the most frequently occurring maintenance problems, in order of occurrence, were

- Incorrect installation of components
- Installation of incorrect parts
- Electrical wiring discrepancies (including cross-connection)
- Loose objects (tools, etc.) left in aircraft
- Inadequate lubrication
- Cowlings, access panels, and fairings not secured
- Fuel/oil caps and refuel panels not secured
- Landing gear ground lock pins not removed before departure

Although these studies tell us little about why individuals made these particular errors, they tell a great deal about the relative error proneness of the different phases of the aircraft maintenance task.

The cure? Perhaps the best overall cure is a well-developed safety culture that places high value on quality work. Other measures include required inspection items, attention to maintenance instructions and procedures, and a supportive management structure.

While business aviation suffers very few accidents due to maintenance error, less serious incidents cause unnecessary aircraft reliability problems and wasted time and money.

Maintenance Manual

- Organization/job descriptions
- Safety
- Technician qualifications
- Training
- Inspection programs
- Quality control
- Work procedures and performance criteria
- Discrepancy management
- Maintenance scheduling
- Unscheduled maintenance away from home base
- Recordkeeping
- Vendor selection
- Spare parts control
- Support equipment
- Facility maintenance

Quality control and maintenance error should never be far from the mind of each technician and manager.

Maintenance Manual

An operations manual was discussed in Chapter 6, and the chapter contains a brief recommendation that a section of it be devoted to maintenance procedures. This brief treatment may be adequate for departments that are small or have most of their maintenance performed by vendors, but large departments possessing capabilities to perform a variety of tasks should consider developing a separate maintenance manual.

The manual is primarily a communications tool to control expectations and provide guidance for all people within the department. Making sure technicians perform actions the same way every time and to a set of standards goes a long way toward ensuring excellence within the maintenance function.

Ideally, this manual should be developed in-house and by the department technicians themselves. If a generic manual is adopted by the department, it probably will not be used as a reference because it does not reflect actual practices used by the department. Developing a manual in-house also serves to clarify and reinforce existing policies and procedures and to indicate the need for additional ones.

Aircraft Handling

The high value of corporate aircraft demands great care during moving and parking. Therefore, the following procedures should be considered for all aircraft movements:

- Training should be provided for all personnel.
- Only qualified personnel will move the aircraft.
- Tugs and towbars should be rated adequately for the gross weight of the aircraft to be handled.
- The aircraft will be chocked whenever parked.
- At least one wing walker, in addition to the tug driver, will be employed to determine adequacy of clearance between an aircraft and surrounding objects during aircraft movement. Additional wing walkers should be employed when warranted.
- Aircraft spotting marks shall be used.
- Hangar doors will be fully opened during operations in which an aircraft is moved into or out of the hangar.

Security

Security is covered more completely in Chap. 6, but some note of security for maintenance needs to be made. Facility security may be of greater importance for maintenance personnel than for others because maintenance controls so many and such large entrances the hangar. Open hangar doors during warm weather are an invitation to thieves, vandals, and the like; a favorite time for leaving everything unattended and open is lunch time. And maintenance personnel are often the last to leave a facility at night. The duty of locking up and setting the intrusion alarms should be spelled out and briefed to all maintenance personnel.

Another problem area arises when leaving aircraft on the ramp unattended with the airstair, equipment bays, and baggage doors open. Even though the aircraft is in front of the hangar or nearby, it does not take long to get in and out.

Maintenance personnel are quite knowledgeable regarding aircraft movements and, possibly, the passengers being moved. This is not only security-related but confidential information, too. This information must be kept close-hold and passed only to those needing to know aircraft movements.

Evaluating Maintenance Performance

Chapter 4 covered evaluation of the flight department's overall performance, looking at measures such as aircraft cost per mile, load factor, and denied trips. Maintenance activity also can be measured and should be to determine its level of performance. Here are some of the measures that can be used:

- *Aircraft availability.*   Answers the question, "How many days was the aircraft available for flight during the month?" The complement of this number is normally the number of days the aircraft was not available due to maintenance, either in-house or contract.

- *Dispatch reliability.*   If an aircraft departs within 15 (or 30) minutes of its scheduled departure time or passenger arrival, a successful flight has commenced. If it does not depart within that time frame, something went wrong—weather, ATC, or a mechanical problem. Therefore, a subdefinition of dispatch reliability is required to adequately define the failure. The value is expressed as a percentage arrived at by dividing the number of successful launches by scheduled flights.

- *Quality.*   How many errors were made in maintenance monitoring/tracking, actual maintenance functions, recordkeeping, and parts ordering/handling? Most departments do not measure these items, but doing so provides useful feedback to all concerned, especially if the data are not used for disciplinary actions.

- *Maintenance hours per flight hour.*   Knowing how much maintenance time, especially nonscheduled time, is spent on a specific aircraft tells a lot about its reliability and general condition. Recording the type of system worked on will help further define issues.

Tracking these values provides information about the effectiveness and efficiency of a maintenance operation, although some issues causing poor numbers may be beyond the control of the maintenance organization. Therefore, each of these measures requires either a footnote or further explanation to describe the cause for either low numbers in the first two measures or a high number in the third.

Recordkeeping

The ICAO and all national regulatory systems require that certain records be kept for each aircraft, powerplant, and appliance to determine whether it has been maintained in an airworthy condition in accordance with approved procedures. This is a critical function because it is usually the only means available for an operator to prove compliance with airworthiness requirements. However, there is an even more important reason to maintain complete and correct maintenance records, that of proving to a prospective buyer of the aircraft that not only has it received the required maintenance but also that it has been done properly and in a high-quality manner. The resale value of a turbine-powered aircraft may be decreased by 20 to 30 percent if its maintenance records are poorly maintained or incomplete.

Inspection status and maintenance performed are important components of recordkeeping but are considered perishable records. This means that they must be maintained for only 1 year after the work has been performed or until the work is superseded or repeated. Perhaps the more important records are those that record the status of life-limited and time-controlled items, airworthiness directives (AD), and records of major alterations or repairs. These are permanent records that stay with the aircraft throughout its life. It is these last records that are most important in proving the status and quality of care of the aircraft. If the accomplishment of an AD or important service bulletin or approval for a major alteration cannot be proved, its reaccomplishment can be very costly.

...2.6.2   Maintenance Records

2.6.2.1   The owner of an aeroplane, or in the case where it is leased, the lessee, shall ensure that the following records are kept for the periods mentioned in 2.6.2.2:

    a. *the total time in service (hours, calendar time and cycles, as appropriate) of the aeroplane and all life limited components;*

    b. *the current status of compliance with all applicable mandatory continuing airworthiness information;*

    c. *appropriate details of modifications and repairs;*

    d. *the time in service (hours, calendar time and cycles, as appropriate) since the last overhaul of the aeroplane or its components subject to a mandatory overhaul life;*

    e. *the current status of the aeroplane's compliance with the maintenance programme; and*

    f. *the detailed maintenance records to show that all requirements for the signing of a maintenance release have been met.*

2.6.2.2   The records in 2.6.2.1 a) to e) shall be kept for a minimum period of 90 days after the unit to which they refer has been permanently withdrawn from service and the records in 2.6.2.1 f) for a minimum period of one year after the signing of the maintenance release.

2.6.2.3   In the event of a temporary change of owner or lessee, the records shall be made available to the new owner or lessee. In the event of any permanent change of owner or lessee, the records shall be transferred to the new owner or lessee.

—ICAO Annex 6, Part II, Operation of Aircraft—International General
Aviation—Aeroplanes

Regulations to Comply With

There are many regulations that must be complied with to ensure that every flight is legal in the eyes of the national regulatory authority. The list shown in Table 7.4 includes just those regulations that form the basis for corporate flight department maintenance issues in the United States and that require some form for proof of compliance. Non-U.S. operators will find similar regulations promulgated by their national regulatory authority, usually based on applicable ICAO annexes.

An individual should be tasked with ensuring that all aircraft records are maintained properly. In larger departments, this may become a supervisory function due to the number of people involved in the recordkeeping process. Regardless of the situation, a single individual should be given the task of ensuring that the recordkeeping job is done correctly. It may be desirable to have all recordkeeping work be subject to an additional supervisory or quality control function to make sure that record entries are complete and correct.

Computerized Record Tracking Systems

The complexities of modern turbine-powered aircraft and their required maintenance actions are such that maintaining a clear picture of aircraft airworthiness status is difficult without some form of computerized tracking system. These systems are commonly offered as services to flight departments, providing a required inspection and time-controlled item action and status structure for the department to use. The service provider reviews inputs from the subscriber prior to applying those actions to the database and creates status reports to be used for

---

Airworthiness Directives

  39.3 General

Maintenance, preventive maintenance, rebuilding, and alteration

  43.5 Approval for return to service after maintenance...

    .7 Persons authorized to perform maintenance...

    .9 Content, form and disposition of maintenance...records

   .11 Content, form, and disposition of records for inspections...

Certification: Airmen other than flight crew members

65.81 General privileges and limitations

   83 Recent experience requirements

General operating and flight rules

  91.7 Civil aircraft airworthiness

    .9 Civil aircraft flight manual, marking and placard requirements

  .203 Civil aircraft: certifications required

  .213 Inoperative instruments and equipment

  .403 General (maintenance)

  .405 Maintenance required

  .409 Inspections

  .411 Altimeter system and altitude reporting equipment tests and inspections

  .413 ATC transponder tests and inspections

  .415 Changes to aircraft inspection programs

  .417 Maintenance records

  .419 Transfer of maintenance records

TABLE 7.4 Regulations for U.S. Flight Department Maintenance Operations

maintenance planning and scheduling. Some services produce maintenance-due lists and work cards to provide the subscriber with inspection and time-controlled item action checklists.

While these services provide a very useful function, they do not take the place of the official recordkeeping function of the department. For instance, actual signatures are required for inspection, maintenance, and AD actions, and AD methods of compliance are required to determine AD status. This function is not normally available on computerized tracking systems. Paper records and actual signatures may no longer be required in the future, but both regulators and prospective buyers want to see them today.

Selecting a computerized recordkeeping system can be a confusing experience because the available systems contain so many features, some useful and some not so useful. The data input mechanism, application of the data to portions of the record, and creation of status and due lists are all fairly standard. The real variables of the systems are the method of portraying airworthiness status and quality control of the data and reports. Some status reports are easier to read and use than others, but the feature that separates the good from the not so good is the vendor's ability to ensure that the data provided by a client make sense (it is up to the operator to ensure their correctness) and are applied properly to the database. Many a frustrating hour has been spent in trying to correct a data-entry mistake or missing entry.

The only way to determine the vendor's ability to perform well in these areas is to ask those who use the systems. A lot is riding on the ability of a computerized recordkeeping system to provide you with accurate, reliable, timely data regarding the airworthiness status of your aircraft; take time and care in selecting a vendor. Most important is to use a system that fits the department's method of operations and that does not require the department to accommodate a feature of the vendor's system that cannot be modified.

Maintenance Records: Fine Points.   Most operators of aircraft, regardless of size, fail to realize and appreciate the fact that the responsibility for maintaining records that accurately reflect the state of airworthiness of an aircraft rests with the owner/operator, not the technician who maintains it. Yes, the technician is responsible for recording the maintenance and inspection work he or she performs on the aircraft, but the final responsibility for an aircraft's records rests with whoever owns up to operating it.

This is important for several reasons. First, the principal method of determining airworthiness, particularly for complex and/or transport-category aircraft, is through an aircraft's status with regard to inspection, required maintenance, compliance with mandatory manufacturer's service bulletins and ADs, and status of life-limited components. Granted, a detailed preflight or postflight inspection will determine obvious airworthiness discrepancies, but the FAA and other national regulators are concerned with whether the aircraft conforms to its original type design and is in a safe condition for flight; the best means of accomplishing this is to comply with the manufacturer's instructions for continued airworthiness (maintenance program), required as a consequence of the certification process.

While many operators (pilots) rely on the integrity of their technicians and maintenance contractors to ensure that all this paperwork is complete, few understand or appreciate the implications of not being in compliance with whatever maintenance and inspection program they use. Other than being in violation of the regulations, the aircraft may not be in a safe condition for flight. Moreover, when it comes time to sell the aircraft, gaps and inconsistencies in the maintenance records may raise questions regarding its true state of airworthiness and value.

Computerized maintenance recordkeeping systems have enabled the corporate operator to plan for upcoming inspections and required maintenance and life-limited items and to track the status of all these requirements. If the combination of source documents (work cards, parts transaction reports, discrepancy actions, etc.) and tracking lists meets regulatory requirements, these systems may constitute the actual aircraft maintenance records. But a word of caution: Computerized maintenance record tracking systems do not automatically provide the operator with a set of records that meet all regulatory requirements for maintenance records.

For instance, if a discrepancy is corrected (the aircraft is repaired), where is this action recorded? Few record tracking systems record repair-only information, merely the fact that a part may have been replaced. Suppose that no part was replaced. Where is this maintenance action recorded?

Regulations require that a current status of ADs be maintained, imposing specific items to be recorded. Most tracking systems do not note the method used to comply with the AD; this is required to be in compliance with the regulations. And merely noting the manufacturer's service bulletin used for compliance may not be adequate either. Some service bulletins offer alternative means of compliance, and these alternatives are seldom noted in computerized systems.

On a procedural note, most tracking systems provide operators with a monthly transaction report, noting the entries made in the previous month. However, many operators fail to check this list against their submissions over the same time period. This is particularly true if a maintenance contractor performs the action, and a transaction report is not submitted to the tracking system service provider. As a consequence, tracking systems may not reflect the actual status of an aircraft due to entry errors and omissions—garbage in, garbage out.

Most flight departments operating turbine-powered aircraft maintain both the required aircraft records and also participate fully in a records tracking system service. The two systems complement one another and act as a system of checks and balances to ensure that all required maintenance is accomplished. This system is particularly impressive when it comes time to sell the aircraft.

Finally, consider the dilemma of losing an aircraft's official maintenance records due to fire, flood, theft, or carelessness. Attempting to reconstruct those records is a time-consuming and imperfect process that may or may not satisfy the regulatory authority or, more important, a prospective buyer. Since a great deal of the value of the aircraft depends on its official records, it makes good sense to back up those records either by photocopying or electronic imaging. The latter is particularly attractive because companies providing this service often certify the authenticity of the images they produce.

Company Private Airline

A telecommunications company bought a competitor whose headquarters were 300 miles away. Company managers foresaw a lengthy transition period in which employees would travel regularly between the two headquarters. Since airline service was spotty and road conditions between the locations were poor, the purchasing company, which already had used corporate aircraft for years, bought a 12-seat turboprop aircraft to provide twice-a-day shuttle service between the two centers.

Over time, the two locations retained their importance, and a third company was acquired, bringing the same requirements for travel as the initial acquisition. As a consequence, a triangle route was established for the turboprop. Now, years after the merger, the small turboprop has given way to a 30-seat jet, providing daily round trips to the same three locations.

Training

Technician training does not end with AMT school, an inspection authorization refresher, or even aircraft initial training. The world of aviation continues to change with increasing rapidity, particularly in manufacturing and maintenance technology. Nondestructive testing, composite repair, data bus technology, and built-in test equipment have created a new world of wonders for the technician within the past decade. A technician who graduated from school more than a decade ago must struggle to stay abreast of the newest hardware and, yes, software.

Like it or not, our aircraft are constantly changing; service bulletins, ADs, alert bulletins, customer bulletins, and new manufacturing methods are literally making our aircraft obsolete, one piece of paper at a time. Manufacturers release a wealth of information concerning aircraft, powerplants, and appliances, but there is often so much information that not much of it gets read and/or absorbed. Whether this flurry of information is absorbed adequately or not does not guarantee that technicians remain knowledgeable of the aircraft in their charge or their craft in general.

Aircraft-specific recurrent training is a must due to these changes and to hear the latest airworthiness problems and cures. How often? Once every 3 years if there is more than one technician working on an aircraft and every 2 years if there is only one. Some of this recurrent training can be obtained at manufacturer's maintenance and operations (M&O) seminars; departments send their technicians to recurrent training one year and to the manufacturer's seminar the next.

Most maintenance training centers around aircraft-specific recurrent courses, but there is much more of value available. Troubleshooting, avionics, and electrical systems basics are quite popular and are credited with extending the basic knowledge of the technician. Some consider these tools that help the productivity of the maintenance department and provide skills that save maintenance hours and lost flights.

Maintenance Training

- Aircraft initial
- Aircraft recurrent
- Selected systems
- Manufacturer M&O
- Troubleshooting
- Electronics/avionics
- Support equipment
- Trade shows
- Maintenance resource management
- Environmental/health/safety
- Personal development/advancement (NBAA PDP)

State and local environmental and safety agencies may require toxic chemical handling, waste disposal, and special vehicle and industrial safety qualifications for hangar personnel. First aid, CPR, fire fighting, blood-borne pathogens, refueling, and airport security courses may be required too.

How about training for advancement and promotion? Management, personnel relations, finances, and leadership courses probably will make your technicians more productive, part of the team, and promotable. It is unfair to expect an expert technician to perform as a supervisor without proper training. Perhaps the best reason for training technicians is to show them that you value their contribution to the department's efforts sufficiently to invest in them, to show them that you consider them to be professionals and want them to be the best qualified and most productive employees. Investing in their future sends them a message that says, "We value your services; you're here for the long haul." Returns on such messages cannot be measured.

Flight departments regularly spend $15,000 to $30,000 per year on pilots to keep them current in the aircraft they are assigned to fly. But how much is spent on those responsible for the maintenance and inspection of the machines they fly? No such numbers exist, but casual observation would place the number at between $2000 and $4000 per year. Does this make sense? Should less be invested in the people who maintain the aircraft than in those who fly them?

As increasing numbers of "electronic" aircraft enter the fleet, training will become not just nice to have, but essential. However, the technology of the 1960s and 1970s continues to change as well and demands constant upgrading of skills and knowledge for the technician force.

The most persuasive reason for training is to ensure that the abilities of the technician are equal to efficiently troubleshooting an aircraft problem to save money and time and perhaps even a flight itself. As with pilots, a training plan for the flight department should include planned technician courses for each individual for the following year. And once the technician has been to school, record the fact in an individual training/qualifications folder.

Maintenance Resource Management[3]

Maintenance resource management (MRM) is a general process for improving communication, effectiveness, and safety in aircraft maintenance operations. Attention should be given specifically to the implementation and evaluation of MRM training. Much as crew resource management (CRM) was created to address safety and teamwork issues in the cockpit, the FAA and other national regulatory bodies, in conjunction with industry partners, developed MRM to address teamwork deficiencies within the aviation maintenance environment.

MRM is a team-based safety behavior. It teaches managers and maintenance personnel skills that enable them to work safely in a complex system. MRM teaches more than just team skills; it teaches and reinforces an organizational philosophy in which all members of the organization are oriented toward error-free performance. This is accomplished by teaching

- How the effects of individual actions ripple throughout organizations
- How to use available resources safely and effectively
- How to propagate a positive culture of safety in the organizations through specific individual actions

The overall goal of MRM is to integrate the technical skills of maintenance personnel with interpersonal skills and basic human factors knowledge in order to improve communication effectiveness and safety in aircraft maintenance operations.

Passenger Handling

The passengers carried in the company aircraft are, first, fellow associates and, second, guests or clients of the company. As such, they deserve courtesy and consideration while they are in and around company aircraft. More important, they often are not familiar with aircraft operations, and their safety and wellbeing should be given special attention any time they are in an aircraft operating area.

Maintenance section personnel are often the first contact that passengers have with corporate aviation. Therefore, maintenance personnel carry a special burden because they must ensure that the passengers are cared for properly. Although passenger service representatives and the flight crew normally are responsible for passenger handling, maintenance personnel are an integral part of their care.

All passengers should be greeted warmly, directed to the aircraft or passenger assembly area, assisted with their luggage, and given answers to their questions. A maintenance segment employee will be available to passengers for every launch and recovery.

Passengers should be escorted at all times on the ramp by a line service technician, mechanic, or flight crew member to ensure their safety and to prevent inadvertent damage to the aircraft. If a qualified person is not available to escort passengers, they may be instructed to remain in a safe location until an escort can be arranged.

The aircraft handler or mechanic should remain available at all times before departure to assist the flight crew with baggage or other passenger concerns.

Safety

Safety is as important for the maintenance department as it is for the flying side of the house. Perhaps it is more important because more personal injury accidents occur in maintenance operations and in hangars than they do in flight operations. Management commitment to a safety policy and to its enforcement is an essential part of any flight department. Expediency and just plain sloppiness are the principal enemies; constantly keep these unwanted "visitors" in check.

Chapter 8 is devoted entirely to safety, but much of what is presented relates specifically to operational safety. Maintenance safety is equally important; in an effort to ensure that

Safety Policy

Safety is the most important operating rule. It is an essential ingredient in all measurements of success. It is the *direct* responsibility of all personnel connected with the aviation department.

To accomplish the above safety mandate, the company shall operate under the following safety policy:

- Safety shall be considered by all personnel to be an integral and vital part of the successful performance of any job.

- Safety is a paramount part of good operating practice and, therefore, a management function that will always be given priority.

- Direct responsibility for the safety of an operation will rest with the person performing that operation.

- Each employee is responsible for the performance of his or her duties, giving primary concern to his or her own safety and that of fellow employees. The company's guests, property, and equipment are entrusted to their care.

- Management at all levels shall initiate procedures for prompt corrective action to eliminate unsafe acts, conditions, equipment, or mechanical hazards.

maintenance personnel do not overlook the safety subject, this special maintenance safety section is inserted into the maintenance chapter. However, make sure that at least the first and last sections of the next chapter are also reviewed because a healthy overview of safety and the accident/incident response section applies to all.

Safe Hangars for All

Consider the following:

- A passenger has to duck under a wing to get to her airplane on the ramp. She doesn't duck far enough and gets a nasty gash in her forehead.
- The aviation department manager, in a rush to get to his aircraft, turns from a conversation and moves forward in a darkened hangar, impaling himself on a trailing edge static discharge wick.
- A flight department secretary, looking for a technician, falls from an airstair and breaks her wrist.
- A manufacturer's customer service representative slips and falls on a wet hangar floor, injuring his back.

What do these accidents have in common? They all occurred in hangars. What do they not have in common? None of these accidents involved an aviation maintenance technician. Why is this important?

Most people in corporate aviation think of the hangar floor, its shops, and its equipment to be the sole domain of the technician and not the concern of others in the department or outsiders. The mishaps just mentioned tend to debunk that myth, underscoring the need to provide a safe operating environment for all who may venture into the hangar area.

The aircraft hangar is principally the domain of the technician, aircraft cleaner, and hangar attendant—it is these employees who must be protected first. Most corporate flight departments are aware, at least to some degree, of the requirements of the Occupational Safety and Health Administration (OSHA), mandated by federal and state law. These codes specify equipment standards, personal protection devices, operating procedures, and information elements for all workplaces that involve industrial activity. Companies engaged in this type of work are subject to inspections to ensure compliance with the regulations and sanctions if the codes are not met. However, just being in compliance with OSHA and company safety standards does not guarantee that safe operations will occur.

Rather than just compliance with a set of rules that sometimes seem to be of little value except to hinder the worker and his or her company, it is the concept of *protection* that must be considered to ensure workplace safety. Consider the following:

- Aircraft check stand safety rails are provided but not used by technicians.
- Material Safety Data Sheets (MSDS) are prominently displayed, but no employee has ever read them or understands their value.
- Eye protection devices are provided, but maintenance supervisors do not enforce their use.
- An eye wash station is available on the hangar floor, but it has never been tested or flushed.
- No technician has ever viewed available hangar safety videos.

While this fictitious company seemingly is in compliance with most OSHA standards, it is probably unsafe. Why? The attitude of management and employees seems to be that safety is a paperwork exercise and not a necessity. Like people sitting on their seat belts, these are accidents waiting to happen.

Perhaps one of the reasons companies are seemingly in compliance with regulations but are not in practice is that they do not consider safety to be a part of their standard operating procedures (SOPs). That is, even though they have safety rails for check stands, the requirement to use the safety equipment is not written into their normal procedures. However, the real reason may be that supervisors do not take the time to enforce the few simple rules that go along with safe operations around aircraft.

Safety pays! Try replacing a disabled employee if you don't believe this.

- Provide a safe working environment—walking surfaces, noise, light, temperature.
- Use personal protective equipment.
- Provide proper tools and equipment.
- Report unsafe conditions and practices—correct them as soon as possible.
- Educate for safety.
- Regularly inspect work areas for unsafe operating conditions.

Basic Hangar Inspection Checklist[4]

- Fire detection and extinguishing equipment
- Personal protective devices—use, maintain
- Work surfaces—floors, stands, ladders, fall hazards
- Housekeeping—spills, trash, clutter, obstacles
- Storage of flammables
- Compressed gasses—storage/use
- Motorized machines—guarding/anchoring
- Environment—ventilation, noise, light, temperature
- Toxic substances—labeling, storage, use, protection
- Electrical, compressed air, water—safety, integrity, availability
- Work procedures

The final item on the list needs some explanation. Work areas and practices are subject to a slow deterioration if they are not constantly looked after. The "we've been doing things this way for a long time and they work OK" syndrome creeps into daily operations, and the downward slide begins. If there are safe working standards in place and they are upheld and enforced on a continuing basis, the battle for safety is more than half won. Take time to inspect the hangar area using a checklist (see the following sidebar) at least quarterly. Better yet, have a company or other safety expert look at your facility annually in addition to your inspection efforts.

There must be buy-in to these tenets from all levels of management. Without management educating, coaching, demonstrating, and enforcing safe practices, the troops will not do what's needed. Management "must walk the talk" if a safe operating environment is to be created.

Technicians and other hangar workers must be guided and supported in the extra effort it takes to properly position the right check stand or prepare a safety harness for a fall hazard. Without a cooperative effort between management and the workers, safety will not occur in the hanger.

And remember, hangar safety is for everyone, not just the technicians. Safe operating practices must be taught to and lived by all personnel who have access to the hangar. Any person with access to the hangar should either be educated in its hazards or continuously escorted. This is why your automobile service center prominently displays signs forbidding entry to shop areas—the workers know that hazards exist in those areas for the uninitiated and realize that they incur a liability by admitting nonemployees to their work areas. Everyone must be involved in safe hangar operations.

Steps to take:

- Assign hangar safety as a collateral duty to an individual.
- Create a hangar safety plan.
- Conduct regular inspections (see the sidebar on page 292).
- Enforce safety regulations.
- Strive for continuous safety improvement.

Thousands of people will be injured in an aviation ground environment this year. Keep your people out of those statistics.

Solo Technicians

Many flight departments have just one AMT. This is the norm for single-aircraft operators and appears to work well in terms of reliability and safety. However, there are significant challenges associated with this work process.

Regardless of the number of technicians assigned, every department is subject to the single-technician phenomenon. Even in large departments, technicians often work alone or with little supervision. And there may not be enough time to allow a second person to review the work of the individual prior to approval for return to service.

Airworthiness.   The name of the game for any AMT is to maintain assigned aircraft in an airworthy condition. In corporate aviation, the pride and personal integrity shown by most technicians make airworthiness a given rather than the ultimate goal. However, having just one brain, one set of eyes, and two hands presents significant barriers to the solo technician.

Regardless of the experience level of the technician, a second opinion from another is always welcome. And a second set of eyes and hands increases the confidence level and speeds completion of the task.

Perhaps the greatest challenge faced by the solo technician is having to be right all the time—a tall order for mere human beings. Using the specified torque value, getting the sequence of wire clips in the correct order, and servicing the hydraulic system to the proper level are all critical items—were they done correctly? Unfortunately, this thought may occur at 2 A.M. when the technician is on the way home from the hangar. Inspection of critical components during an inspection or verifying one's own work on these systems may be the most difficult part of the AMT's job; on some systems, you only get one chance to get it right. Airlines and repair stations have mandatory inspections imposed on their critical systems/component work via required inspection items (RII). This system is a luxury the single technician can seldom afford.

However, the individual may ask a fellow solo practitioner, a pilot experienced in maintenance issues, or an A&P from a maintenance vendor to perform a quality check on his or her work. Yes, even pilots can be trained to look for errors if they know what to look for.

Personal Safety.   The AMT works in an industrial setting that contains many hazards. The technician often works in an isolated area, surrounded by hazards, where a serious injury may go undetected for a significant period of time. Some companies do not allow technicians to work alone, instead assigning another individual who may not be a technician to work with them. Alternatively, the solo technician may be required to check in with company security at regular intervals or to wear a wireless alerting device to ensure safety.

Administration.   Some estimates claim that 40 percent of an AMT's time is spent on paperwork. While not seemingly productive time, this function is an essential part of the technician's job. This paperwork includes maintenance recordkeeping; keeping manufacturer's directives current; ordering, receiving, and shipping parts; and vendor management. The company probably adds its own administrative requirements to the aviation-related ones, further compounding the problem. The volume of paperwork requiring the technician's attention and the accuracy of the completed work are both problems for a one-person maintenance operation. Any clerical assistance is appreciated.

Help for Singles.   Both the single technician and every member of the flight department must realize the difficulties and hazards that confront these individuals—the concept of looking out for one another is especially important when the maintenance department only has one name associated with it. Judicious amounts of assistance, professional and otherwise, will help to ensure quality work and safe technicians.

Upstairs, Downstairs

Keeping all members of the flight department informed and up to speed on current and future operations becomes a challenge when the department consists of more than one person. Communications and coordination tasks of the manager's job begin to multiply rapidly as the number and type of personnel grow. Keeping two pilots in the loop is hard enough in small departments, but when AMTs, schedulers, administrators, and purchasing agents are brought into the picture, more of the manager's time becomes devoted to keeping these diverse elements of the team moving in the same direction.

One of the greatest communications gaps observed in flight departments occurs between operations personnel (pilots) and maintenance personnel (AMTs). These two groups are highly specialized and usually work well within their respective departments but may not work well at the interface between their specialities. The "they fix 'em, we fly 'em" and "they break 'em, we fix 'em" mind sets describe the gap that may exist between the pilot and AMT. And because pilots traditionally have offices on the second floor of the hangar and AMTs on the first, this rift has been described as the "upstairs, downstairs" phenomenon, a title borrowed from the PBS Masterpiece Theater series that described the interaction of a nineteenth-century British noble family and their servants.

This communications gap that may exist between the two groups can affect the efficient operation of the department in a variety of ways. Misunderstood aircraft discrepancies, poorly scheduled maintenance planning, handling repeat discrepancies, and just being unaware of either the operations or maintenance schedules can cause increased expenses, lowered efficiency, and flights that do not go. If permitted to exist for long, these conditions may grow into feelings of mistrust and hostility between the two departments.

Maintenance and Ops on the Same Team

- Encourage frequent visits between the two camps.
- Hold joint pilot and AMT meetings to discuss areas of mutual interest.
- Locate the pilots and AMTs in the same area of the hangar.
- Create quality action teams that consist of people drawn from all departments to solve specific problems.
- Designate an AMT to be present at the launch and recovery of all flights to discuss the airworthiness of the aircraft with the pilots.
- Take AMTs along on flights as observers.
- Designate pilots to assist/observe with scheduled inspections of the aircraft they normally fly.
- Create pilot/AMT teams to manage each aircraft.
- Use AMTs and pilots to conduct in-house aircraft systems and operations training for their counterparts.
- Provide lunch for all hands on a regular basis to encourage interaction between the groups.

Just think how the two groups might better operate as a team if there were no "stairs" separating them.

Know the Regulators

Horror stories abound regarding interaction with the local aviation regulatory representatives, especially with FAA Flight Standards District Offices (FSDO). Red-tagging aircraft for minor airworthiness items, unreasonable ramp checks, unannounced visits that result in fines and airman certificate actions, and general unreasonableness about interactions with the dreaded airworthiness or operations inspector are common aviation lore. Are they true? Some undoubtedly are. How pervasive are these unjustified federal actions? Probably not very.

Personal and business turbojet operations enjoy one of the best accident records and operational reputations in civil aviation. The insurance industry knows this, so it places few restriction on those operations. Airport operators know this and essentially leave corporate operators alone. And most airworthiness and avionics inspectors know this and do the same.

Most airworthiness inspectors have a massive workload placed on them every year through their national work plan. FSDOs are tasked with visiting and surveilling all air agency certificate holders within their geographic area not once but several times during the fiscal year. Scheduled air carriers, air taxi operators, repair stations, instrument shops, and FBOs are all targets of their scrutiny on a continuing basis. Inspectors will tell you that working with some of these operators can be a complex and challenging task. Keeping these companies in compliance with the myriad of FARs and FAA orders is more than a full-time job.

So how much time do they have to monitor and inspect general aviation? Not much. And for the corporate operator who enjoys such a sterling reputation, not much indeed.

Does this mean that you are home free, that you never have to deal with the feds? Probably. Is this a good thing for you and your operation? Probably not.

While you may consider it dancing with the devil, getting to know at least one contact at your local FSDO will benefit you and your operation someday. An aircraft ferry permit, a field sign-off for a 337, and an interpretation of a regulation or AD are all reasons to know a local FAA inspector. And being able to associate a name with a voice or face makes the transaction more friendly.

If you are really feeling brave, why not invite your local airworthiness inspector to visit your facility? While there is an actual Part 91 maintenance inspection procedure listed in the inspector's handbook, chances are that the inspector will be delighted at your hospitality and limit his or her surveillance activities to a detached admiration of your facilities.

Speaking of "the handbook," FAA Order 8900.1 which incorporates the *Airworthiness Inspectors Handbook,* is an invaluable compendium of the way the FAA expects its inspectors to carry out their activities. This tome (two large loose-leaf binders) is a treasure trove of information about acceptable means of compliance, standards, and regulatory processes. And when an inspector hears you referring to "the handbook," the conversation is automatically elevated to a higher plane.

While it is difficult to have a perfect relationship with those who are paid to ensure compliance with the welter of regulations and orders that govern our activities, these people joined the FAA for the same reason we started in our jobs, a love of aviation. They are kindred spirits in many ways. Get to know your fed.

Selling Maintenance

Every day you tackle the monthly due list, work off discrepancies, and maintain the hangar. Keeping your aircraft airworthy and in good condition is the reason for your existence, so you do it cheerfully and with satisfaction.

Long hours, knotty problems, bad parts, obscure manuals, and sometimes unappreciative pilots do not make the job any easier, but it is your chosen profession, so you enjoy doing it most of the time. It is what is expected of you, so you do your job with little expectation of recognition other than an occasional "Good job" or "Well done" from those who count.

However, did you ever think of the significant contribution the maintenance section provides to the corporate aviation effort? The name of the game is airworthiness—without it, nothing happens. The complexities and number of items that go into the concept of airworthiness are not well understood outside the AMT fraternity. Pilots and schedulers wonder what takes so long to approve an aircraft return to service, managers wonder why you need another set of hands, and the accountants always ask why windshields cost so much. The truth is that most people, even aviation people, do not understand what it takes to maintain a modern turbine-powered aircraft to the standards required by the FAA, the manufacturer, and most important, you. Aviation maintenance technicians have been doing their job quietly and efficiently for so long that everyone has come to expect miracles from their technicians. And like Rodney Dangerfield, they may not be getting the respect they deserve.

Why not tell them more about what is involved in the maintenance process? Why not let the rest of the department and the folks downtown in on what it takes to keep the aircraft in excellent condition?

The immediate effect probably will be greater respect from all concerned; the long-term effects may be more significant. If the department manager understands how many individual actions and work cards it takes to perform a phase inspection, including those requiring two sets of hands, he or she may be more willing to provide temporary or permanent help or to send the aircraft out to a maintenance contractor to complete the work. If the accounting department understands how hard you try to get the best deal for rotables and replacement parts, there may be fewer questions about the invoices. And if senior company management understands some of your money-saving successes or notable airworthiness safety finds you make in the course of your work, the entire flight department's image will benefit.

By nature, most technicians are not predisposed to blowing their own horn or selling their virtues to anyone. However, a little low-level selling may bring big benefits.

The suggestions shown in the sidebar may help get more recognition for the maintenance function and its personnel. The most important lesson: Don't hide your light under a basket.

Good maintenance is not an option; my airplanes don't fly without it.   —AIRLINE CEO

Techniques for Promoting Maintenance

- Keep records of your searches for major parts and services, recording the quotes for each. When significant differences exist, share these with your manager and those within accounting. A formal memo may be appropriate for major finds.

- When an inspection turns up a significant fault or defect, make a photographic record of it and share it with the pilots and company management. If you can save the part or component, use it as a dramatic visual aid.

- Take the time to show schedulers, dispatchers, and pilots what is involved in a major inspection. Invite them to the hangar deck to observe notable procedures and inspection techniques, e.g., landing gear swings, nondestructive testing, electronic BITE procedures, etc.

- Become involved in a company nonaviation activity such as an employee research team, sports team, or charitable activity. Most people in the company do not know the aviation maintenance section exists. Let them know it does.

References

1. *NBAA Management Guide.* Washington: NBAA, 2012.
2. Spencer, F. W., and Schurman, D. L., *Reliability Assessment at Airline Inspection Facilities,* Vol. III: *Results of an Eddy Current Inspection Reliability Experiment.* DOT/FAA/CT-92-12. National Technical Information Service, 1992.
3. From FAA Advisory Circular 120-72, *Maintenance Resource Management Training.*
4. US 26 CFR 1910 and NBAA OSHA Compliance Basic Compliance Checklist.

CHAPTER 8

Safety

Safety Programs

Corporate aviation enjoys an enviable safety record, maintaining an accident rate equivalent to the major scheduled airlines. In fact, corporate aviation's record compares favorably with that of the airlines in a number of recent years according to the data released at the end of 2011 (see Table 8.1). How does this happen?

Dedicated, professional pilots and mechanics, good management, excellent equipment, standards that are set and adhered to, and perhaps, most importantly, a corporate and departmental culture of conservatism are the causal factors in a safe operation. This last item is most important because safety as a concept and safe operations specifically are largely intangible; safety is an *attitude*, a mind-set established at the highest levels of the organization and projected downward through the ranks. While it is possible to run a safe flight department without a high-level commitment to safety in the organization, this is a rarity and a credit to the professionalism of the individuals within the department.

However, since safety is essentially an attitude and a way of life, it must constantly be demonstrated and reinforced within the organization. Merely telling professionals to be safe will not work. Senior pilots and mechanics who go by the book, take the conservative approach, and comment on unsafe and potentially unsafe practices that they observe are the backbone of a safe operation. Other practices reinforce the safety message as well:

- Recurring references to standards
- Constant training
- Required reading of safety materials
- Reviewing standardization, regulatory, and procedural materials
- Engaging in spontaneous discussions of aircraft systems, procedures, and regulations
- Initiating spot inspections of facilities and records

Recognizing the execution of safe practices helps, too. Company and departmental recognition serve as local reinforcement, while the NBAA Corporate Flying Safety Awards Program serves as a national incentive.

Safety as a concept does not sell well, but daily practice of an ingrained safety culture does sell.

Operational versus Safety Tensions

The company purchased its aircraft to realize the benefits of on-demand air transportation: to go when and where its personnel want to go, do it expeditiously, and, most importantly, do it safely. At times, the when, where, and expeditious qualities may be in conflict.

The boss wants to go to his lakeside home in northern Michigan in February for a hunting trip with his friends. The narrow, 4000-foot, single strip serving the destination has only a VOR approach to 600-2 minimums, and it is often unattended during the winter. At departure time, light snow and strong crosswinds are forecast for the destination; arrival will be just after dark with the prospect of a circling approach to minimums. The nearest decent alternate is 100 kilometers (km) away, a 2.5-hour drive away from the boss's house; and fuel could be tight on a divert. It is clear that the boss wants to get there as soon as possible to get the weekend started. What should you do?

Option A: go take a look-see at the local airport. Option B: go directly to the alternate and put the party in a prearranged limo. What do you tell the boss? After a quick risk assessment using your flight risk analysis tool, the red numbers direct you to option B. Since you have previously instructed the boss regarding your use of operational standards and risk assessment techniques, he readily agrees with your choice and tells his party that just to be on the safe side, we are going to take a "scenic tour" of the Michigan woods after landing.

But if the boss were uninformed and an impulsive sort, what would have happened if option A had been his choice? Suppose the concepts of hazard identification, risk averseness, and operational standards had not become a part of the boss's vocabulary? What then? Anger the boss, and start looking for a new job? While extreme, worse cases have happened.

The best course of action is to make the boss (the *big* boss, preferably) a part of the safety program:

- Relate how the flight department treats safety and standards.

- Show how department personnel identify hazards and act through risk assessment techniques.

- Demonstrate how the entire department participates in the safety process.

- Provide evidence of how SOP and policies have been modified as a consequence of the above processes.

- Best method: co-opt the boss into the department's safety program, ideally a comprehensive safety management system, by having him sign a comprehensive safety statement at the beginning of your safety documents. More on this later in the chapter.

What Is the Problem?

Why all the emphasis on safety? Business aviation is relatively safe, so why bother with safety programs and the like? The reason the system is relatively safe is the emphasis on safety issues for the past 20 years. Efforts by the International Business Aviation Council (IBAC), National Business Aviation Association (NBAA), European Business Aviation Association (EBAA), International Civil Aviation Organization (ICAO), Flight Safety Foundation (FSF), FAA, aircraft manufacturers, and others have provided research-supported evidence of the need for safety programs suited to the business aviation community. Take a look at Table 8.1 to see how safe business aviation is, and where its weak spots are.

In the table, it is gratifying to note that, over the most recent five-year period of available data, the business and corporate aviation accident rates have continued to be relatively stable and compare favorably with airline mishap rates. The corporate rates are a testament to

Operator Type	Departures (5 yrs)	Total Accidents	Fatal Accidents	Total Accident Rate	Fatal Accident Rate
Commercial (air taxi)	10,626,845	248	92	2.33	0.87
Corporate	19,331,071	71	12	0.37	0.06
Owner-operated	4,998,812	110	35	2.20	0.70
All business aircraft	34,956,730	563	150	1.61	0.43
Airline	~30,000,000	126	16	0.42	0.05

TABLE 8.1 2011 Business Aircraft Accident Rates by Operator Type (Extrapolated) (per 100,000 departures) (*Courtesy IBAC and Boeing*)

the care and management of those operations from both the flight department and the companies that own and operate the aircraft. Yet we are still plagued with the fact that more than *two-thirds* of business aircraft accidents occur in the final minutes of the flight, i.e., the approach and landing phases. Similarly, another 15 percent of accidents happen during the takeoff and climb phases, which take up approximately three percent of flight time. While it may seem logical that many accidents would occur during the time the aircraft is closest to the ground, concentrated focus on these percentages has yielded little change in the rate of occurrences. These are clearly areas that we must work on if improvements in the overall accident record are to be made.

Notably, loss of control, controlled flight into terrain, and increasingly, runway excursions are the most frequently cited accident causes for commercial aviation. Business aviation's difficulties seem to concentrate on the approach and landing phase. A number of specific programs, presented by ICAO, the Flight Safety Foundation, and various worldwide regulatory agencies, are all excellent programs designed to curtail these persistent types of accidents.

Additionally, non-flight related accidents are often overlooked in a safety overview of business aviation operations. These occur mostly to individuals in maintenance, fueling, aircraft ground movements, and hazardous materials handling. While much emphasis is placed on protecting aircraft, passengers, and crew in the flight regime, less attention may be focused on ground operations, which often feature very damaging and debilitating personnel injuries. Many of these mishaps are also classified as incidents, not qualifying for either the ICAO or US NTSB definitions for accidents; but incidents can often be serious to the point of being as bad as an accident.

Safety Imperatives

- Identify hazards.
- Evaluate risks.
- Communicate safety concerns and ideas.
- Train to understand hazards.
- Set high operating standards.
- Make safe operations a collective effort.
- Emphasize curiosity and inquiry about risks as important elements.

Therefore, we should not concentrate on just one or two obvious and recurring faults. Rather, we should concern ourselves with an overall program that will make operators more aware of the need for a *universal safety program* that will encompass all of the potential hazards operators face in both routine and unusual operational environments. This should lead us to a *systematic* approach to evaluating and mitigating the risks generated as a result of the hazards each operator may face. But first, some useful background information.

Where Accidents Happen

In many ways, business aircraft are at a disadvantage when compared to their airline counterparts. They have fewer support functions—dispatch, crew scheduling, en route assistance, and the use of a wide variety of airports, which may have lesser advantages for them than for the airlines. So, while we must credit the majority of business aviation operators with a very good safety record when their situation is compared to the advantages held by the airlines, accidents do happen. It is clear when viewing Fig. 8.1 that these accidents occur predominately during the approach and landing phases.

In 2012, IBAC performed an analysis of these accidents and discovered the following:

- Overall, turbojets have fewer accidents than turboprops, but more than half (55%) occur during the landing phase of flight.

- A significant number of turboprop accidents involve single-pilot operations.

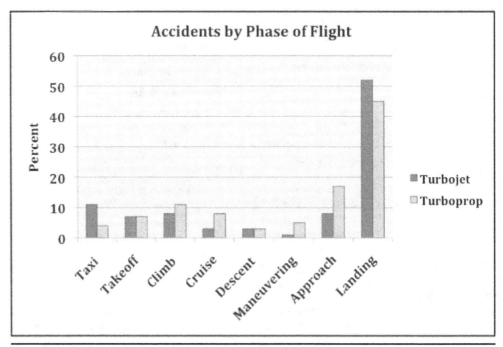

FIGURE 8.1   Business aviation accidents by phase of flight. (*IBAC.*)

- Both turbojet and turboprop landing accidents involve:

  - Poor speed control and unstable approaches
  - Incorrect or lack of runway condition information
  - Crosswinds and gusts as frequent factors
  - Poor runway conditions and snow clearance as frequent factors

- Overall

  - *Runway was seldom a factor.*
  - Fatigue was *not* a factor.
  - Pilot experience was not an evident problem.
  - Low ceilings and visibilities were *not* a problem.
  - Either day or night was not a factor.

Additionally, the Flight Safety Foundation cites the major risk factors in landing excursions as:

- Go-around not conducted
- Landing long
- Ineffective braking (contaminated runways)
- Gear malfunctions
- Fast approaches and landings

Recommended mitigation factors include:

- Adherence to operations manual and aircraft flight manual
- Safety management systems and flight data analysis data
- Improved runway condition reporting
- Accelerated implementation of vertical guidance approaches

To these I would strongly add strict adherence to stabilized approach parameters.

The main point is that if more than half of all business aviation accidents occur during the final few minutes of flight, during the approach and landing phase, then all aspects of the SOP, limitations, training, and risk assessment programs should concentrate on making these flight phases as risk-free as possible. In addition, note from the above recommendations that added attention should be given to the "soft factors," such as adherence to SOP and use of all parts of the SMS.

Hangar Safety

Hangar safety is for the technicians, right? So, the pilots can skip this part? Well, no…*the number of pilots injured in industrial-type accidents often exceed the number of technicians injured.* The next most likely personnel to be injured are passengers passing through the hangar en route to their aircraft. So, hangar safety is everyone's job, especially if you are likely to be injured.

Examples:

- A pilot preflighting an aircraft in a confined corner of the hangar is impaled on a static discharge wick.
- A passenger is forced to duck under a wing tip en route to an aircraft, straightens up too soon, and receives a nasty head wound.
- A technician running to meet an arriving aircraft slips and falls on a wet hangar floor, severely injuring his back.
- A receptionist looking for a pilot falls from an airstair, breaking an arm.
- A technician falls from a high check stand while inspecting the tail of a Gulfstream, suffering a mild concussion and broken ankle.
- A lineman's foot is caught under a moving hangar door and loses two toes.

So there are inherent hazards within the confines of the hangar. But awareness of these potential problems may be a greater problem than the existence of the hazard. No one wants to think about injuries, hazards, and the like. Everybody would rather think about the task at hand, to be goal directed and operationally minded.

If your company has a company-wide safety program, there are probably scheduled inspections conducted and/or checklists available to help with this essential safety task. If your company does not have any of this available, you can conduct your own safety inspection using readily available checklists and reference documents (see documents mentioned at the bottom of page 306).

But even more important is to ensure that everyone in the department is "sold" on the concept of hangar safety, and subscribes to an ongoing need for attention in this area.

Therefore, gather a representative sample of people from all occupations within the hangar and start talking about a few of the following issues:

- Is our hangar safe? How do we know?
- If not, what is unsafe about it? What can be done to improve it?
- Have we had any hangar safety accidents in the recent past? Have the causes of these accidents been addressed and remedied?
- What can we do to make our hangar safer?
- What level of attention and emphasis should be used to ensure hangar safety on an on-going basis? How can we get everyone involved in safety awareness?

For more information about hangar safety, see the "Safety" section of Chapter 7.

Safety Culture

Over the past 50 years, organizations involved in critical operations, such as nuclear power, human surgery, and oil exploration and production, have discovered that merely having a series of unrelated and poorly coordinated processes and procedures do not make an effective safety program. Rather, a series of mature, coordinated, and fully integrated measures all designed to create a risk-averse environment provides superior safety within complex organizations. During the past half-century, the previously mentioned fields of specialization and aviation interests have developed a series of policies, processes, and procedures known as a safety management system (SMS) to provide a positive safety environment.

Much of this chapter will be devoted to a discussion of these systems and how they operate. But first, before any consideration of safety systems can be made there must be a foundation for SMS implementation.

Each organization develops a distinct set of behaviors over time that gives it a unique identity. This *organizational culture* consists of its values, visions, norms, working language, systems, and symbols; it includes beliefs and habits. It is also the pattern of these collective behaviors and assumptions that are taught to new organizational members as a way of perceiving, and even thinking and feeling. Organizational culture affects the way people and groups interact with each other, and with others outside the group. Organizational culture is an important aspect of any organization, since it provides the group's identity and justification for existence.

Safety is an important component of culture, since it encompasses a number of organizational norms including risk perception, adherence to standards and procedures, interaction with others in critical situations, a cautious view of goal direction and respect for other team members' opinions, and views of critical situations. Safety culture has several components, according to Dr. James Reason, a leading proponent of culture-based safety activities, including:

- An *informed culture* in which the organization collects and analyzes relevant data and actively disseminates safety information.

- A *reporting culture* that cultivates an atmosphere where people have the confidence to report safety concerns without fear of blame. Employees must know that confidentiality will be maintained and that the information they submit will be acted upon; otherwise, they will decide that there is no benefit in their reporting.

- A *learning culture* in which an organization is able to learn from its mistakes and make changes. It will also ensure that people understand the SMS processes at a personal level.

- A *flexible culture* in which the organization and the people in it are capable of adapting effectively to changing demands.

- A *just culture* in which errors and inconsequential unsafe acts will not be punished if they were unintentional. However, those who act recklessly or take deliberate and unjustifiable risks will still be subject to disciplinary action.

Note that the first four items above form the core of an SMS, which we will discuss at length later. First the concept of the just culture will be discussed since, without it, there can be no true, productive safety culture within an organization.

Just Culture

Dr. Reason describes the just culture as an atmosphere of trust in which people are encouraged, even rewarded, for providing essential safety-related information, but in which they are also clear about where the line must be drawn between acceptable and unacceptable behavior. So it's just not a free ride or a get-out-of-jail-free card when an individual crosses the line between behavior that is good, reasonable, and acceptable and that which is irresponsible, contrary to accepted procedures, and potentially damaging.

Just culture refers to a set of organizational behaviors that promotes a questioning attitude, is resistant to complacency, is committed to excellence, and fosters both personal accountability and corporate self-regulation in safety matters. This series of policies and behaviors is attitudinal as well as structural and relates to both individuals and organizations. Personal attitudes and corporate style can enable or facilitate the unsafe acts and conditions that are the precursors to accidents and incidents. A just culture requires not only actively identifying safety issues, but also responding with appropriate action.

Most of all, the just culture centers and is based on *total mutual trust within the organization*. Without trust, there can be no just culture and, unfortunately, no real safety culture. The non-trustful organization will be reduced to safety slogans, empty pronouncements, non-standard behaviors, and a reluctance to become truly risk averse. I hate to make it sound so grim and unremitting, but without trust at *all* levels, the organization reverts to a series of individuals marching to their inner safety drummer.

The benefits of a just culture include:

- *Increased reporting*—It not only can lead to increased event reporting, but also can aid in identifying trends that will provide opportunities to address latent safety problems.
- *Trust building*—Trust comes from a series of organizational experiences that reinforce the validity of working together towards a common goal. The lack of trust builds walls and creates resentment.
- *More effective safety and operations management*—It permits an organization to better determine whether violations are occurring infrequently or if deviation from established procedures has become normalized among its front-line employees and supervisors.

But how do you establish this type of culture in a flight department? First, it is not something that can be created instantly; you cannot install it and expect it to perform well immediately, as you would in replacing an aircraft part. This is a program that requires constant work and attention after it has been installed; attitudes and behaviors are being changed. This requires a demonstration of working methods, especially by management, and a process for nurturing mutual trust within the organization. Given that, here are some ways of getting started:

- Establish a legal framework to support no-fault hazard reporting.
- Initiate policies and procedures that encourage reporting.
- Clearly define the roles and responsibilities of the people required to implement and maintain a just culture reporting system.
- Create feedback to people within the organization. Rapid, useful, accessible, and intelligible feedback to the reporting community will prove the system's value to all hands.
- Educate the users with regard to the changes and motives of the new system.
- Promulgate methods for developing and maintaining a safety culture.

Sounds easy, doesn't it? In reality, this is like growing a garden; it takes a long time and much care to get from planting the seed to enjoying the fruits of your labor. However, whether it be growing tomatoes or realizing a mature just culture within your organization, it is worth the effort.

Take a look at your organization's safety by using Dr. Reason's *Checklist for Assessing Institutional Resilience* at www.tc.gc.ca/eng/civilaviation/publications/tp13844-menu-275.htm. Then see *A Roadmap to a Just Culture: Enhancing the Safety Environment* at www.flightsafety.org/files/just_culture.pdf

Sample Flight Department Safety Program Goals

- Safety mission/task profiles reviewed quarterly and used to modify policies and risk assessment measures
- New personnel fully trained in SMS/emergency issues prior to commencing productive work
- Internal evaluation program conducted quarterly and deficiencies handled expeditiously
- Emergency training courses evaluated and modified semiannually, as necessary
- Hazard identification and risk mitigation system regularly used to evaluate work processes/procedures by personnel from outside the area of interest
- Management level effectiveness discussions immediately following quarterly safety meetings
- Annual environmental, health, and safety evaluations conducted by company personnel or consultant
- Semiannual effectiveness evaluation of risk mitigation measures taken

Note that all goals are measurable

Managing Safety

Most pilots and mechanics who have been in the business for a while are familiar with safety programs. In fact, they are familiar with a variety of programs from the past that were tried and discarded with such regularity that they were known as the "program of the week." This was the case because most people and organizations focused on hit-or-miss accident prevention and causes rather than on a comprehensive, fully integrated set of policies, processes, and standards. Some of the causes focused on were

- outcomes and causes,
- unsafe acts by operational personnel,
- assigning blame or punishment for failure to "perform safely", and
- addressing identified safety concerns exclusively.

Little was considered regarding organizational or cultural aspects of a flight department. All efforts were focused on foreseeing the most common accident types and building barriers to prevent their occurrence. In reality, accidents never are caused by a single failure such as lack of training, poor judgment, or inadequate supervision; accidents are caused by lack of adequate policies, failure of management to enforce existing standards and norms, poor supervision, a desire to accomplish assigned duties and missions without the benefit of assessing the attendant risks, and inadequate operational policies and standards.

This piecemeal and fragmented approach was only marginally effective because it neglected the need to address proximate *and* underlying causes of an accident and to achieve an all-encompassing view of the entire operation. Most operators believed that with a little

Safety Management Formula

- Identify the hazards.
- Assess and measure the associated safety risks.
- Mitigate to eliminate the hazards or reduce the risks to an acceptable level.
- Track and evaluate safety management as appropriate and effective.
- Modify safety management activities and policies and SOP as required.

bit of training, a few policies, some standard operating procedures (SOPs), and occasional reviews of selected operational risks all would be OK. Yet, these items were not well balanced or coordinated—the trainers did their thing, policy makers worked in an organizational vacuum, SOP designers did what they thought best, and management tried to keep up with it all. What was really required was a comprehensive, flight-level 410 view of the *entire* operation and its environment. Moreover, all of this needed to be tied together with well-integrated policies, processes, procedures, personnel accountabilities, and a feedback mechanism to tell all concerned whether the organization was meeting the goals it had devised for evaluating the entire safety picture of the operation. Enter the SMS.

Safety Management Systems

Over the past 20 years, a systems approach to safety emerged from several industries and activities involving critical operational elements. Among these are nuclear power, medical surgery, and oil and gas exploration and production. Military aviation and the airlines were the first to adopt the principles laid down by the original critical occupational specialties. All, to one degree or another, use the same principles:

- *Senior management's commitment to the management of safety.* This includes a high-level commitment to the concept and the allocation of resources to ensure project viability.
- *Effective safety reporting.* Most of such data will be acquired through voluntary reporting and self-reporting. Organizations must develop environments in which safety reporting takes place freely.
- *Continuous monitoring* of safety data and the resulting analyses of operational hazards that occur during normal operations. Inherent in this process is the sharing of this information with all operational personnel.
- *Investigation of safety occurrences* to identify systemic safety deficiencies rather than to assign blame.
- *Sharing of safety lessons learned and best practices* through the active exchange of safety information.
- *Integration of safety training for operational personnel.* There is an urgent need to include dedicated training that addresses the basics of safety management at all levels of operational personnel training.
- *Effective implementation of SOPs,* including the use of checklists and briefings. Realistic, properly written, and constantly adhered to SOPs; checklists; and briefings are excellent safety tools.
- *Continuous improvement of the overall level of safety.* Safety management is an ongoing activity that can be successful only through continuous improvement.

From these basic principles, the International Civil Aviation Organization (ICAO) has devised aviation-organization-specific recommended practices that provide a comprehensive, fully integrated SMS. It consists of four major components, divided into twelve elements, intended as a guide for the development and implementation of a service provider's SMS, as follows:

1. Safety policy and objectives

    1.1 Management commitment and responsibility

    1.2 Safety accountabilities

    1.3 Appointment of key safety personnel

    1.4 Coordination of emergency response planning

    1.5 SMS documentation

2. Safety risk management

    2.1 Hazard identification

    2.2 Risk assessment and mitigation

3. Safety assurance

    3.1 Safety performance monitoring and measurement

    3.2 The management of change

    3.3 Continuous improvement of the SMS

4. Safety promotion

    4.1 Training and education

    4.2 Safety communication

This basic outline was devised and published in 2006 in the *ICAO Safety Management Manual*. This has become the industry standard and has been adopted by many ICAO States. All States, air operator certificate holders, and large and/or turbojet-powered aircraft must have an SMS appropriate to their individual operations.

In 2002, the International Business Aviation Council (IBAC) developed the International Standard for Business Aircraft Operations (IS-BAO), which is built around the SMS concept and incorporates a comprehensive set of operational and administrative standards. When this book was printed, IS-BAO had been adopted by more than 700 business aircraft and charter operators worldwide through an auditing and registration process. For more information, see "IS-BAO" section in Chapter 6 and www.ibac.org/is_bao.

In flying I have learned that carelessness and overconfidence are usually far more dangerous than deliberately accepted risks   —WILBUR WRIGHT IN A LETTER TO HIS FATHER

Why an SMS?

Most corporate and charter operators are aware of the existence of SMS, but not all may fully appreciate the rationale for its use and the advantages it brings to an operator. They practice some form of overall safety program, especially if they have an up-to-date flight operations manual that contains policies, standards, and procedures that apply directly to their operation. What is missing is the integration of these elements into a system of interlocking policies and procedures that considers all elements in concert.

More importantly, the SMS forces an operator to actively consider potential hazards, analyze them, and create measures that will minimize the risks involved with the hazards. Further, the system provides for participation of all members of the flight operation in the SMS; teamwork is a welcome result of this action in most operations. The concept of constant improvement through a series of regular reviews of the operation's activities and compliance with its own standards completes the action loop of hazard identification and mitigation, active risk assessment, management of organizational and environmental change, internal evaluation, and program revision.

The advantages that result from being able to actively measure operational and organizational risks (and mitigate them), better management practices, increased customer confidence, loss prevention, and preferred insurance rates, as well as full integration of a team within the flight operation provide a rewarding return on the investment of time and effort spent in implementing and maintaining an SMS.

For the naysayers who contend that their seat-of-the-pants brand of safety management is just fine without the advantages of an SMS, they should ask this question: Really, how safe is my operation? How can I know if I have no means of measuring the risks we encounter on a daily basis?

The ICAO definition of an SMS is a *systematic approach to managing safety, including the necessary organization, structures, accountabilities, policies, and procedures.*

IBAC defines an SMS as *the systematic and comprehensive process for the proactive management of safety risks that integrates the management of operations and technical systems with financial and human resource management.* This definition goes a bit beyond the more fundamental ICAO and FAA definitions, but the concepts of *risks, comprehensiveness, proactiveness,* and *integration* are very important. The most important advantages of an SMS over conventional, more fragmented safety programs are the full integration of supporting programs and the ability to measure the degree of risk exposure.

Implementing SMS

Now that ICAO has made it mandatory for large and turbojet-powered aircraft operating internationally to "establish and maintain a safety management system (SMS) appropriate to the size and complexity of the operation" after November 18, 2010, implementation of the system is in full swing.

Experience has shown that both grasping the SMS concept and implementing it are not particularly easy tasks for corporate operators. The fully integrated approach to safety, and the diversity of missions and wide range of flight department sizes make a one-size-fits-all approach to SMS impossible. As a consequence, IBAC through its IS-BAO program (see "IS-BAO" section in Chapter 6) has over the years provided increasingly detailed implementation guidance for SMS, culminating with the IBAC *SMS Toolkit.*

The SMS Toolkit was developed by a working group comprising members from the IBAC staff and a number of its 15-member associations to assist noncommercial fixed- and rotary-wing aircraft operators, on-demand charter operators, and aerial work interests in developing and implementing safety management systems that meet the ICAO Standards and Recommended Practices (SARPs) and the regulatory requirements of major aviation regulatory authorities.

The *Toolkit* consists of a 57-page hard-copy *SMS Tools* booklet that provides a step-by-step process to develop and implement an SMS, a CD that contains an electronic copy of the booklet, plus 18 tools and 6 reference documents in electronic format. There are also hyperlinks to numerous other related sources. This is one of the best guides to SMS implementation, but there are aids available from other vendors as well.

The components of an SMS include the following methods of creating and sustaining safety:

- Policies and objectives
- Risk management processes
- Assurance elements (is it getting done?)
- Education and promotion within the organization

The basic formula for achieving these objectives is to:

- Identify hazards.
- Assess and measure risks created by the hazards.
- Eliminate the hazards or reduce risks to an acceptable level.
- Track and evaluate safety management activities for effectiveness.
- Modify safety management activities as required.

Putting It Together

A comprehensive implementation plan must be employed to assure that the resulting SMS meets all program goals and objectives. Not starting with a detailed plan is like launching on an international flight without flight planning.

An effective plan will ensure that

- management is committed to its success;
- required resources are allocated;
- responsibilities are assigned;
- milestones are established and tracked;
- existing policies, programs, systems, and procedures are integrated with "the new"; and
- linkages are maintained.

Failure to ensure that the first item on this list is truly present will jeopardize the success of any SMS. Without the boss's support, making it work will be difficult. Like other means of curing an addiction, we advocate a 12-step implementation program. The addiction we are trying to cure is reliance on weak or poorly integrated safety programs.

1. *Study the SMS concept*—Read as much as possible on the subject to gain a comprehensive understanding of what the program is supposed to accomplish, how the parts fit together, and what level of effort is required for success. Talking to others who have instituted an SMS in your type of operation will prove quite useful. See the sidebar on page 312 for sources.

2. *Obtain senior management commitment*—This means selling the flight department manager or chief pilot on the merits of the program. However, whoever runs the flight department must also gain support from corporate executives at the highest levels for the program to work. Management personnel at all levels must stay *actively involved*.

3. *Establish an SMS team*—A project leader and representatives who come from the pilots, schedulers, and maintenance personnel are essential. This is truly a team effort; without a development team, there may not be overall buy-in from the entire organization. Many organizations also note the improved overall teamwork within the organization as a consequence of implementing an SMS.

4. *Conduct a Gap Analysis*—This means taking one of the SMS checklists (see sidebar below) and conducting an internal audit of the organization to see what you already have that meets the criteria for an SMS. The deficiencies noted will provide a road-map for your implementation program.

5. *Conduct initial hazard identification and risk assessment and develop a safety risk profile*—This aspects asks the organization to identify the "standard" hazards they face on a daily basis and how they handle them. It also asks what other hazard they may face based on the variety of operations they occasionally perform. Information on how to accomplish this is available in most of the sidebar publications. The IBAC *SMS Tools* is a good starting point.

6. *Develop safety management strategy and safety assurance processes*—This means that the organization must have a well-integrated plan to implement, sustain, and measure the overall SMS effort. Measuring the relative risk involved with hazards is one of the most important features of the SMS concept and should be used extensively. It relies heavily on checklists and processes to ensure compliance.

7. *Identify safety accountabilities*—Who is in charge of the overall SMS program, but more importantly, who is in charge of each element of the program? Without designating a responsible individual for each program task, the likelihood of any of them being done may be in doubt. In addition, do not forget to designate a due date or time interval for each required action.

SMS Implementation and Reference Tools

- International Business Aviation Council (IBAC)—www.ibac.org/safety-management

- Flight Safety Foundation—http://flightsafety.org/

- National Business Aviation Association—www.nbaa.org/ops/safety/

- European Business Aviation Association—http://www.ebaa.org/content/dsp_page/pagec/Safety_1

- International Civil Aviation Organization
  - ICAO Safety Management Manual—: http://www.icao.int/safety/ism/Guidance Materials/DOC_9859_FULL_EN.pdf
  - ICAO Safety website—www.icao.int/FSIX/

- Federal Aviation Administration—http://www.faa.gov/about/initiatives/sms/

- Transport Canada—www.tc.gc.ca/eng/civilaviation/standards/sms-menu-618.htm

- CASA Australia—www.casa.gov.au/scripts/nc.dll?WCMS:STANDARD::pc=PC_91430

8. *Develop ongoing hazard identification and tracking system and risk assessment procedures*—This is a key feature of the program. SMS depends on a constant flow of information regarding actual and anticipated hazards and the methods for dealing with them to be successful. Each primary document in the sidebar contains one or more methods for accomplishing this critical element of the program.

9. *Develop an emergency preparedness plan*—While we hope that any person or aircraft within the organization will never be involved in an incident or accident, having a plan for dealing with the many consequences of these events is essential. Without it, the aftermath of one of these events is often chaotic and confusing. It is essential to match the flight operation's activities with those of the main company or client to ensure comprehensive coverage.

10. *Amend programs, procedures, and documents as required*—This is the feedback loop for the SMS. Once hazards and consequent risks are identified, how they are handled should be incorporated into the program to ensure that similar events will not reoccur.

11. *Conduct staff training and education*—Without training and constant hazard and risk education, the SMS will probably not be effective or even survive. Again, this is a team effort, and the team must be kept in the loop. If they do not see both activity and results from the program, it will be less effective.

12. *Track and evaluate safety management activities*—This aspect asks the questions: "How are we doing? Are we achieving our goals and objectives? Are we becoming more risk aware? What can we do better?"

These items form a *process*—one that will permit your organization to recognize potential hazards, evaluate and mitigate the ensuing risks, and measure progress and effectiveness of these elements. This is not another manual designed to meet a requirement imposed by someone else. A well-developed and integrated SMS is a continuing *process* that is designed to reduce your organization's risk to the lowest possible levels commensurate with your type of operation. Ideally, practicing the *process* will lead to a positive change in the organization's safety culture.

Even better, the *process* will lead to a more effective, efficient, and productive organization.

Bringing SMS Down to Earth

Safety is paramount; Safety is everyone's job; Safety before all else.... They sound pretty good, don't they? I often see pledges to safety in the safety managements system chapter of an aviation operations manual, but wonder how the flight department will actually implement these glittering generalities.

Too often, we are caught up in safety strategy and neglect to ensure that the big picture becomes sufficiently digestible and action oriented to make a difference to *each* member of the flight department. Risk profiling, assessment, and mitigation are good examples of this phenomenon. These are obviously interesting exercises for a flight department safety manager, but how are the troops expected to use these tools on a daily basis?

The goal of any safety program should be to provide each member of the organization with tangible tools and techniques that enable each of them to perform their daily tasks as safely as possible. Or, in SMS terms, to manage safety risks as effectively as possible. To do so requires that specific procedures be implemented that will both remind and enable individuals to "do the right thing."

The Big Picture

Risk should be viewed in two perspectives: big and little. The big means: what is the overall risk profile of the organization? What are the broad issues that face operations most of the time? What has the organization done to accommodate or prevent the "normal" risks? Then, what are the situational risks associated with the various missions the organization undertakes? Winter ramp operations, flights to mountainous airports, critical system maintenance, and catering safety are examples of situations that occur either frequently or rarely but are associated with unique operations. Whether frequent or rare occurrences, they must be evaluated for known, expected, or anticipated hazards and the probability of their occurrence.

An overview of the organizational risk profile follows:

- Corporate safety education and operational expectations
- Management view of risk issues
- Adequate accountabilities and responsibilities
- Detailed and comprehensive policies, standards, procedures
- Training adequacy for all hands
- Aircraft types versus potential missions
- Aircraft equipment
- Maintenance availability and expertise
- Internal evaluation of critical factors

Detailed Operational Risk Assessment:

Management

- What is different about what we are about to do? How are our risk profiles affected?
- What future changes bring added risk?

Flight crews

- Perform a flight risk assessment (FRAT) for every flight.
- Address risk factors during crew brief.
- Perform abbreviated risk assessments during pre-takeoff and pre-approach briefings.
- Assess risk when unanticipated events occur—runway change, holding, system malfunction, etc.

Maintenance personnel

- Explore risks involved with *all* significant maintenance procedures; explore risks associated with maintenance error, hazards to maintenance personnel, and environmental hazards.

Scheduler

- Evaluate risk associated with every trip—the operational, personnel, and environmental aspects.

An overall organizational risk assessment profile should be developed by examining the items above and any other factors that have a direct effect on overall risk. This assessment will address relative risk levels and mitigating factors designed to ensure risk is reduced to as low a level as possible, given the missions involved.

While good pilots, mechanics, schedulers, and managers have traditionally looked after the above items on an intuitive basis, they have rarely had the quantitative risk assessment tools available to them that will permit concrete evaluation of risk and enable them to *communicate* risk to others and devise reasonable mitigation scenarios. The dual ingredients of probability and severity of occurrence contained in the risk equation will better arm us to assess, mitigate, and communicate about the bad stuff.

The end game for safety in flight departments is to get everyone both thinking and communicating in terms of risk on a daily basis. When you hear the troops actively discussing risk factors and ways to reduce them, you will know the system is working.

If one took no chances, one would not fly at all. Safety lies in the judgment of the chances one takes.
— CHARLES LINDBERGH

Risky Business

Flying into Teterboro or Van Nuys on a bright sunny day is a different experience from flying into Hilton Head or Aspen on a dark and stormy night. While each of these flights contains certain potential hazards, the former may contain more than the latter. Recognizing and dealing with these hazards is a must before and during flight. If done effectively, we go home satisfied that we met the day's challenges; if not, we may be the first at the scene of the accident or incident.

More to the point, each flight contains risks; the risks are measurable; they can be controlled. Why risks? The term often gets confused with the concept of hazards, but risk involves two factors: hazards and probability. Therefore, a risk is the combination of a specific hazard and the likelihood that the hazard occurs.

$$\text{Severity of Hazard} \times \text{Probably of Occurrence} = \text{Risk}$$

While some use this equation mathematically to arrive at a rate, this method is less useful than a more simple method: mentally combining the two factors to arrive at a subjective risk index. (See Fig. 8.2.) While this matrix does not provide an absolute value, it does put things into perspective. For instance, the consequences arising from an engine failure after V_1 under normal circumstances may contain a substantial hazard, but the probability of occurrence may be so low that a risk index of only 3 would be assigned. However, a TCAS Traffic or Resolution Advisory shortly after takeoff is a great hazard with a high probability of occurrence, warranting a risk index of 5.

You may have arrived at a different index than I did, simply because our values are different, we engage in different types of operations, or we come from different backgrounds. Thus, it is important to evaluate the relative risks within your operation as a collaborative effort, involving all members of the flight department in the exercise.

Big Picture

Before we get too far into the business of assessing and controlling risk, realize that we should consider risks from several perspectives. The overall view and attitude of both the company and the flight department management determine the primary frame of reference

Severity of Occurrence / Occurrence Probability	Highly likely to occur in the immediate future (80%–100%)	Probably will occur within the next year (50%–80%)	Possibly will occur within the next 5 years (10%–50%)	Unlikely to occur in the foreseeable future (0%–10%)
May cause death; permanent or total disability; critical damage to facility or property; or critical compromise of safety	5	5	4	3
May cause major injury; permanent, partial, or temporary total disability; severe occupational illness; major damage to facility or property; or major compromise of safety	5	4	3	2
May cause minor injury or occupational illness; minor damage to facility or property; or minor compromise of safety	4	3	2	1
May cause negligible compromise of safety of employees or property	3	2	1	1

FIGURE 8.2   Risk assessment matrix.

for risk. Then a variety of specific, detailed areas bring risk into focus on a flight-to-flight basis. First, the big picture.

Realize that the company and flight department organization may present risks that might affect us even before we get close to an aircraft. These are broad-brush systemic issues that color our entire working environment. Look at the following factors and see if you think that any of them might affect the way we do business:

- Safety culture
- Policies, standards, and procedures
- Management commitment to policies, standards, and procedures
- Organizational stability and personnel turnover
- Operating environment
- Aircraft properly equipped for required trips

I can almost hear the flickers of recognition occurring as you read the list. Sure, each of these factors can present risks to a greater or lesser degree if they are not recognized and *mitigated*. Good word, *mitigated*; we will return to it.

If we work for an entrepreneur who has little regard for rules and limits, we could be in trouble. By their very nature, entrepreneurs are risk-takers; that is how they get ahead of the competition. However, taking risks with an airplane and its passengers may be the wrong way to get ahead, or anywhere, for that matter.

If the boss cannot buy into the basic limitations specified by the FAA, aircraft manufacturer, and insurer, you may be working for the wrong company. Most flight departments go beyond the basic FAA regulations in an effort to reduce risk and improve safety. Items, such as takeoff weather minimums, semiannual pilot training, and landing fuel minimums, are policies and procedures practiced by many high quality flight departments; if the boss can't accept the wisdom of these, the entire risk index shifts toward higher numbers.

Likewise, flight departments with few policies and standards tend to leave too much to chance or to individual interpretation. Good standards and operational guidance keep everyone on the same sheet of music, operating in harmony. The lack of harmony begins to shift the risk index in the wrong direction.

In addition, rapid turnover of personnel within the department indicates that something is wrong with the way the place is being run. Stability is requisite to low risk.

Narrowing the Focus

Now that we have a feel for the big picture, let us look at the details that may present risk. These fall into several categories: operational, technical, and human.

- Operational factors include the environmental items that come with every flight: weather, airport facilities, en route structure, approach aids, and available information about all the foregoing.

- Technical factors include the type of aircraft, its operating limitations and shortcomings, and the quality and amount of its maintenance.

- Human factors include personnel supervision and pilot and technician qualifications, currency, personal fitness, and mental state.

All of these "little picture" issues have the potential for affecting each flight and each operational evolution within the department. While most operations are routine, it is the *seemingly routine* matters that can create risky situations.

When considering these issues, we must think in terms of severity and probability, the two components of risk. Stopping on a short runway is not normally considered a hazard unless other factors decrease the probability of a safe stop. Wet runway, tailwind component, inoperative anti-skid, and a relatively new pilot all conspire to decrease the probability of a safe stop. Risk puts the effects of relatively benign hazards into a new light.

The cumulative effect of multiple hazards is also worth noting. For instance, suppose the PIC for a flight scheduled to arrive at a relatively short airfield after dark had not landed at night within the past 90 days. Is this a high risk item? Similarly, suppose this airfield had only a nonprecision approach available; also, taken by itself, not a particularly high-risk item. But just suppose that the VASI was out of service at the airport, again not a high-risk issue if taken in isolation. The point being, that taken by themselves, the potential hazards mentioned are relatively insignificant, but together they spell a high degree of risk.

Mitigation

Once an increased level of risk has been discovered, two choices are available to remain safe: do not do it or take steps to *mitigate* the risk. Mitigate is a fancy word for lessen, reduce, or eliminate. Increased risks may be reduced by applying mitigating factors. In the example of the short field landing scenario mentioned previously, the risk of an unplanned departure from the runway surface during landing might be reduced by using a pilot with more current experience, landing only if an additional landing distance safety factor were added to the AFM runway performance data, or just not landing at that airport. The last mitigating factor, not attempting a risky event, is often overlooked in the rush to complete the assigned mission.

CFIT/ALAR

Flight Safety Foundation Controlled Flight into Terrain/Approach and Landing Accident Reduction problems and solutions.

Problems

- Nonprecision instrument approaches
- Lack of altitude call-outs
- Lack or nonuse of standard operating procedures
- Failure to execute a missed approach
- Unstabilized or rushed approaches
- Approaches in low light or darkness
- Approaches in poor visibility
- Wet or contaminated runway operations
- Optical illusions associated with the above items
- Disregard of radio altimeter

Antidotes

- Standard operating procedures
- Approach briefings
- Altitude call-outs
- Stabilized approach parameters
- Sterile cockpit
- Automation awareness
- Runway safety factor
- Contaminated runway factors
- Define nonprecision approach procedures and limitations
- Use night approach visual vertical guidance lighting
- Crew resource management emphasis

From the broader perspective of the company supporting its flight operations, additional training, better equipment, additional standards and limitations and educating company personnel regarding the limitations of the aircraft and its personnel are all important mitigating factors. These big picture items are really the first line of defense in alleviating risk and must be worked at constantly to achieve a safe operating environment. Without this environment the flight crew assigned a sticky mission may have the deck stacked against them even before they begin the flight planning process.

Risk comes from not knowing what you're doing   —WARREN BUFFETT

Changes

Safety experts have found that a major source of risk within organizations results from changes to the routine. A new aircraft, pilot, technician, destination, airport, or procedure can easily alter the smooth operation of the flight department to the point where an unnecessarily high degree of risk is introduced.

An unfamiliar aircraft by itself is not necessarily a high-risk item if certain mitigating steps are taken: training the pilot and technician, assigning an experienced shotgun rider for the first 25 hours, and imposing more conservative operating limitations are the normal means of ensuring that the new toy is treated properly. In a similar case, a new pilot, although quite experienced in your type of aircraft, may not understand the SOP required to operate into some of your "normal" airports and may raise the risk for what had been considered routine operations. A suitable amount of supervised initial operational experience, detailed discussions about special airports, and in-depth training regarding your flight operations manual and SOP should adequately reduce the risk to acceptable levels.

The important point is to recognize changes when they occur, define the potential hazards associated with the changes, determine the risks, and apply mitigating factors as necessary.

From the operational standpoint, changes are quite important. A last minute change of destination or addition of passengers, a deferred maintenance discrepancy, unexpected headwinds, a runway change while approaching the destination, and a wind shift in close are all examples of events that may raise the risks for the flight crews without them realizing that anything is happening. Therefore, every change before and during flight must be carefully evaluated for potential hazards that may contain the probability for an unacceptably high risk. This is especially true when cockpit workload is high and changes are coming rapidly.

Perspective

Risks must be considered from two standpoints:

1. Does the organization adequately support the mission of the flight department to provide a low-risk environment?

2. Is every flight and maintenance event evaluated for risks associated with it? The first factor is an ongoing task overseen by management. The second becomes everybody's business.

Ultimately, the responsibility for a safe flight rests with the PIC, but management must create a sufficiently risk-free environment to enable each flight, at least, to begin with a reasonable expectation of a successful conclusion.

Process

Every flight department should have a safety program that incorporates a risk assessment and mitigation process. While it does not have to be a very formal process, procedures should be published that involve every member of the department in the process. A culture should emerge that, when each activity is planned, a risk assessment is conducted to ensure that the associated risk is acceptable.

The International Standard for Business Aviation Operations (IS-BAO) developed by the International Business Aviation Council (IBAC) contains a comprehensive safety management system for corporate flight departments that incorporates risk profiling techniques and associated checklists designed to facilitate the risk assessment and mitigation process;

Potential Hazards Checklist

Organization

- Safety culture
- Policies, standards, and procedures
- Management commitment to policies, standards, and procedures
- Organizational stability and personnel turnover
- Aircraft properly equipped for required trips

Operations

- Weather
- Airport facilities
- En route structure
- Approach aids
- Information on the above

Technical

- Aircraft limitations and shortcomings
- Maintenance

Human Factors

- Personnel supervision
- Personnel maturity, skill, and integrity
- Pilot and technician qualifications and currency
- Personal fitness and mental state
- Teamwork and crew resource management

using them will assist in avoiding risk. Even better, implementing their safety management system will provide an in-depth safety program supplying additional layers of protection against accidents.

The ultimate goal of such a risk management program is to get everyone to automatically think, "What is the risk?" They should be thinking this whenever any activity, no matter how slight, is mentioned. The concept of risk assessment must pervade all tasks within the organization, whether big or small. Ideally, risk assessments should be applied to every part of the organization's operations to include planning, anticipated changes, and flight and maintenance operations by using checklists designed for the purpose. It is important to remember that risk management is a never-ending process that must be revisited as conditions change.

If done correctly, your business does not have to be risky.

Operational Risk Examples (Derived from Flight Safety Foundation CFIT/ALAR Studies)

High Risk

- Nonprecision/circling approach
- Terminal radar/air traffic control service not available
- EGPWS or radar altimeter failure
- Minimum runway length

Medium Risk

- Long duty period
- Unfamiliar airport
- International operations
- Darkness—especially with visual approach
- Limited lighting system—low intensity only, no VASI or approach lights
- Mountainous/hilly terrain
- Reduced visibility
- Contaminated runway
- Equipment failures, especially multiple
- No weather reporting available
- Crew experience/currency

Combine these factors with their probability of occurrence to arrive at a risk factor

SMS for the Long Term

Once the basic structure of SMS is in place—standards, processes, procedures—performance comes next. Rather than a static, non-functioning system, SMS is meant to be a dynamic, constantly improving, daily habit that all members of the organization use to continually assess risks and ensure that goals are being met. Performance is the key to an effective SMS. This is the time when you need to shift gears from the "we can implement it" to the "now that we're there, how do we make it work for us on a daily basis?"

Using the following questions as guidelines, go for *continuous performance* and *sustainability* with your SMS:

- Is the CEO and/or responsible person *still aware* of the organization's safety policy statement? Has that person continued to support the intent and spirit of the statement? How do you know?

- Have those with designated safety-related authorities and responsibilities continued to carry out their duties effectively and conscientiously? That question is directed to more than just the safety officer/manager.

- Are all (a majority, some, a few) employees continuing to *actively participate* in the SMS by submitting and tracking hazard reports, risk management techniques, and safety awareness exercises? Do management-level employees *actively* participate, too?

- Is progress toward stated safety goals being periodically measured? Do the safety goals still measure relevant issues? Are they truly measurable?

- Is the system designed to track regulatory standards and guidelines compliance effective?

- Are employees aware of and using the contents of the operations manual and SOP? How do you know? Moreover, are these living documents?

- Is safety (risk mitigation?) performance being regularly *measured* and *assessed*?

- Are hazard identification and tracking techniques being applied to human factor issues, third party vendors, and known or anticipated organizational changes?

- Are risk assessment and mitigation techniques being used on a daily basis by *all* (most, a majority, a few?) employees?

- How often is the internal evaluation program being employed? How long does it take to get through the entire checklist? Are resulting deficiencies being mitigated and fed back into the standards?

All of this information should be evidence based—show me the data! Repeat quarterly and track your improvements!

Passenger Safety

What do your Very Important Passengers want from corporate on-demand air transportation? While no definitive polls are available on this issue, most would bring it down to a very few wishes: safety, reliability, and good service—in that order of importance: Get me there in one piece; do it on time, every time; and make it a pleasant experience. Not surprising or unreasonable.

Most flight departments go to great lengths to ensure that the pilots and technicians are well-qualified and trained, that the aircraft are airworthy, and appropriate policies and procedures are in place to ensure safety, effectiveness, and efficiency of the entire operation. Understandably, these measures are principally oriented toward operational considerations for pilots, technicians, and aircraft. But what about keeping the precious cargo, the passengers, safe? What policies and procedures are in place to ensure their health and well-being? In most flight departments the answer is, not many.

This omission is not intentional but stems from the belief that passengers will be safe if the rest of the operation is. If airworthy aircraft are provided, aircrews are properly qualified and trained, and the SOP is appropriate, then passenger safety can be assumed, right? Maybe not, since there are many nonaircraft, non-SOP items that can reach out and grab an unsuspecting passenger.

The sidebar contains just a few of the many different ways passengers inadvertently expose themselves to the potential hazards associated with flight. The point is that unless the subject of passenger safety is focused upon, too much will be left to chance. In addition, given the importance of keeping them safe and pleased, leaving this subject to chance is an unacceptable risk for the health and well being of the flight department.

Sadly, every year a number of unfortunate and avoidable passenger-related accidents occur in on-demand air transportation. Death from being struck by a moving tail rotor, broken limbs and sprains from airstair falls, slip-and-fall accidents in hangars and on ramps, burns from galleys, head gashes from baggage doors, and eye injuries from blowing ramp debris are a few examples of these personal injuries. Are they avoidable? Certainly, but passengers are quite ingenious in their ability to harm themselves, so precautions must be equally thorough and ingenious.

Create Expectations

Flight departments should give significant attention to passenger injury prevention if a relatively hazard-free environment is to be achieved for its prized customers. A reasonable means of accomplishing this is to select several flight department members and at least one relatively inexperienced passenger to explore the potential hazards associated with passenger-carrying operations. The passenger's participation is essential to provide department personnel with a nonaviation perspective.

This group should step their way through the entire air transportation process, looking at common (and uncommon) hazards, and using the passenger's perspective to formulate policies and procedures designed to protect passengers and prevent accidents. Then publish the committee's work and ensure that passengers know what to expect when they fly.

Many potentially hazardous situations may be avoided if the passengers know what to expect in the aviation environment and are apprised of policies and procedures designed to protect them from harm. Therefore, creating expectations regarding preflight through postflight activities will set the scene for a safe and enjoyable flight.

Ways to Injure

- A passenger runs back across a crowded ramp to retrieve a forgotten article of baggage and is blown to the ground by a large jet at breakaway thrust.

- Passengers refuse to wear their seat belts or ignore the seat belt sign when illuminated.

- Individuals use galley facilities without being properly instructed and are burned.

- A passenger's child darts dangerously close to a running turboprop.

- An unescorted passenger disembarks and walks aft of a helicopter that has engaged rotors.

- Individuals under the influence of alcohol or drugs become unruly or refuse to take directions from the flight crew.

- A passenger attempts to carry a toxic chemical or loaded weapon on board.

Preflight.   Upon arrival at the departure facility, be it the hangar at home base or a lean-to at a small airport, passengers should know the scheduled departure time, have a clear idea of where to meet the flight crew, and what to do with their baggage. Additionally, they should understand that they will *never* be allowed to move about the ramp area unless they are escorted by competent personnel, be it corporate staff or a qualified FBO employee.

Boarding.   Again, no unescorted passengers are allowed on the ramp. When approaching the aircraft, the route should be one that does not require walking under wings, engines, or baggage doors. A crewmember should stand by the door/airstair to ensure that passengers are not injured while boarding. Once aboard, each passenger should be shown to a seat and have his or her carry-on items stowed by a crewmember.

Briefing.   In addition to the required safety briefing, passengers should be made familiar with *all* cabin safety features, not just how to operate the airstair and use the seatbelt.

Passenger Safety Checklist

- Passenger policy and procedures documents
- Ramp access and control
- Activities in the hangar and around the aircraft
- Baggage (including carry-on) control
- Cabin briefing—safety, cabin features, flight activities
- Seatbelt policy—takeoff, en route, landing, taxiing
- Handling unruly passengers
- Cabin emergency procedures
- Passenger safety training
- Reporting safety hazards

Primary and secondary emergency exit operation, fire extinguisher location, and emergency oxygen usage should be covered thoroughly (comply with FAR 91.519 as a starting point). This is also a good time to brief passengers regarding anticipated en route weather conditions, point out the fasten seat belt sign, and advise them about the desirability of keeping their seat belt fastened at all times.

Comprehensive knowledge of galley, refreshment center, and lavatory features and operation will go a long way in preventing injuries associated with these items. Additionally, this briefing will allow the passengers to fend for themselves during flight and thus allow the cockpit crew to remain at their stations, as required by law.

En route.   Perhaps the greatest hazard during the cruise phase of flight is unexpected turbulence. A policy should be developed to ensure the seat belt sign is turned on at appropriate times with accompanying cabin public address announcements alerting passengers to the possibility or presence of turbulence.

Descent/Landing.   Again, perhaps the most hazardous condition associated with this phase of flight is the possibility of a passenger not being seated with his/her seat belt fastened. Visual inspection of the cabin by flight crewmembers and cabin PA announcements should preclude this possibility.

Postflight.   If passengers are properly briefed and controlled, they will remain in their seats until a crewmember has opened the exit door and announced the possibility of deplaning.

Emergencies.   While emergencies directly involving passenger participation are rare in corporate air transportation, procedures for cabin fires, passenger medical emergency, cabin decompression, and emergency evacuation should be developed. These procedures do not necessarily have to be briefed to the passengers prior to flight, but flight crewmembers should be very familiar with them. Practicing them, as indicated below, will prepare the crew for these events, which we hope are as rare as commonly simulated (and practiced) emergencies like engine fires or emergency descents.

Practice Makes Perfect

While the passenger participation in emergencies mentioned above is quite rare, informal surveys indicate that many passengers would like to take part in training sessions in which these events are practiced. Some flight departments do so using their own resources, allowing passengers to remove emergency exits, exit the aircraft through those openings, discover the contents of the first aid kit, practice CPR, and extinguish a fire under controlled conditions. Others contract for this service to be done on a regular basis. Either way, both passengers and crew will learn from the experience. Moreover, the passengers will appreciate the additional level of care and attention provided to them.

The Name of the Game

Pleasing passengers is a highly desirable goal in corporate aviation. However, keeping them safe is essential. Months or years may pass in peace and harmony, with little of note occurring regarding passenger safety. Nevertheless, when least expected, passengers will unwittingly attempt to do harm to themselves.

Remember, safety, reliability, and good service are what passengers want. Make sure those goals are prioritized in the order shown.

Flight Attendants

It may seem strange to put this subject here, but I feel that the realization of the need for a flight attendant flows naturally from the subject of risk assessment. The principal reason for the use of flight attendants is safety; secondary reasons have to do with customer service issues. But if corporate aircraft accidents requiring an emergency evacuation are extremely rare, why bother? Further, the cockpit crew is often trained to direct and assist with emergency evacuations, right? Not necessarily, for two reasons. First, you may not be considering all types of emergencies, and second, do not be so sure that most cockpit crews have received recent evacuation training.

Most passenger emergencies do not involve an emergency evacuation. Emergencies requiring a trained and qualified *cabin safety attendant* fall into three main categories: passenger illness, self-inflicted injuries, and cabin equipment safety.

Passenger illnesses come in many varieties: heart attack, choking, breathing difficulty, intestinal distress, anaphylactic shock, stroke, seizures, drug overdose, fainting, and diabetic shock. All of these events have occurred in corporate aircraft to varying degrees of severity and consequence. In a number of cases, flight attendants were able to assist the passengers through first aid and telephone consultations with doctors. Significantly, in a number of these instances, suitable divert fields were unavailable due to distance/time considerations, placing responsibility for primary care on the flight attendant.

Flight attendants all have a variety of humorous and not-so-humorous anecdotes regarding the ingenious ways passengers have of inflicting injuries on themselves. Burns in the galley, mashed and broken fingers, falls in the cabin and on the air stair, bumped heads, and injuries due to turbulence. The latter can be deadly, and sometimes the only way to prevent it is by having an assertive flight attendant on board who insists that the passenger remain seated.

Cabin equipment safety refers to a potential hazard to the aircraft due to an improper passenger action. Most of these involve such mishaps as spilling liquid into entertainment consoles resulting in overheated, smoking wiring; galley misadventures including fires, overflowing sinks, and overloaded circuitry; and lavatory fires caused by clandestine smoking.

How would passengers cope with these "little" emergencies? Are they capable of saving a life or preventing damage to the aircraft? Would the cockpit crew be able to cope with these events and still safely fly the airplane? The answer to all of these is a qualified "possibly."

And how about the big one—the evacuation of the cabin after a takeoff or landing accident or other ground emergency? Simulated and real evacuations reveal that

- getting passengers organized in the first few moments after the event is essential;
- passengers are often confused and undirected immediately following an accident;

- passengers are hesitant to act in the face of an emergency;
- passenger unfamiliarity with escape routes causes significant delays;
- unfamiliarity with escape routes causes significant delays;
- failure to see alternative escape exits wastes valuable time; and finally,
- assertive and organized cabin crewmembers facilitate exit success.

So, despite repeated passenger safety briefings and the availability of passenger safety briefing cards, passengers may not prove to be effective during self-evacuations. Moreover, frequent flyers may not have paid attention to the briefing in some time, especially given the permissiveness of the regulation requiring safety briefings for frequent flyers.

While the cockpit crew may be able to help with some of these issues, their duties following an event may keep them from assisting the passengers in the first critical moments. In addition, one or more of the cockpit crew may be rendered ineffective due to shock, injury, or incapacitation. The flight attendant, on the other hand, has been specifically trained for this moment.

Does every operation or corporate aircraft warrant a flight attendant? Obviously not for aircraft with six or fewer passenger seats (although a few operators choose to). However, once you get up into the Challenger/Gulfstream class of aircraft, employing flight attendants makes sense because of cabin size, complexity, and the number of passengers. Although not safety related, the image and level of service provided by the flight attendant goes a long way to enhancing the passengers' view and appreciation of the flight department. Think about it.

Airplanes are near perfect, all they lack is the ability to forgive.   — RICHARD COLLINS

Emergency Response Planning

Business aviation is safe; perhaps the safest form of transportation available on the face of the earth. But accidents happen. While no one ever anticipates an aircraft accident or incident, each flight department should have a plan for such an eventuality. The two key factors in pre-accident planning are education and communication.

Educating all flight department personnel *and* key company personnel about what to expect in case of an emergency will ease the confusion and uncertainty of the event. Connecting the proper people with adequate information after an accident or incident will allow established procedures to be set in motion with a minimum of confusion. Experience has shown that gathering and disseminating information concerning the accident in the few hours after its occurrence are critical to controlling the situation and minimizing the trauma associated with these events. If the company has an overall crisis or disaster plan, the aircraft accident/incident plan should conform to and be integrated with it.

Accident/Incident Flow

- Receive information
- Verify information
- Notify crisis management team
- Log all telephone calls
- Initiate crisis management plan
- Communicate frequently with team members
- Limit and *control* contact with media
- Each team member maintains a chronological record of events

The Plan

A section in the operations manual should be devoted to the subject of providing references, procedures, and lists of people to contact. Key elements of this section are the procedures taken to obtain essential information and then to disseminate it to the proper individuals within the company.

Having a plan is the key to coping with an aircraft accident or incident. A flood of human emotions comes into play in the aftermath of such a traumatic event—disbelief, denial, shock, fear. Couple these with the confusion, conflicting reports, inability to communicate with key persons, and the rumors and lack of verifiable information that accompany such events, and the flight department's tasks become more than difficult.

As in any emergency event, having developed a plan, practiced it, and communicated its contents are the keys to getting through the first days following an accident. Pilots have extensive emergency and abnormal procedures checklists, regularly practice them during recurrent training, and work with other crewmembers to successfully handle the emergencies. Why not such a system for accident/incident planning? See Fig. 8.3.

If there is a single goal for those who may be directly involved in post-accident matters, it would be this: control chaos. The minutes and hours following an accident are fraught with disorder and competing demands on one's time; controlling these events and demands will have long-term implications.

- **RECEIVE INFORMATION**—Use the Aviation Accident/Incident Reporting form.

- **VERIFY INFORMATION**—Obtain an independent second report.

- **NOTIFY CRISIS MANAGEMENT TEAM**—Always notify *the Manager of Corporate Aviation* first—Use the emergency response telephone numbers.

- **LOG ALL TELEPHONE CALLS**—Use a permanent log.

- **INITIATE CRISIS MANAGEMENT PROCEDURE**—See the plan.

- **COMMUNICATE FREQUENTLY WITH TEAM MEMBERS.**

- **LIMIT AND *CONTROL* CONTACT WITH MEDIA**—Work with Corporate Communications.

- **EACH TEAM MEMBER MAINTAINS A CHRONOLOGICAL RECORD OF EVENTS.**

Figure 8.3   Aircraft accident plan action items.

The best way to start the process of planning for the worst is to engage in a role-playing exercise with your peers. Pilots, flight attendants, technicians, schedulers, and receptionists all have unique perspectives that will prove useful as you develop the plan. Also, seek the advice of members of your aviation network; they will provide still other perspectives and experiences relative to your plan. The object of the role-playing is to explore all probable scenarios of how notification could take place, what support may be needed, who should be notified, how information should be disseminated, and who can properly represent the company to members of the media, government investigators, and employees.

If the company has a disaster or crisis response plan, the aviation emergency response plan should follow its guidelines as closely as possible. In fact, the aviation plan should be integrated into the company plan.

Elements of the Plan

Here are some elements of an aviation crisis response plan that should be considered:

- *Initial notification.*   Word of an accident could come from a variety of sources: media, law enforcement, airport, FAA or NTSB. Each will have a different level of information and expertise. How will you handle them? What information elements do you need to get from them?

- *Anybody can get the word.*   Company security guards, public relations personnel, and receptionists may receive word of the event prior to an aviation-literate person. How should those people be prepared to receive the information?

- *Verification*
    - Never accept a single report from an unknown source regarding an accident/ incident, no matter how authoritative it sounds. Get the basic facts before the word is spread: The what, where, when, and who and the damage/injury assessment—see Fig. 8.4.
    - Ensure that the crew and passenger manifest is accurate. Once verified, release the manifests only to corporate human resources or corporate counsel for further dissemination.

- *Communications*
    - After informing necessary flight department personnel, notify essential corporate managers—see the sidebar on page 330. Key personnel should have up-to-date telephone lists for these people to facilitate the communications process. Company receptionists, security guards, and all flight department personnel should have these lists available and be trained in collecting and transmitting the information to the proper people. This process should be tested periodically to ensure its validity.
    - The NTSB must be notified immediately of a reportable accident/incident. While someone else may have done this, verify by calling a NTSB field office.
    - Notification/release of crew and passenger names should be done only by corporate human resources or other designated and experienced persons.
    - No, *absolutely no,* communication regarding the event should be attempted with media, press, insurance companies, attorneys, and curiosity seekers. Any outside communications must be accomplished by corporate communications or legal

Date _____   Time _____
Name of person taking the report _____
Person providing the report: Name _____
                                Location/Phone _____
Aircraft involved: Type _____ Registration _____
Crew: Pilot-in-Command _____ Condition _____
         Second-in-Command _____ Condition _____
Location of occurrence _____
Time of occurrence _____
Details of the occurrence _____
_____
Injuries _____
_____
Damage to aircraft _____
_____
Damage to property _____
_____
Passenger names/condition_____
_____
_____
Names/phone of available witnesses _____
_____

Other pertinent information _____
                                        (continue on back)

Reminders at the scene: **Reminders at home base:**
• Care for injured   • Notify corporate officers
• Protect wreckage   • Notify NTSB
• **No contact with media**   • **No contact with media**
• Only minimum facts to FAA/NTSB   • Collect/impound crew,
• Photograph the scene   flight, and maintenance
• Physical examinations for crew/passengers   records

FIGURE 8.4   Incident/accident reporting form.

counsel; aviation personnel are too close to the problem to provide objective answers about accidents/incidents, particularly in the first hours following the event.

• Log all telephone contacts and significant events for post-event reconstruction purposes.

• *Preservation*

  ○ Provide assistance to the injured.

  ○ Protect the aircraft from further damage or vandalism.

  ○ Protect crewmembers from providing possibly self-incriminating answers to NTSB/FAA officials—those involved in an accident/incident must always be represented by counsel.

  ○ Obtain photographs of the event scene.

• *Participation.*   Owner/operators are normally afforded official status as parties to the accident investigation by the NTSB/FAA. An operator may wish to participate to protect his/her interests and to provide necessary background information for the event. However, participants should be familiar with aircraft accident investigations and the NTSB rules and rights regarding investigations.

People to Notify within the Company

Companies vary, but here is a basic list:

- Flight department manager or chief pilot
- Chairman and/or president
- Legal counsel
- Risk manager
- Public affairs/corporate communications
- Human resources

- *Data collection.* As an investigation unfolds, large amounts of data will be required. It should be collected expeditiously, impounded, and released only upon advice of counsel. Such information may include:

  o Flight crew employment, operational, and training records

  o Flight schedule and passenger manifests

  o Maintenance discrepancy, inspection, and airframe/powerplant/accessory maintenance records

  o Flight plan and weather briefing information

  o Fueling records

- *Control.*   A single individual within the flight department should be designated as the central control point for all communications and actions relating to the event. This is normally the flight department manager or director of operations. While this individual may, and should, delegate certain tasks, the responsibility for the overall control of the communications and investigation must lie with this individual for purposes of continuity.

- *Format.*   While nicely crafted prose and attractive formatting are expected for most corporate communications, these may not be useful for the accident plan. Events will unfold rapidly after initial notification—there will be little time to pour through detailed directives and instructions. Therefore, checklist and brief outline formats will be useful for all concerned.

More information regarding plans and their structure is available at the NBAA website, nbaa.org/safety.

Implement It

Once a draft plan has been devised, it should be tested by conducting at least a notification exercise to ensure that essential individuals or their surrogates can be contacted at any time. Note how long it takes to complete the notification process and if the individuals involved received the notification information in a timely and accurate manner. Then, revise the plan to accommodate its shortcomings.

Before publishing the plan, obtain the blessing of applicable company personnel, especially human resources, communications, risk management, and legal.

Once the plan is published, your job is not finished. Training company personnel to use it properly and react to its contents is essential. If the security guard or receptionist who receives the initial call doesn't know how to complete the initial notification form or who should be called, then the plan will be only marginally effective. All flight department personnel must be thoroughly familiar with all aspects of the plan, since any one of them may be the person who has to "hold down the fort" until designated department managers arrive.

Anyone who may possibly become involved in an aircraft accident/incident should receive applicable portions of the plan. Each aircraft; all flight department personnel; human resources, risk management, and communications departments; and corporate counsel should be provided with the complete plan. Security guards, receptionists, and others who may be involved in the initial notification process only need portions of the plan that apply to their level of involvement.

Just to make sure the plan works, conduct at least one communications drill annually to ensure that it is still operational and effective. People, places, and missions change; so should the plan.

Once a basic plan is in effect, you may want to consider enhancing it to include items such as passenger/crew incapacitation, passenger/crew ground injury, vandalism, terrorism, or hijacking.

But because the principal purpose of the plan is to control the confusion that inevitably follows an aircraft-related mishap and to provide organization and structure to a highly disturbing and unstructured event, realize that simple is better. It should serve as a guide to those who will become involved in the notification, communication, and investigation processes. Without a plan, chaos may occur.

No job is so important and no task so urgent that we cannot take time to perform our work safely. The safety of people must come first.   —ENDURING SAFETY SLOGAN

CHAPTER 9

Putting It All Together

Institutional Flight Departments

Taken by themselves, the rules, suggestions, and guidelines presented in this book will create pockets of excellence within flight departments. However, there must be a more cohesive, centering force that will enable an on-demand flight department to succeed. This largely undefined force comes from both the experience and insight of those managing the operation.

Most flight departments seek to and succeed at operating safely, keeping up with the flight schedule, and keeping the budget under control. In short, they do the technical stuff fairly well. But many fail at being truly superior service providers.

Call it the big picture, the ability to accurately read the needs of the parent organization, or merely organizational talent, but the best flight department and all successful organizations seem to have it. They are constantly seeking to make themselves better, to exceed the needs of their customers, and to achieve excellence for the sake of being excellent. All this takes good management and leadership skills, but it also requires the realization within the flight department and company/principal that on-demand air transportation is an essential part of the larger organization. In such an organization, aircraft are not just limousines with wings, and the people at the airport are not just taxi service order-takers. The flight department is an integral part of the organization.

Few business organizations could do without their accounting, human resources, or information technology departments; these are essential support functions of the company. Although these groups do not produce, market, or sell the company's goods and services, they nonetheless perform critical functions within the organization. It is this revered status that flight departments should be striving for.

Most Admired

The list is in: Apple, Amazon, Coca-Cola, IBM, Starbucks—all traditionally are in the upper strata of *Fortune*'s annual "Most Admired Companies" list. Each year *Fortune* surveys 4000 businesspeople in all industries to see which companies they admire the most—all very subjective, yet with structure.

"Tireless innovation. Robust financials. The ability to lure and keep the smartest people." These are but a few of the features of the much admired. The following equally weighted criteria are used to arrive at which companies enjoy the finest reputations:

- Innovativeness
- Quality of management

- Employee talent
- Financial soundness
- Use of corporate assets
- Long-term investment value
- Social responsibility
- Quality of products/services

Great. But why is it important to be admired? Isn't a profit enough? Well, profitability isn't specifically mentioned, but "financial soundness" is close enough for government work. No, if you cannot attract and retain talent, stay ahead of the pack in developing new products and services, and convince investment bankers and fund managers that your stock has merit, your financials will not be sound for long. The same is true for flight departments: Sheer numbers of passengers flown does not a success make.

While we have a difficult time showing financial soundness in a profit-and-loss sense, the ability to show that we are good stewards of the company's funds through a healthy budget variance report and financial goals met should mean a lot to your company. Real profit for the flight department is measured in broad-based company support, passengers who are raving fans, and people clamoring to use the aircraft.

The other issues mentioned earlier are applicable to any business venture, be it selling lemonade on the sidewalk or producing world-class, on-demand air transportation services. For instance:

- *Innovation* means providing transportation solutions for the movers and shakers within your company. Finding these solutions requires cleverness, forward thinking, creativity, and sensitivity to corporate needs. Innovation is a natural product of smart people working to please customers.

- *Quality of management* means people having the best interests of the company at heart and working to fit people to tasks with the goal of winning situations for both the company and individuals. *Quality* means adherence to standards, consistency, and good leadership traits.

- *Talent* means the best people being used for jobs in which they can excel. Knowing raw talent when first seen and then developing it is an art form. The real key to the talent issue is working every day to help your employees realize their personal and corporate goals.

- *Exemplary use of corporate assets* means using the aircraft effectively and efficiently to accomplish stated goals. This means that you have to measure results, compare them with benchmarks or self-trends, and evaluate the result.

- *Long-term investment value* is important to the flight department because it addresses the issue of whether the capital or lease payments invested in the aircraft result in reasonable returns on investment. This is measured in terms of *value created* for the company through use of its aircraft. (*Hint:* Use Travel$ense.)

- *Social responsibility* relates to participation in larger corporate programs that demonstrate that the company cares about the people within its community more than just self-centered profitability. Your company does not have such a program? Start your own—it may become a trend.

- The name of the game—*quality products and services*—if you do not make this the "holy grail" of your operation, the flight department may soon be found only as a chapter in the corporate history book. It is *not* about flying; it is *not* about producing handsome reports; it is *not* about making the budget numbers to the penny—*it is about the customers.* How we serve their needs and treat them while we are serving those needs is the fulcrum around which we leverage our department. Ask them if you are meeting (exceeding?) their expectations on a regular, structured basis. This should be the department's true measure of success.

Staying ahead of these issues not only will make your flight department admired but it also should ensure your place as a significant contributor to the success of the company. In reality, *admired* really means "successful."

Profit And Loss

Companies thrive on profit and avoid loss. The concept of keeping one's corporate head above a sea of red ink drives everything the company does. One quarter of decreased profits or (shudder) losses is OK; two consecutive poor quarters is grounds for drastic action. It is a basic fact of the business world: No profit, no business; contain and minimize the losses.

Since we all realize that the flight department is a business unit unto itself (we do, don't we?), we should explore the profit-and-loss (P&L) concept as a means of determining our relative level of success. Unfortunately, we do not have the convenient performance measure our parent company does—dollars—but it is possible to assign some value to our performance.

First, understand what business you are in: the transportation solutions business, not the airplane business. The flight department is there to help individuals within the company achieve success by moving them to their desired destinations in a timely manner. Realize that this may not always involve company aircraft; alternate transportation resources outside the company occasionally may be necessary in achieving this goal.

The market for this mission is literally everyone within the company *and* all of its existing and potential customers. While corporate aircraft traditionally have been reserved for a select group of employees, new utilization strategies have opened the door to a wide variety of uses and groups within the company (see the excellent National Business Aviation Association [NBAA] publication, *Business Aviation Fact Book.*).

This market must be researched and cultivated constantly to gain maximum effect from existing and potential customers. The

Profit

- Excellent *relationship* with passengers
- Repeat business
- Respect from all
- Increasing demand
- Broad-based support from the company

Loss

- Unexplained decreasing demand
- Budget under fire
- Increasing micromanagement from above
- Constant review of department activities
- Poor reliability record
- Apprehensive passengers

movers and shakers in the company are constantly on the move, exploring new markets and revisiting old ones. How can you help them with their quest? This process requires some marketing of your own—ask your marketing department how it does it.

At the end of every month, sit down and assess that month's performance. Conventional measures of dispatch reliability, number of denied trip requests, and cost per hour all provide *quantitative* measures, but the *qualitative* measures may be more important. Check the sidebar on the previous page to see how well you did last month.

Is it possible to assign absolute values to these qualities? No, but it is possible to assess the department's relative and ongoing position within a spectrum of these items. Start with a baseline at the beginning of the year, and chart your performance throughout the year using anecdotes, random comments, and actual numbers (such as demand). Ask other members of the department to help with the process, and do not forget to include the boss in your calculations; he or she will be glad to know that you are concerned and attempting to *qualify* your performance.

The flight department is a service organization, one that supports the company mission. All service organizations within the company—human resources, information services, planning, legal, etc.—must measure their value to the company. If that value sinks below a predetermined level, the service organization's utility to the company is brought into question, and its viability may be jeopardized; personnel changes, reorganization, and outsourcing are all options for the ailing service organization.

How does your service organization measure up? Remember, it is *the little things that count*.

What Business Are You In?

If the flight department is to succeed, it must actively support the corporate mission and objectives.

Ask an airline pilot what business he's in, and he'll tell you, "The airline business." The same is true of the airline mechanic: "The airline business." But, ask the airline CEO or the director of marketing, and you will probably get a different answer. They may say that they're in the transportation business or that they move people. Ideally, they will answer that they are in the customer service business and, incidentally, use airplanes to serve their customers.

It's all a matter of point of view. But, that point of view may make the difference between the options of failure, success, or astounding success.

Ask a corporate pilot what business he/she is in, and you will probably hear, "We fly airplanes," or "We fly executive (business) aircraft." While this is true, it is only part of the story. Sure, corporate pilots fly aircraft; it's implied in the job title. But, bush pilots, agricultural pilots, EMS pilots, and night cargo pilots fly aircraft, too. They are all pilots but for widely different purposes. And, it's how they approach that purpose that may make the difference between failure, success, or astounding success.

Name of the Game

If we start from the focal point of all company operations, we will get some insight into the reason for its existence. Most companies are in business to produce a product or service but, more importantly, they must make a profit to survive. Breakeven is OK for a while, but investors and stockholders become increasingly impatient with companies that do not turn a profit within a predetermined time period. So, regardless of the contents of the neatly framed corporate mission statement, production and profit are the bottom line for virtually all companies.

As with any support organization within the company, if the flight department is to succeed, it must actively support the corporate mission and objectives. While this may seem an intuitive truth, this very important fact is often lost in the day-to-day chaos that marks most flight departments.

Juggling flight crews, planning upgrades, completing scheduled maintenance, choosing vendors, training, reconciling expenditures, and numerous other items often interrupt our view of the *big picture*. Although it may be a natural consequence of dealing with details to lose sight of the big picture, doing so obscures the real purpose for the department, namely, that of *serving customers*.

Yes, *customer service* is job one in running a flight department. Flying, maintaining, and scheduling aircraft are the means to the end of supplying customer service, but the needs and concerns of the customer should always come before the means of accomplishment.

Focus

In the maelstrom of the 1992 presidential race, the Clinton campaign had lost its focus and was trying to be all things to all people and to react and respond to every need, allegation, and demand placed on the beleaguered candidate. The campaign manager finally realized that Clinton could not be all things to all people and correctly surmised that the central issue of the campaign was the economic future of the country. The hastily printed sign, "It's the economy, stupid," prominently placed in campaign headquarters became the rallying point that focused the campaign's efforts and contributed significantly to Clinton's victory.

While a sign in the flight department stating, "It's the customers, stupid," may be overly dramatic and even counter-productive, the essence of this message should drive and qualify every action taken by the department. If a proposed action does not support the delivery of high quality customer service, it may not be necessary.

Should a flight be deadheaded to pick up a passenger? Should we standardize FMSs within the fleet? Is a standby flightcrew required for the weekend? All tough questions, but how do they relate to superior customer service?

It is easy to say that, since we're in the customer service business, no service is too small not to provide to our passengers. Therefore, we will provide backup aircraft for all flights, repaint and refurbish every three years, and equip the aircraft with the finest avionics available, right? Obviously few flight departments can afford the luxury of these actions. Therefore, some additional criteria should be included in the decision-making process.

Safety, Service, Value

An informal survey of a number of flight departments reveals that the following decision-making criteria are often used: *safety, service, value*. That is, when considering whether to provide a backup aircraft to an important flight, determine whether it will increase safety, provide better service, and/or improve the value of the services offered to the customer. How do these criteria balance among themselves, given their relative importance? Naturally, you must have buy-in to this process from the flight department's reporting senior to make it work. This may take some education of the corporate hierarchy to accomplish the ultimate goal of excellent and consistent customer service.

Hard Sell

Now that you've placed customer service signs all over the hangar, articulated this mission to all hands, and told the boss of your actions, how will this sell within your department? Realize that employees see new management, service, and administrative plans come and go, often with lightning rapidity. Therefore, any new program is viewed with some degree of skepticism, or even cynicism. The real test of the new program is whether management uses and upholds it. If management gives the new program only lip service or continually makes exceptions to a stated goal or policy, the rank-and-file soon see whether it's a valid program or merely management posturing.

If every flight department action focuses on the concept of serving the company and the flight department's customers day-after-day, month-after-month, the troops will soon get the word and act accordingly. Those who refuse to buy into this concept will begin to stand apart from the rest and become candidates for corrective action.

The Mission

Most companies widely distribute copies of their mission statements to employees. These handsomely framed statements of purpose find their way onto the walls within the flight department, too. Sadly, they are rarely internalized or serve as guidance for employees; they often serve only as monuments to executive posturing.

Similarly, the flight department mission statement found at the beginning of the *Flight Operations Manual* may not capture the essence of the department or be used for practical purpose. For instance, the statement, *The Acme Flight Department exists to provide safe, reliable and high-quality on-demand air transportation to our employees and customers*, somehow lacks the focus and punch of, *It's the passengers, stupid!* Yet, neither extreme captures the flight department's reason for existence.

Here's how the big guys express their passion

- Singapore Airlines: "Singapore Airlines is a global company dedicated to providing air transportation services of the highest quality and to maximizing returns for the benefit of its shareholders and employees."

- Southwest Airlines: "The mission of Southwest Airlines is dedication to the highest quality of Customer *Service* delivered with a sense of warmth, friendliness, individual pride, and Company Spirit."

- Delta Airlines: "We—Delta's employees, customers, and community partners together—form a force for positive local and global change, dedicated to bettering standards of living and the environment where we and our customers live and work."

- United Airlines: "To be recognized worldwide as the airline of choice."

- Virgin Atlantic: "To grow a profitable airline where people love to fly and people love to work."

But consider simplification:

- We move people

or

- We help people succeed through transportation

or

- We provide transportation solutions

The point being that whatever you choose must be simple, direct, and catchy— the statement must be capable of being easily imprinted in the minds of all employees. It must serve as a focal point for every action taken within the department. To do less is to squander valuable resources.

Once you have figured out what business you're in, putting flesh on the bones of the idea takes time and effort.

Doing It

Establishing a mission statement is important, as is defining goals and objectives for the long and short term. But, the real trick is getting your people to believe in the aforementioned items. There is no easy answer to getting people to believe that they aren't really in the airplane business, but the people business. Let's face it, your employees signed on to fly, maintain, and schedule airplanes; the people were secondary interests at the time of employment.

Too many flight departments operate as if they ran a large flying club. The consequence of this attitude is that they will fly the aircraft, and if passengers wish to join them, so much the better. Somehow, this seems to be looking through the wrong end of the telescope.

Perhaps the attitude should be: "A customer wants to go somewhere—how can we help him/her achieve that goal?" With this focus, you start with an authorized request and work your way up to the technical side, that of dealing with aircraft. Doing this requires that flight department personnel be sensitized to the needs and preferences of their passengers (see customer service sidebar on page 344).

All of this may seem a bit overwhelming at first, but realize that virtually every company has some service or support functions it maintains. Human resources, information systems, finance, and risk management are a few of the normal corporate service organizations. Find out how they go about running their *businesses*. There will certainly be parallels between the flight department and their organizations. See how others deliver support services.

Internalizing It

It is up to every manager and supervisor to internalize and promote the customer service and personal aspects of providing transportation services for their company. Only if the troops see management living the message—*walking the talk*—will they become a part of the business. This process should be reinforced with written service standards and methods of measuring performance in this critical work area.

Your employees need to be constantly reminded of which aspects of your business are important. If management places customer service on a par with safety, the message will be accepted and *lived* by them. Then there will be no doubt about what business the department is in.

The Excellent Flight Department

Most corporate flight operations regularly deliver the boss to his or her destination and make it back to home base safely—and do it on time. However, some operations do it better than others, and the ones that do it really well gain the admiration and respect of the boss *and* ensure the longevity of the flight operation. These are important considerations in these lean times, well deserving of the attention of the entire flight department, from the manager to the hangar sweeper. What makes some departments better than others? Consider the following scenarios.

The boss shows up at Frantic Industries' hangar for a scheduled flight only to find that the assigned aircraft is not ready because the ordered fuel load had not been added. While the top man cools his heels for 20 irritating minutes, the flight department rushes to fix the problem. Much shouting and running about ensues, capped off by the chief of maintenance being chewed out by the department manager, unfortunately, within earshot of the boss. Eventually, the flight departs, yet no follow-up action is taken by either supervisor concerning the incident.

Coincidentally, a very similar incident occurs at the Frantic flight department's neighbor, Squared Away, Inc., on the same day. However, during the delay, the flight department uses the moment to discuss the proposed acquisition of a new aircraft with the boss. The maintenance manager quietly tells the lineman to bring the fuel load up to the proper amount and actually assists the fueler with the task.

Immediately after seeing the boss off, the department manager launches a quiet but thorough investigation into why the fuel load was incorrect. In so doing, a procedural glitch in the way fuel orders are filled is discovered. The department manager asks the maintenance manager to find a better way to accomplish the procedure. The maintenance manager then involved all his people in the process of finding a simple yet effective solution.

Obviously, Squared Away had a better day than did Frantic. Frantic's boss is left wondering whether his department manager is rehearsing for a TV sitcom or running a multi-million-dollar flight department, to say nothing about the departure delay. Further, all the flight department personnel involved in this episode are left in an unhappy mood.

Squared Away fared somewhat better. The boss got to discuss a topic of interest with a trusted manager, while the head of maintenance showed that he is a team member. A hidden problem was uncovered and presented to the people who were most involved in the procedure. The solution to the problem involved everyone; they were all given ownership of the new procedure.

Here's another set of contrasting scenarios: The Stackum and Packum (S&P) Company flight department knew it had to put some procedures and standards on paper because it had just taken delivery of its second aircraft and was in the process of negotiating for a third. New people were rapidly being hired and had to be indoctrinated and trained into the S&P way of running the "airline." The department manager gave the responsibility to the chief pilot.

Being a hands-on person, the chief pilot worked long hours and weekends, and during her scheduled vacation period, to complete a master training plan and an operations manual. As she tried to implement these directives, major flaws and oversights were discovered, to the point where training courses were considered meaningless, and most of the department ignored the new operations manual.

Meanwhile, in a flight department several hangars down, Paradigm Enterprises, an equally fast-growing company, was faced with a similar situation. At Paradigm, however, the department manager sat down with his chief pilot and maintenance manager and told them he wanted a comprehensive training plan and operations manual, stating that he wanted every manager and supervisor to have a part in their development.

Because the department manager selected effective project leaders, all the members of the department were soon energized. Within a week, everybody in the hangar was working on some aspect of the projects. The best news was that because all members of the organization had worked on the projects, both directives were embraced and followed by all concerned.

Ways and Means

Attitude, attention to detail, leadership, good communications, customer orientation, personnel empowerment, and a willingness to go the "extra mile" made the difference in these little morality plays. It is these few management and behavioral traits that spell the difference between "merely adequate" and "excellent" in any organization, corporate flight department's included.

Sounds easy, doesn't it? Just put those magic words to work, and excellence will soon follow. Not so! Throughout many flight department evaluations and consultations, the preceding characteristics stand out as common threads of excellence. But just talking about them in the abstract will not make them so. Lots of hard work and constant attention to making them all work together just may yield the desired superior performance. Let's take them one at a time.

Attitude.   Does the department manager think that he or she has reached the pinnacle of his or her profession? Does he or she believe that the department personnel are the best available? Is providing on-time, smooth, and safe air transportation that exceeds airline standards his or her top priority? Does all this flow to every subordinate? If not, then excellence for the department is doubtful.

Conversely, if the manager views his or her job as a stepping stone to an airline cockpit or a larger flight department, he or she may not be putting his or her best into the job. If he or she thinks that some of his or her people are not good at their jobs and chooses to accept this, how can he or she expect superior performance? In addition, if better than airline standards are not set and compliance not demanded, then mediocre or uneven performance is the inevitable consequence.

Details.   This is why we delegate, right? Yes, but if attention to detail is not demanded on a continuing basis and followed up on, delegation is a shallow exercise of avoiding responsibility. But micromanagement is a danger if supervisors go overboard in their attention to detail. This is a trait best passed onto the troops by example rather than by oversight and second-guessing.

Leadership.   Unfortunately, we spend too much time concentrating on building managers and not enough time on growing leaders. *Growing* is the operative term here; all the training and authority in the world will not create a leader. It is often said that the leader gains his or her authority from his or her subordinates and not his or her supervisor. This is certainly true of the flight department, since most of its occupants are highly trained professionals who expect to be treated as such and look to their supervisor for intelligent direction and mutual respect.

However, John Wayne types cannot run the flight department by sheer force of will and charisma. Good management and planning are essential adjuncts to the total view of running the operation. The good leader must provide firm and consistent direction while giving his or her people the opportunity to make self-realized decisions of their own.

Communications.   Perhaps the most important failing of flight departments is their failure to adequately communicate within their organizational boundaries. This is especially true of the often not-so-thin line that separates the operations types from the maintenance types. Up, down, and crossways are all essential interpersonal communications models that must be emphasized and fostered, whether it be intradepartmental or interdepartmental.

If there is a simple way to foster the communications habit, it is to ask lots of questions. Research has shown that only when we are actively listening do we understand problems and form solutions to be transmitted eventually to the sender. Too many of us try to transmit simultaneously; all we get is a squeal as a result.

Customer orientation.   Too often flight departments get the mistaken notion that they are in the flying business and not the customer satisfaction business. True, flight departments do a lot of flying in the pursuit of pleasing the customer, but this is only a part of it. Is the transportation service done pleasantly and on time? Moreover, does the service meet the expectations of the customers? When was the last time you asked your customers what they thought of or expected of the service you provide?

Customer orientation means selling the product, too. Even if you are flying all your aircraft more than 700 hours per month with a 50 percent load factor, you must be constantly marketing the flight department to the entire company. Divisions come and go, plant locations come and go, and chief executive officers (CEOs) come and go; constant marketing to all parts of the company will ensure that the flight department will not fall victim to the impermanence of the business world.

Personnel empowerment. A new touchy-feelie buzzword? Not really, if used properly. This is the concept that people, especially subordinates, often have good, creative, and useful ideas that will benefit the organization, if those people are presented with the proper environment in which to implement those ideas. The subsets of empowerment are delegation, trust, and support for subordinates. Inherent in the concept is that people can be more productive if they work in an environment that recognizes their value to the organization. It does not mean that the inmates are in control of the asylum, however. See "Leadership" on the previous page.

The Extra Mile. Adequate service is the goal of the mediocre provider. Excellent service is the goal of the organization that wants to be held in high esteem and to be considered invaluable. Don't you want your department to be considered at least invaluable, or preferably, indispensable? If you don't, your company may not be able to justify just an adequate operation for very long.

When you go the extra mile, there is little traffic to contend with.　　　　　—JOHN SHEEHAN

Attainment

So how does one attain this high level of management purity and excellence? Unfortunately, there is no right or single answer. There are just too many variables involved. Good people, a clear sense of mission, and a good product to sell make the job easier, but the basic tenets remain: Attitude, leadership, good communications, customer orientation, and extra effort make it happen every time.

However, since most of us have not attended the Peter Drucker or Tom Peters schools of management, how does one get the ball rolling on the road to an excellent flight department? Fancy techniques and programs with impressive-sounding acronyms do not make it happen. The attributes I have mentioned involve commitment, dedication, and hard work by all managers associated with the enterprise. Here are some starter ideas:

- Establish/reestablish contact with key company personnel in planning, accounting, human resources, production, etc. Know what the company needs.
- Start a "Market the Flight Department" campaign within the department. Ensure that all department members realize the value of the flight operation to the company and "sell" the concept whenever possible.
- Get the pilots talking to the maintenance crew. Form a safety committee, a training committee, a service-upgrade committee, anything to get all the players talking.
- Have an independent party perform a management and safety evaluation of the department and make specific recommendations to improve all aspects of the operation.

- Hold a planning retreat for the department to find out where you want to be in the next 3 to 5 years.

- Hold brief in-house training sessions each month. While department personnel can act as instructors in some cases, draw on company, airport, service representative, and FAA personnel for these functions.

- Ensure that your routine reports and budgeting documents are adequately supporting your needs. Are you reporting information that is no longer relevant? Does corporate headquarters have a full appreciation for your efforts and problems?

These are just a few ideas on how to get started on the road to a better running department. Want more ideas? Ask your people.

Safety, Service, Value

We all pride ourselves on providing good service to our passengers. Show up on time with a shiny airplane, be nice to the people in back, get them to the destination with minimum hassle, say "Please" and "Thank you," and smile as they deplane. What more do they want? Well, maybe a bit more.

While most chairpersons or CEOs will not articulate the reasons they have a corporate aircraft, experience has shown that they are looking for an alternative to airline travel. This alternative must be as safe as, or safer than, the airlines, provide high-quality service to the passengers, and ensure that good value is received for the company in the process. With this in mind, the corporate flight department can structure its organization and operations to incorporate features that will focus on *safety*, *service*, and *value*.

Safety First, Last, and Always

I have never met a flight department manager or chief pilot who said that the boss ever grabbed him or her by the lapels, pulled him or her close, and said, "You'd better be safe!" Hardly ever is the subject of safety raised with regard to corporate aircraft operations. Why? It is taken for granted that all aspects of the flight department will be conducted as safely as possible. This is the most important tenet that underlies corporate aviation: We are as safe (hopefully safer) as the airlines. Without this essential foundation, we might as well pack up and go home.

Fortunately, turbine-powered corporate aviation regularly betters the airlines accident rate. This segment of corporate aviation has had a lower accident rate than the airlines for the past seven years and should continue to do so, given the current culture within the community. And culture is the key.

You cannot go to the corner store and buy a six-pack or half-gallon of safety; it is not something that can be readily defined or obtained. It is a *culture* or state of being that provides us with the safety we seek. However, culture is one of those Psych 101 terms that normally eludes definition. The best definition of a safe aviation operation I ever heard was one in which every decision started with the subject of safety, i.e., an assessment of the risks involved with any and every decision made. Doing so is not all that difficult; assessing relative risk sometimes can be tricky. Yet practice should make the act second nature.

Make sure *everybody* in the company knows about this decision process, too, especially the boss and all flight department members. If everybody understands that safety is not only first but also last and always, decisions will be accepted more readily and less subject to challenge.

Service with a Smile

Most flight department members will tell you that good service for their passengers goes without saying; they always give good service. And therein lies the problem; it usually goes without saying.

Improving Customer Service

- Make all personnel aware, on a continuing basis, that the department is customer oriented.
- Know where the company is headed—what's the plan?
- Conduct a face-to-face survey of all of your regular passengers to determine their desires and perceptions of flight department services. Include the passengers' secretaries and administrative assistants in the survey, too.
- Provide customer service training for flight department personnel.
- Institute a standardized flight crew flight briefing and a debriefing procedure for passengers.
- Provide incentives for flight department personnel to find ways to improve service.
- Discuss customer service at all departmental meetings.
- Evaluate department personnel on their level of customer service.
- Provide several briefings each year regarding flight department mission, operations, and scheduling procedures for passengers, potential passengers, and their assistants.
- Encourage informal contact between flight department and corporate support personnel to find out what the company thinks about your service and to communicate your service orientation to them.
- Find ways to measure passenger satisfaction.

Few, if any, flight departments abide by written passenger service standards; yet they maintain volumes of administrative, financial, and operational standards. The problem is that every flight crew member thinks that he or she knows what constitutes good service. However, this is exactly the reason that standard operating procedures are published in such great detail—everybody has a different opinion about how things should be done.

Flight crews take great pride in exhibiting their professionalism in how they train and plan for and fly their flights. This professionalism involves both ongoing training and adherence to standards. Why not for the customer, too?

Every time you walk up to the check-in counter of a major hotel chain or call the help line of a major company, you should witness the results of a carefully crafted customer standards and training program. These people are trained to help, to be pleasant, and to provide *results.* Why not your flight department, too?

How and by whom are passengers greeted prior to a flight at home and on the road? Which crew member provides both the required safety brief and the "soft" trip brief prior to flight? What are the contents of the soft briefing? How is catering handled before and during flight? These may seem to be small items, but the passengers *always* notice how well they are done.

Good service provides your flight department with its reputation. That reputation should be impeccable. It should generate raving fans. It should make the flight department an indispensable part of the corporate culture. Work at it.

Good Value Creates Job Security

A common expression within the community says that a sustained high level of demand for the flight department's services creates job security. This may or may not be true. If you fly the aircraft 800 hours per year, you may just be creating wear and tear on the aircraft, if the company does not perceive that it is receiving good value for its transportation dollar.

Companies often confuse the concept of *cost versus value.* Just about everything costs something, but many things may not provide a value that makes the cost worthwhile. For example, your aircraft may cost $1000 per hour to operate, but can you prove that the company gets at least $1000 in value per hour out of that expenditure? Well, it depends on how you define value.

In this case, *value* refers to the usefulness or importance of a product or service. In the case of corporate aviation, just the availability of rapid and reliable on-demand air transportation creates usefulness for the company. That is, the ability to visit a customer, plant in trouble, or supplier at will should *create value* for the company. Then individual trips themselves create value, depending on the mission and desired outcome of the trip. Creating one's own schedule, security, confidentiality, and lack of the hassles associated with airline travel complete the value equation.

Much of this is measurable via the NBAA Travel$ense software program. Being able to measure the *value* of a single flight, series of flights, or an entire year's worth creates a powerful impression on the corporate aviation detractors within the company.

Hierarchies

Safety, service, and value always should be prioritized in the order stated. That is, safety first (naturally), service next, and finally, value. If taken out of order (i.e., if value is sought before either safety or service), the equation breaks down. If safe, service-oriented operations are not desired before calculating the value, the company and, likely, the passenger will be disappointed in the long run. Similarly, if a disproportionate amount of emphasis is placed on value or service over safety, warning flags should be raised.

All this revolves around the reason the company got an airplane in the first place. This reason, hopefully, was to provide safe, reliable, on-demand air transportation to serve company interests in achieving its basic mission. *Safety, service,* and *value* always should support this basic thought.

When in Doubt

Flight department personnel are faced with decisions, great and small, every day. If they are taught to always consider *safety, service,* and *value* in the decision-making process, they probably will choose wisely most of the time. Finally, they must see this process being used by both flight department and company management on a daily basis. Management by example is a powerful tool.

Use your own best judgment at all times.
—ENTIRE CONTENTS OF NORDSTROM CORPORATION POLICY MANUAL

APPENDIX **A**

Business Aviation Background

Organizations

Aircraft Owners and Pilots Association (AOPA)
421 Aviation Way
Frederick, MD 21701
Tel: +1 301 695 2000
www.aopa.org
Represents the interests of aircraft owners and pilots in the U.S. For foreign affiliates, see IAOPA.

Airports Council International—ACI World (ACI)
Suite 1810
800 Rue du Square Victoria
Montreal, Quebec H4Z 1G8
Canada
Tel: +1 514 373 1200
aci@aci.aero
Represents major airports—regional branches are shown on the website.

American Association of Airport Executives (AAAE)
601 Madison St., Suite 400
Alexandria, VA 22314
Tel: +1 703 824 0500
aaae.org
Represents airport managers/executives.

Flight Safety Foundation (FSF)
801 N. Fairfax Street, Suite 400
Alexandria, VA 22314-1774
Tel: +1 703 739 6700
flightsafety.org
Promotes civil aviation safety worldwide.

International Council of Aircraft Owner and Pilot Associations (IAOPA)
421 Aviation Way
Frederick, MD 21701
Tel: +1 301 695 2220
iaopa.org
Represents interests of aircraft pilots and owners in 71 countries (see website for a list of countries represented).

International Business Aviation Council (IBAC)
999 Rue University, Room 16.35
Montreal, Quebec H3C 5J9
Canada
Tel: +1 514 954 8054
ibac.org
Represents a number of national business aviation organizations—see website for listing.

International Civil Aviation Organization (ICAO)
999 Rue University
Montreal, Quebec H3C 5J7
Canada
Tel: +1 514 954 8219
icao.int

National Business Aviation Association
1200 8th St NW
Washington, DC 20036
Tel: +1 202 783 9000
nbaa.org
Represents the interests of U.S. business aviation operators—member of IBAC.

Periodicals

AOPA Pilot
421 Aviation Way
Frederick, MD 22153
Tel: +1 301 695 2000 Fax 2375
aopa.org/pilot

Aviation International News
214 Franklin Ave.
Midland Park, NJ 07432
Tel: +1 201 444 5075
ainonline.com

Aviation Week and Space Technology
2 Penn Plaza
New York, NY 10121
Tel: +1 212 904 2000
aviationweek.com

AvWebBiz
avweb.com/register/

Business and Commercial Aviation
54 Danbury Rd
Ridgefield, CT 06877
Tel: +1 914 939 0300
www.aviationweek.com/publications/bca.aspx

BART International
(Business Aviation Real Tool)
20 rue de l'Industrie
B1400 Nivelles
Belgium
Tel: +326 788 3603
www.bartintl.com

Flight International
Quadrant House
Sutton, Surrey
SM2 5AS
United Kingdom
Tel: +44 (0) 20 8652 3315
www.flightglobal.com

Flying
460 N. Orlando Ave.
Winter Park, FL 32789
Tel: +1 407 628 4802
www.flyingmag.com

Fly Corporate
Mach Media
Kortrijksesteenweg 62
9830 Sint-Martens-Latem
Belgium
Tel: +32 9 262 0330
fly-corporate.com

Weekly of Business Aviation
1200 G Street N.W. Suite 200
Washington, DC 20005
Tel: +1 202 383 2350
Fax: +1 202 383 2438
http://www.aviationweek.com/publications/weeklyofbizaviation.aspx

World Aircraft Sales
1210 W 11th St
Wichita, KS 67203
Tel: +1 800 620 8801
avbuyer.com

Air Transportation Requirements

Air Transportation Needs Analysis Checklist

The ability of the aviation department to successfully support the air transportation needs of the company depends heavily on its ability to provide support *when* and *where necessary*. A clear understanding of company strategy and *anticipated* air transportation requirements is essential to being able to provide transportation services when and where needed.

Historical Research

Airline Activity.   Obtain the following data for the previous 12 months for key executives/ personnel from corporate travel department:

- City-pair analysis (note multiple-leg trips)
- Most popular destination frequency
- Total and average fares
- Average trip length in days
- Include or factor out international trips, as applicable

On-Demand Air Transportation.   Develop data similar to airline activity (above) for corporate, charter, and fractional operations. Record positioning (deadhead) trip frequency, if applicable.

- Executive use—individual/average
- Popular destinations
- Costs versus *value*
- *Evidence of value created* (reasons for trips)

Company Strategy.   Obtain access to company strategic plans and/or interview key employees to gain knowledge of company plans (see below).

- Overall change in goals/mission
- Mergers/acquisitions
- New products/territories

- Customer targets
- Unique transportation requirements

Analyze Data

Develop tables showing:

- Top 10 airline destinations, with frequency
- Top 10 corporate/charter destinations, with frequency
- Average time away from home base per trip
- Costs per trip/leg or other relevant measure

Create maps portraying frequent trips/destinations for both airline and on-demand air transportation operations.

Develop Interview Information

Schedule 30-minute private interviews with key employees/air transportation users. These should be done in the subject's office to put him or her at ease and to imprint you into his or her territory.

Create Baseline Information

- Understand the interviewee's area of responsibility.
- Individual estimates of
  - Airline travel frequency.
  - On-demand travel usage (breakout by type).
  - Load factors.
- Why these trips are critical to accomplishing individual/company goals.
- Opinions regarding quality/efficiency of current/previous service
  - Airline travel.
  - On-demand travel.

Develop Projected Usage (18-Month Time Horizon)

- Ask interviewee to explain planned new products/markets for his or her organization.
- List estimates for interviewees and their direct reports of anticipated
  - New destinations.
  - Airline travel.
  - On-demand travel by type.
  - Special needs.
  - Load factors.
- Why new trips are required and/or their effect on individual/company goals.

Create Projection

Alternatives

- Tabular data by individual or organizational subdivision
- Overall narrative estimates of change
- Graphic presentation of old/new destinations/frequencies
- Maps showing current and projected destinations

Analyze Data.   Create an aggregate of:

- Current uses, destinations, frequencies
- Projected uses, destination, frequencies
- Special needs
- Innovative uses
- Impact of anticipated travel on company goals/objectives

Analysis

Aircraft.   Requirements:

- Average and longest trip lengths
- Load factors
- Cabin comfort requirements and amenities
- Airport/range-payload performance requirements
- Special equipment (international)
- Trip frequency, RONs, simultaneous trip requirements (how many aircraft?)

People.   Requirements:

- Type
- Skills
- Number

Facilities/Infrastructure.   Requirements:

- Hangar
- Office
- Equipment
- Services

Financial.   Resources:

- Capital expenses
  - Current
  - Projected

- Operational expenses (present as direct and indirect)
  - Current
  - Projected

Alternatives

- Corporate aviation
- Joint ownership/interchange/time share
- Charter
- Fractional ownership
- Airlines
- Combinations of the above

Impact

- Competitive advantage. Beat the competition.
- Time saved. Time away from office, nights away from home.
- Value provided. *Competitive advantage*, productivity, force multiplier, schedule prioritization, customer orientation.
- Intangibles. Safety, security, confidentiality, lifestyle, image, hassle factor.

Presentation

Use maps, graphics, table, narrative.

- Current situation
- Future needs
- *Transportation solutions*
- Options/alternatives

Specifics

- Methods
- Aircraft
- People
- Facilities
- Finances
  - Capital
  - Operating expenses

Choosing an Aircraft Checklist

Background

- Who uses the aircraft?
- How much did they use it last year?
- How much did they use the airlines?
- Where did they go?
- How long did they stay?
- What are the shortcomings, if any?

Analysis

- Number of trips
- Popular destinations
- Load factor
- Leg length
- Trip length (days)
- Charter used
- Costs involved − capital + operational

Future (18-Month Time Horizon)

- Determine strategic direction of the company.
- Who could/should use the aircraft?
- Interview movers and shakers:
  - How do they travel now?
  - What will their part of the company be doing next year?
  - How will this affect their and their direct reports' travel needs?
  - What unique travel needs will they have?

Analysis

- Summarize company strategic direction.
- Compile travel requirements.
- Note changes from past.
- Consider preferences of senior staff.
- Build travel resource alternatives—value, suitability, utility, flexibility, cost.
- Choose one alternative.

Presentation

- Build a story.
- Use charts and graphs rather than tables and data.
- Present strategic direction for company departments.
- Provide resource alternatives.
- Use anecdotes to describe value of on-demand air transportation to company.
- Recommend one alternative.
- Provide cost impacts as long-term (life cycle) values.

Reference http://www.nbaa.org/admin/management-guide/ for the *NBAA Management Guide,* Appendix B.

Aircraft Purchase Checklist

Background

- Perform air transportation analysis (see checklist).
- Choose likely aircraft types.
- Query personal network regarding likely types and sources.
- Research aircraft choices in detail.
- Research market.
- Determine company acquisition format and financing preferences.
- Decide on broker/dealer, if appropriate:
  - Reputation
  - Years in business
  - Expertise/experience in your area of the market

Purchase Considerations

- New/used:
  - Resale value
  - Warranty provisions
  - Maintenance costs, special maintenance items
  - Customer support
  - Insurance restrictions
- Installed equipment
- Parts availability
- Repair station/service center availability
- Cabin appointments
- Pilot/maintenance training available
- Demonstration flight for principal passengers

Purchase Specifics

- Insurance coverage desired and operating restrictions imposed (see checklist)
- Inspection:
  - Qualifications of inspector
  - Statistics on airframe, powerplants, accessories—time in service, cycles
  - General condition—paint, interior, hidden areas
  - Major repair/damage history
  - Modification/alteration status/documentation (Form 337)
  - Equipment list and parts authenticity validation (Form 8130/yellow tags)
  - Discrepancy list/deferred items review
  - Logbooks/records—status of inspections, life-limited items, AD/SB, required overhauls, and accessory/appliance items
- Specific inspections/tests:
  - Engine borescope, hot section, APU, accessories, oil analysis
  - Test flight/functional checks—all systems, cabin features, avionics, flight envelope
- Legal review of sale/lease documents

Documentation

- Purchase contract/lease
- Bill of sale
- Aircraft registration certificate/application
- Certificate of airworthiness
- Title/title search (release of lien, if necessary)
- Certificate of insurance
- Aircraft records/logbooks (FAR Part 91.417)

Reference: Manufacturers, brokers, dealers: www.nbaa.org/emarkets and www.nbaa.org/prodsvcs.

On-Demand Air Transportation Policy and Procedures (Example)

Purpose

The Ace Widget Corporation operates its own aircraft and contracts for charter air transportation for use as a management tool in improving its efficiency and effectiveness in the marketplace. The rationale for use of on-demand air transportation must be consistent with the overall management policies and practices of the corporation.

Use of on-demand air transportation should be considered when significant advantages or savings over airline travel may be realized in terms of time, money, security, or productivity.

Methods

Ace Widget maintains a Cessna Citation aircraft in Anytown to provide air transportation services for its employees and customers. This aircraft is capable of carrying up to seven passengers for distances of up to 1200 statute miles. Certain other restrictions such as minimum runway length and maximum weight apply to the aircraft.

Additional on-demand air transportation is available via aircraft charter. A wide range of aircraft types are available, priced at an equally wide variety of hourly rates. In using these services, the type of aircraft chartered must fit the proposed travel mission and not unduly exceed the desired level of capability.

Only the company Citation and authorized charter aircraft will be used for on-demand air transportation. The standards set for these aircraft operations will ensure that the highest levels of safety and efficiency are received by corporate travelers. *All flights, regardless of supplier, will be scheduled through the Corporate Aircraft Coordinator.*

Authorization

Individuals occupying the following positions within Ace Widget may authorize the use of on-demand air transportation:

Chairman	Strategic Planning
Chief Executive Officer	Corporate Counsel
Chief Financial Officer	Administration
Group Presidents	Treasurer
Government Affairs	

These individuals may authorize on-demand air transportation for use by their direct reports and for other activities directly related to essential business purposes, when required.

Scheduling Procedure

Priority for the use of the Citation aircraft will be determined by the Office of the Chairman and will be chosen on the basis of the business purpose containing the greatest benefit for the company.

Those authorized to use on-demand air transportation will supply the following information via e-mail/fax to the corporate aircraft coordinator as far as possible in advance of the requested use date:

Dates desired

Business purpose(s)

Type of aircraft desired

Detailed itinerary, including specific times and locations

Passenger manifest, including applicable cost centers

Special requirements

Ground transportation required

Catering requests

Travelers will make their own lodging arrangements.

The aircraft coordinator will make appropriate arrangements and provide an e-mail/fax copy of the proposed trip sheet to the traveler, requesting confirmation of the travel details shown. Once confirmed, a final trip sheet will be transmitted to the requester. Any desired changes to this document must be communicated to the aircraft coordinator as soon as possible; direct communications between a requester and the charter broker or charter provider are discouraged because this procedure will deny the aircraft coordinator needed information.

Charges

At the completion of each trip, the flight department/charter company will supply the aircraft coordinator with a final manifest and flight times/charges accruing to the trip. The aircraft coordinator will verify the data received and provide the accounting department with appropriate charges by cost center and number of the travelers.

Charges for the Citation will be computed at the direct operating cost for the aircraft, currently $650 per flight hour. Charter charges will be actual costs billed by the charter company.

Restrictions

The knowledge and expertise of senior corporate executives are such that the loss of more than one of them may jeopardize the continuity and smooth operation of the company. Therefore, in an effort to manage the risk associated with such losses, the following travel restrictions apply:

1. No two persons in the Office of the Chairman may travel together.
2. Not more than five corporate officers may travel together.
3. Not more than one direct report may travel with a group president.

(This list needs a detailed review prior to implementation; keep it simple.)

It shall be the responsibility of the executive authorizing the flight to ensure the restrictions above are not violated.

Security

A principal reason for using on-demand air transportation is to protect the well-being of individual travelers and to ensure privileged information regarding corporate activities is not compromised. Therefore, information regarding travel plans, potential or actual, will be treated as corporate confidential information. Specific details of trips will be made known only to those requiring that information.

International Travel

International travel requires an additional level of planning and detail due to the customs, immigration, and aircraft operating facilities requirements; adequate time to plan and make arrangements for these items is essential for a trouble-free trip. Therefore, each executive authorizing an international trip will appoint a contact person to work with the aircraft coordinator to gather the information and documentation required for the requested trip.

Special Uses

The Federal Aviation Administration (FAA), Federal Election Commission, and Internal Revenue Service (IRS) regulate the operational and business use of on-demand air transportation.

Compliance with applicable regulations is essential for Ace Widget aircraft operations. Since each of the uses listed below may be subject to several regulations and interpretations, further guidance from the corporate aircraft coordinator and corporate counsel should be sought when seeking definitive answers for specific usage situations.

Elected Officials

In general, carriage of candidates for elected office or their agents must be paid for by the candidate *prior to the flight*. Carriage of office holders is generally prohibited except in cases where fact-finding trips or activities in which substantial participation by the official is involved. Obtain clearance from corporate counsel prior to carrying any candidate or office holder.

Empty Seat Provisions

Each employee traveling on Ace Widget aircraft or aircraft charters must do so for a stated business purpose. However, if more than 50 percent of the aircraft seats are filled on a single flight by those with a valid business purpose, any remaining seats may be used by employees, spouses, and children traveling for nonbusiness reasons without assigning a value to that transportation.

Personal Use

Personal and family use of Ace Widget on-demand air transportation may be available to employees and their guests only with the express permission of the Office of the Chairman. Such use is subject to IRS rules that impose imputed income for the value of the flight to the employee(s) using the service, normally at a standard industry fare level (a per-mile rate issued periodically by the Department of Transportation) and a multiplier factor based on aircraft size and employee position. Note that some spousal services may qualify as business purposes.

Aircraft Operating Standards

Although both corporate and charter aircraft activity are regulated by the FAA, those regulations provide only minimum operating standards. It is the desire and policy of the Ace Widget Corporation to provide the safest possible air transportation, offering the highest service standards to its employees and customers.

Therefore, corporate aircraft operations will be conducted in accordance with the Ace Widget Flight Department Flight Operations Manual. All aircraft charter operations provided by the charter broker will incorporate the provisions of the Standards for Qualifying Outside Charter Companies attached to this directive.

In all cases, the pilot in command of the aircraft has full authority over the conduct of that flight and shall not be interfered with.

Owner Flown Procedures

Owner/Employee-Pilot Operating Limitations

Aircraft

- No operations in crosswinds that are in excess of 75 percent of the maximum demonstrated crosswind component should be attempted.
- All aircraft will be maintained in accordance with the manufacturer's inspection program or 100-hour inspection cycle, as applicable.
- Each aircraft will be equipped with appropriate emergency and survival equipment.

Pilots

- No pilot will be allowed to fly the aircraft unless he or she has received a formal checkout from a certificated flight instructor and is listed on the insurance policy.
- No pilot having less than 25 hours in type will be allowed to act as PIC.
- An instrument competency check is required of each pilot annually.
- Annual recurrent training from a training vendor employing classroom instruction and a simulator/flight training device is required of all pilots.
- Flights operating with a single-pilot crew shall plan to arrive at the final daily destination not later than 10 P.M.
- All instrument approach weather minimums will be increased by 200 feet and ½ mile and takeoff minimums increased by ¼ mile until the PIC has logged 100 hours in type.
- Single-pilot IFR flights must have a functioning autopilot available that provides at least a heading-hold function.

Operations

- Only airports having at least 50 percent more landing and takeoff runway available than required by the airplane flight manual performance charts will be used.
- An approved checklist will be used for *all* aircraft operations.
- Takeoff minimums for all flights require a minimum prevailing visibility of ¼ mile or 1200 RVR, or as published. Takeoffs when ceiling and visibility are less than specified

for an available instrument approach at the airport of departure require a takeoff alternate less that 30 minutes flying time from the departure point. This takeoff alternate must be forecast to have at least alternate weather minimums for the instrument approach providing the lowest approach minimums at the proposed time of arrival.

- Only one instrument approach should be attempted when weather for the approach is reported to be less than published minimum values.
- No night circling instrument approaches are allowed, unless at home base.
- No aircraft will land with less than one hour of fuel on board, computed at normal cruising speed consumption.
- Weather: Takeoff is not permitted when a thunderstorm is located within 5 nautical miles of the airport and within 60 degrees of the departure course.
- Takeoff is not permitted when wind shear warnings of 15 knots airspeed or wind speed are reported.
- Flights will not be operated into areas of known or forecast thunderstorms without an operational weather radar or other thunderstorm detection device.
- No flight will be flown into areas of known or forecast moderate icing or greater or severe turbulence.
- Takeoff or landing is not permitted when winds or wind gusts exceed 40 knots.
- Aircraft are not allowed to take off when weather conditions are such that frost, ice, or snow may be expected to adhere to the aircraft unless it is deiced and operated in accordance with AC 135-17, Appendix A.
- The PIC will always observe the fueling of the aircraft.

Owner/Employee-Pilot Standard Operating Procedures

Note: The SOPs listed here are only suggested and should not be adopted without a thorough evaluation of their usefulness and effect on other operations. They have been compiled from a variety of sources, including air carriers, corporate flight departments, training vendors, and other good operating practices.

Preflight Preparation

Preflight planning should commence not less than 30 minutes prior to desired departure time. Preflight preparation duties shall consist of the following:

- Ensuring that the appropriate flight information publications are available for the flight
- Preparing a full flight log, listing all locations that will define the route of flight, times en route, and total fuel required
- Obtaining a full weather and NOTAM briefing, either from a computerized briefing service or from a flight service station
- Filing a flight plan for all flights, except local flights
- Computing aircraft weight and balance for the load to be carried
- Reviewing the airworthiness status of the aircraft

- Ensuring that the proper amount of fuel is on board the aircraft
- Completing a walk-around inspection of the aircraft as described in the pilot's operating handbook or airplane flight manual
- Ensuring that passengers are properly briefed concerning the flight including normal and emergency procedures. The briefing will cover the following items:
  - The duration of and flight conditions expected on the flight
  - Smoking
  - Use of safety belts and shoulder harnesses
  - Location and operation of normal and emergency exits
  - Location and use of safety and survival equipment
  - Use of normal and emergency oxygen equipment

Engine Start and Taxi

- All crew members and passengers must be on board, briefed, and seated with seat belts fastened prior to engine start.
- Airport Traffic Information Service (ATIS), ATC clearance, and taxi clearance will be received prior to commencing taxi.
- Taxi at a reasonable and safe speed.
- Check gyro instruments, including horizontal situation indicator (HSI), alignment with magnetic compass during taxi.

Takeoff

- Have departure, initial segment, and departure airport approach charts readily available.
- Compute takeoff safety data (use Takeoff and Landing Data Card).
- Ensure the chosen/assigned runway is suitable regarding length, width, and surface. Note any climb-out limitations regarding required climb gradient or controlling obstructions prior to departure.
- Conduct a takeoff briefing or self-briefing to include the following:
  - Speeds, power settings, aircraft limitations, and special conditions
  - Communications and navigational frequencies for normal departure and emergency return
  - Departure procedures
  - Initial assigned/chosen altitude
  - Emergency/contingency procedures
- Use an appropriate aircraft lighting configuration to ensure conspicuity.
- Select go-around mode on flight director (if installed) for initial pitch attitude guidance.
- Cross-check pitch and roll indications with all available instrumentation.

Climb-Out

- Engage autopilot only after reaching 500 feet AGL or twice the altitude loss associated with autopilot failure as specified in the AFM/POH, whichever is higher.
- Maintain a constant lookout for other aircraft.
- Note/callout assigned altitudes 1000 feet prior to reaching each altitude change; use an altitude alerter if installed.
- Cross-check instruments for pitch and roll information.

Cruise

- Recompute fuel required to reach destination at least hourly.
- Obtain an updated destination weather forecast hourly.
- Cross-check navigational accuracy by all available means.
- Cross-check instruments for pitch and roll information.
- Cross-check and note engine instruments at least every quarter hour.

Descent/Approach

- Plan/conduct descents to maintain a cabin rate of descent not greater than 1000 feet per minute.
- Note/callout assigned altitudes 1000 feet prior to reaching for each altitude change; use an altitude alerter, if installed.
- Conduct an approach briefing to include the following:
  - Destination airport features, including approach and runway to be used and lighting to be anticipated
  - Destination weather
  - Navaids setup
  - Field elevation
  - Initial approach altitude
  - Final approach course
  - MAP/timing required
  - DH/MDA
  - Missed-approach procedure
- Note/callout the following during approach:
  - Localized/final approach course capture
  - 500 and 100 feet above DH/MDA
  - DH/MDA
  - Runway in sight/no contact
  - Landing/go-around decision

- Stabilize approach at least 1000 feet above DH/MDA (aircraft in landing configuration and rate of descent stabilized).
- Maintain approach deviation from course/glideslope within 2 dots deflection and $V_{ref} \pm 10$ knots during an instrument approach.
- Place aircraft in landing configuration no later than the final approach fix during an instrument approach.
- Complete landing checklist prior to reaching 500 feet above the runway.
- Execute missed approach at the DH/MAP if visual reference requirements of FAR Part 91.175(c) are not met.
- Configure aircraft lights to provide maximum conspicuity.
- Maintain continuous lookout for other aircraft.
- Cross-check instruments for attitude and position.

Landing

- Maintain V_{ref} (adjusted for condition correction factors) to -5 and $+10$ knots limitations until over the runway threshold.
- Leave configuration of aircraft unaltered until after clearing the runway, unless required by the POH/AFM.

Postflight

- Complete aircraft log and flight record prior to deplaning.
- Conduct brief postflight inspection to determine aircraft's airworthiness.

These standard practices should be reviewed regularly to ensure their validity and conformance with other procedures and limitations used for a specific aircraft operation. At a minimum, they should be reviewed annually.

Budget Justification

Replacement Aircraft Tug (Example)

Background

The aviation department uses a special motorized vehicle known as an aircraft tug to move aircraft in and out of the aircraft hangar at the Orville Wright Airport. The weight and size of the department's two aircraft make their movement impossible to accomplish by hand and dangerous to do so except with a specially designed vehicle.

The current tug used by the department, a Fleetway Model 102, has been in service for 12 years. The commonly expected service life for this type of vehicle is 8 years. While the tug has been well maintained, it has reached a point beyond which economic repair is not possible. This conclusion was reached by the director of aviation maintenance in consultation with the Fleetway service representative.

Prior Expenses

The current tug was purchased from the Fleetway Corporation in 1984 for $23,000. The following major expenses have been incurred in the recent past:

1993: Transmission overhaul	$1100
1991: Engine overhaul	$1600
1989: Steering mechanism repair	$900
1989: Electrical repairs (3)	$650

Current Situation

The tug has become unreliable for its use of moving company aircraft around the hangar and airport ramp areas. It requires large amounts of aviation department maintenance technician time and outside mechanic assistance to keep it in running order.

Engine starting in cold weather is very difficult even when the tug is fresh out of preventative maintenance. Four times during the past winter, aircraft launches have been delayed due to an inability to start the tug, necessitating the use of a commercial aircraft handler's tug. Additionally, the tug's transmission and braking systems have chronic difficulties that have caused a potential safety problem during aircraft movements.

Since the initial purchase of the tug, the company has upgraded its aircraft fleet, moving to the larger Gulfstream III aircraft. The weight of this aircraft exceeds the Fleetway maximum aircraft weight recommendations, making handling of the Gulfstream aircraft difficult and potentially unsafe during wet weather and on the sloped portions of the airport ramp.

Assumptions

1. The current aircraft tug has exceeded its service life.
2. The current tug is incapable of moving some company aircraft safely and effectively.
3. A replacement tug will have a service life of 8 years (manufacturer's data).
4. The most suitable replacement for the current tug is a Tractionmaster 204. Purchase price: $34,000.
5. Value of the current tug (Fleetway 102) is $600.

Proposal

Purchase a new Tractionmaster Model 204 aircraft tug to replace the existing tug. [This should be accomplished prior to the wet spring weather (May).]

Justification

The current aircraft tug moves the two company aircraft approximately four times each per day in preparation for flight operations, for maintenance purposes, and to store them after completion of flight operations. Additionally, the tug is used for trips to local parts suppliers and other corporate aviation departments located on the airport ramp area. Occasionally, the tug is used to move heavy equipment within the hangar.

When the tug is unserviceable, either flight or maintenance operations must be halted for as long as it takes to repair the tug. Given the on-demand nature of the company's flight operations, such delays affect critical operations. Reliable, safe aircraft movements are essential to the aviation department's operations.

The current state of repair of the existing aircraft tug is such that its reliability is never assured. Excessive amounts of maintenance technician and management time are spent on this support device. Repeated visits by the factory service representative and other automotive mechanics have not improved its reliability appreciably.

The current tug is barely adequate to move the Gulfstream aircraft under ideal circumstances. During wet weather and for operations on sloped portions of the ramp, it is sometimes impossible to safely control the movement of the tug-aircraft combination. During these times, a rental tug from a nearby aircraft service company is engaged to ensure the safe movement of the aircraft. Renting a tug is often difficult due to its high demand and limited hours of availability.

Used tugs of the proper size and state of repair are not in ready supply within the aviation community. A number of types of tugs have been investigated, both new and used; a new Tractionmaster 204 is the best replacement choice.

Benefits

- Reliable and safe aircraft ground movements
- High-quality, new-technology, low-maintenance vehicle
- Eight-year service life is high, considering the type and amount of usage

Employee Jobs and Performance

Flight Department Collateral Duty Job Descriptions

Safety.   Responsible for ensuring that the flight department maintains a high level of safety consciousness and active safety program promotion. Advises management on safety issues.

- Helps formulate/reviews all operational procedures and policies to ensure safety is included in the process.
- Conducts a semiannual safety review of the entire department.
- Maintains the flight department hazard reporting system.
- Works with flight department management to ensure that safety-related training is accomplished.
- Provides or arranges for monthly safety training for pilots and technicians.
- Maintains a pilot and technician safety reading file.
- Conducts quarterly safety inspections of the hangar.

Training.   Responsible for ensuring flight department personnel receive appropriate training.

- Devises and maintains a long-range training plan for the department.
- Arranges for and schedules pilot training and oversees technician training.
- Devises/reviews pilot training curriculum and syllabuses.
- Maintains pilot training records.
- Regularly reports to management on the state of department training.

Standardization.   Responsible for maintaining a high level of flight procedure standardization among department pilots.

- Devises and maintains standard operating procedures for all aircraft types.
- Coordinates SOP training with training vendors.
- Conducts and/or supervises regular standardization checks for flight personnel.
- Maintains standardization evaluation records for flight department personnel.

- Works closely with the aircraft manufacturer, training vendors, and the FAA to ensure that SOPs are correct and current.
- Regularly advises department management regarding the state of department flight standardization.

Flight Information Publications/Planning.   Responsible for ensuring that current flight information publications and flight planning devices (FLIPPs) are available to department personnel.

- Devises and maintains a FLIPP plan for the department.
- Orders and ensures timely delivery of required FLIPPs.

Automation Support.   Responsible for ensuring that appropriate automation support is available to the department.

- Devises and maintains an automation support plan for the department.
- Evaluates additional automation support/upgrade needs for the department.
- Arranges for computer/software training for department personnel.
- Ensures automation devices are properly maintained.

Catering.   Responsible for providing consumable products for use by passengers.

- Maintains a standard inventory of consumable materials/food for the aircraft.
- Evaluates and contracts with catering services to provide food service for passengers at home base and at normally used en route stops.
- Regularly evaluates the quality and adequacy of aircraft consumable products and food service and takes action to ensure a high quality of service to the passengers.

Records/Reports.   Responsible for maintaining adequate flight operations and financial records and producing relevant management reports from these sources.

- Devises and maintains appropriate data recording and storage systems for flight department operations.
- Devises and produces operational and management-level reports that adequately describe department operations to the levels needed.

Employee Performance Measures

Writing skills:

- Writes clearly and informatively.
- Edits work for spelling and grammar.
- Varies style to meet the needs of the audience.

Team participation:

- Balances team and individual responsibilities.
- Exhibits objectivity and openness to other views.
- Gives and welcomes feedback.
- Contributes to building a positive team spirit.

Quantity:

- Completes work in a timely manner.
- Achieves established goals.

Quality:

- Demonstrates accuracy and thoroughness.
- Displays a commitment to excellence.
- Looks for ways to improve and promote quality.
- Applies feedback to improve performance.
- Monitors own work to ensure quality.

Problem solving:

- Identifies problems in a timely manner.
- Gathers and analyzes information skillfully.
- Develops alternative solutions.
- Resolves problems in early stages.
- Works well in group problem-solving situations.

Planning and organization:

- Prioritizes and plans work activities.
- Uses time efficiently.
- Plans for additional resources.
- Integrates changes smoothly.
- Sets goals and objectives.

Organization support:

- Follows policies and procedures.
- Completes administrative tasks correctly and on time.
- Supports the organization goals and values.
- Benefits the organization through outside activities.

Oral communication:

- Speaks clearly and persuasively.
- Listens and obtains clarification.
- Responds well to questions.
- Demonstrates group presentation skills.
- Participates in meetings.

Judgment:

- Displays a willingness to make decisions.
- Exhibits sound and accurate judgment.
- Includes the appropriate people in the decision-making process.

Job knowledge:

- Competent in the required job skills and knowledge.
- Exhibits ability to learn and apply new skills.
- Keeps abreast of current industry developments.
- Requires minimal supervision.
- Displays an understanding of how the individual's job relates to other team members' jobs.

Initiative:

- Volunteers readily.
- Undertakes self-development activities.

Dependability:

- Responds to requests for service and assistance.
- Follows instructions and responds to management direction.
- Takes responsibility for own actions.
- Commits to doing the best job possible.
- Meets attendance and punctuality guidelines.

Customer service:

- Displays courtesy and sensitivity.
- Meets commitments.
- Responds to customer needs.
- Manages difficult or emotional customer situations.

Cooperation:

- Establishes and maintains effective relations.
- Displays positive outlook and pleasant manner.

- Offers assistance and support to coworkers.
- Works actively to resolve conflicts.
- Works cooperatively in group situations.

Communications:

- Expresses ideas and thoughts verbally.
- Expresses ideas and thoughts in written form.
- Exhibits good listening and comprehension.
- Selects and uses the appropriate communication methods.

Analytical skills:

- Synthesizes complex and diverse information.
- Collects and uses data.
- Uses intuition and experience to compliment data.
- Identifies data relationships and dependencies.

Glossary

Abort   To terminate a preplanned aircraft maneuver, e.g., an aborted takeoff.

ACAS   Airborne Collision and Avoidance System. See also TCAS.

ADF   Automatic Direction Finder—Aircraft radio receiver that will point to a ground-based radio transmitter. See also NDB.

Aerial work   An aircraft operation in which an aircraft is used for specialized services such as agriculture, construction, photography, surveying, observation and patrol, search and rescue, aerial advertisement, etc.

Aeronautical Information Publication (AIP)   A publication issued by or with the authority of a state and containing aeronautical information of a lasting character essential to air navigation.

Air taxi operator   Company providing on-demand (unscheduled) commercial air transportation. Also known as *aircraft charter operator*.

Air transportation   An aircraft operation involving the transport of passengers, cargo, or mail.

Airworthy (aircraft)   Conforms to the type design specifications and is in a safe condition to fly.

ALAR   Approach and Landing Accident Reduction. An extensive study resulting in a series of recommendations to reduce accidents in the stated regime. The Flight Safety Foundation has promoted the work.

Annex (ICAO)   Method of promulgating ICAO standards and recommended practices (SARPS).

AOC   Air Operator Certificate. National Aviation Authority authorization for an operator to use aircraft or a repair station for commercial purposes.

AOG   Aircraft on ground. Parts needed to repair an aircraft, without which the aircraft remains unflyable.

ASOS/AWOS   Automated Surface/Weather Observation System. Remote sensors that record and broadcast weather conditions at an airport.

ATC   Air traffic control. A service operated by an appropriate authority to promote the safe, orderly, and expeditious flow of air traffic.

ATIS   Automated Terminal Information Service. Recorded message broadcast by air traffic control towers providing weather and operational conditions at an airport.

Block time   See Flight time (airplane).

Business aviation   On-demand air transportation used in support of a business.

CAA   Civil Aviation Authority (United Kingdom and other countries).

Cabotage   Situation where aircraft picks up passengers, cargo, and mail at one point in a State, other than the State of its own registry, and discharges same at another point in the grantor's State.

Category I/II/III operations   Precision instrument approach operations providing landing minimums as follows:

- Cat. I. Decision height (DH) 200 feet and runway visual range (RVR) 2600 ft (these are the normal precision approach minimums)
- Cat. II. DH 100, RVR 1200
- Cat. IIIa. DH (N/R), RVR 700
- Cat. IIIb. DH (N/R), RVR 150
- Cat. IIIc. DH (N/R), RVR (N/R)

Note: Special ground and aircraft equipment, aircraft maintenance, and pilot training required for operations below Cat. I.

CFIT   Controlled flight into terrain. Causal factor in a significant number of aircraft accidents. CFIT awareness and education programs have been developed by the Flight Safety Foundation.

Chargeback   An internal charge levied on the user of a business/corporate aircraft.

Commercial air transportation   An aircraft operation involving the transport of passengers, cargo, or mail for remuneration or hire.

Contaminated runway   A runway is considered contaminated whenever standing water, ice, snow, slush, frost in any form, heavy rubber, or other substances are present.

Corporate aviation   Noncommercial on-demand air transportation used in support of a business, employing professional pilots (normally two) compensated specifically for their piloting duties.

CRM   Cockpit resource management. A series of recommended practices designed to promote harmony, safety, and efficiency among flight crewmembers.

CVR   Cockpit voice recorder. Device installed in most turbojet aircraft that records the most recent 30 minutes of intracockpit conversation and radio transmissions/reception.

DH   Decision height. Height above the terrain during a precision instrument approach at which a decision must be made to either continue the approach or execute a missed approach (go-around).

Direct operating costs   Operating costs applicable only when an aircraft flies. Also known as *variable costs*.

Dispatch rate   Percent of time flights depart within a specified time frame (normally 15 minutes) relative to the scheduled time or time passengers are available for flight. Failure to meet this criterion may be due to weather, air traffic control, or aircraft malfunction.

EASA   European Aviation Safety Agency.

EFIS   Electronic flight instrument system. CRT/LED-based flight and engine instrument.

EGPWS   Enhanced ground proximity warning system. Electronic system installed in an aircraft to warn of approaching terrain that may endanger the aircraft.

ELT   Emergency locator transmitter. An aircraft radio transmitter designed to activate automatically on impact; it aids in locating downed aircraft by transmitting signals to a locating satellite after an accident.

Employee (owner)-flown business aviation   Aircraft operated in support of a business and using a pilot or pilots not specifically employed or compensated to fly the aircraft.

ETA Estimated time of arrival.

ETE Estimated time en route.

ETOPS Extended-range operations. Flight operations in which a suitable emergency airport is more than 90 minutes distant during some portion of the flight.

FAA Federal Aviation Administration. U.S. aviation regulatory body.

FAR Federal Aviation Regulations. U.S. federal rules controlling aviation operations.

FBO Fixed-base operator. Aircraft service company providing fuel, transient parking, passenger waiting areas, and often aircraft hangars and maintenance services.

FDR Flight data recorder. Electronic device installed in some larger turbojets that records a number of aircraft flight and systems operating parameters.

Fixed costs Operating costs applicable whether an aircraft flies or not. Also known as *indirect costs.*

Flight plan Routing, aircraft, passenger, and pilot information relating to a specific flight, usually filed with a government agency; used for air traffic control and search and rescue purposes.

Flight time (airplane) The total time from the moment an airplane first moves for the purpose of taking off until the moment it finally comes to rest at the end of the flight. Also known as *block time.*

Flight time (helicopter) The total time from the moment a helicopter's rotor blades start turning until the moment the helicopter finally comes to rest and the blades are stopped.

FMS Flight management system. A computer system that uses a large database to allow routes to be preprogrammed and fed into the system by means of a data loader. The system is constantly updated with respect to position accuracy by reference to a variety of navigation aids.

FOQA Flight operational quality assurance. Program for the routine collection and analysis of digital flight data gathered during aircraft operations; intended to amass data to be used in assessing operational quality and safety.

Fractional ownership An arrangement whereby an individual or company purchases a share of an airplane (normally not less than one-sixteenth) that is placed in a pool of similar aircraft to be shared with other similar owners. The pooled aircraft are operated and managed by a company that provides all necessary personnel and services.

FSF Flight Safety Foundation. Organization promoting safe flight operations worldwide.

GA General aviation.

General Aviation (ICAO) An aircraft operation other than a commercial air transport operation or an aerial work operation.

GMT Greenwich Mean Time. Now correctly called *Coordinated Universal Time (UTC).* Time at the prime meridian. Also known as *Zulu time.*

GNSS Global navigation satellite system. Generic term for any satellite-based navigation system.

GPS Global Positioning System. A space-based radio positioning, navigation, and time-reference system. The system provides highly accurate position and velocity information and precise time, on a continuous global basis, to an unlimited number of properly equipped users.

GPWS Ground proximity warning system. Aircraft electronic system that warns of inadequate separation from ground and excessive sink rate close to ground.

IAOPA International Council of Aircraft Owner and Pilot Associations.

IATA International Air Transport Association.

IBAC International Business Aviation Council. An association representing business aviation associations around the world.

ICAO International Civil Aviation Organization. A specialized agency of the United Nations whose objective is to develop the principles and techniques of international air navigation and to foster planning and development of international civil air transport.

IFR Instrument flight rules. Regulations designed to provide safe operating procedures for aircraft operating in weather conditions less that the minimums required for flight under visual flight rules (VFR). IFR flights usually receive a route clearance and flight monitoring by air traffic control agencies.

ILS Instrument landing system. Short-range navigation system that uses precision localizer and glide-slope radio transmitters near a runway to provide landing approach guidance. See also Cat. I, II, etc.

IMC Instrument meteorological conditions. Meteorological conditions expressed in terms of visibility, distance from cloud, and ceiling less than the minimums specified for visual meteorologic conditions.

INS Inertial navigation system.

Instrument approach procedure A series of predetermined maneuvers for the orderly transfer of an aircraft under instrument flight conditions from the beginning of the initial approach to a landing or to a point from which a landing may be made visually.

Interchange agreement Method of mutual leasing of aircraft between parties in exchange for equal time in the other's aircraft. No charge or fee may be made except for the difference in aircraft operating costs.

JAA Joint Aviation Authorities. European-based regulatory body that generates aviation standards.

JAR Joint Aviation Requirements. Standards produced by the JAA.

Joint ownership Operating agreement under which two or more entities own an aircraft, one party provides the flight crew, and each party pays a share of charges specified in a legal agreement.

Kts Knots. Nautical miles per hour. Speed measurement, which is equal to 1.15 miles per hour.

Large Aircraft An aircraft with a maximum gross weight of more than 12,500 lb (5700 kg).

Life-limited items Aircraft parts that must be overhauled or discarded after a certain number of hours or a calendar interval.

LOA Letter of agreement/authority. Letter from a state regulatory authority permitting a specific type of operation, e.g., MNPS, RVSM, etc.

LORAN Long Range Navigation. An electronic navigational system by which hyperbolic lines of position are determined by measuring the difference in the time of reception of synchronized pulse signals from two fixed transmitters.

Maintenance Aircraft inspection, overhaul, repair, preservation, and replacement of parts; excludes preventative maintenance.

Manifest A document that lists the number/names of passengers and weight of baggage and/or freight carried on an aircraft.

MDA Minimum descent altitude. The lowest altitude, expressed in feet above mean sea level, to which descent is authorized on final approach or during circle-to-land maneuvering in execution of a standard instrument approach procedure where no electronic glideslope is provided.

MEA Minimum en route altitude. The lowest published altitude between radio fixes that ensures acceptable navigational signal coverage and meets obstacle clearance requirements between those fixes.

MEL Minimum equipment list. A document issued by the country of registry's regulatory authority that permits operation of a specific aircraft with certain installed systems inoperative or removed.

MLS Microwave landing system. Precision instrument approach system used primarily in Europe.

MMO Maximum operating Mach number.

MNPS Minimum Navigation Performance Specification. Standards that require aircraft to have a minimum navigation performance capability in order to operate in MNPS-designated airspace; regulatory authority certification required.

MOA Memorandum of agreement (also military operating area).

MOCA Minimum obstruction clearance altitude. The lowest published altitude in effect between radio fixes on VOR airways, off-airway routes, or route segments that meet obstacle clearance requirements for the entire route segment

Missed approach A maneuver conducted by a pilot to pull up and go around when an instrument approach cannot be completed, either due to weather being less than published minimums or the aircraft not being in a position that will permit a safe landing.

Mode S Type of secondary surveillance radar (SSR) equipment that provides Mode A and Mode C interrogations, discrete address (Mode S) interrogations from the ground or air, and a data link capability.

Nav Navigation.

Navaid Navigational aid.

NBAA National Business Aviation Association.

NDB Nondirectional beacon. Ground-based radio transmitter designed to be located by an aircraft ADF.

NM Nautical mile. 1 NM = 1.15 SM (statute miles).

NMC National Meteorological Center.

NOAA National Oceanic and Atmospheric Administration.

Nonprecision approach procedure A standard instrument approach procedure in which no electronic glide slope is provided.

NOTAM Notice for airman. Advisories to pilots regarding inoperative features of airports, navigation aids, or air traffic control facilities.

NTSB National Transportation Safety Board. Independent federal agency responsible for investigating transportation accidents and incidents.

NWS National Weather Service.

On-demand air transportation The transportation of passengers and cargo by aircraft from one point to another in a manner and at a time designated by the person exercising operational control.

Operational control The exercise of authority over initiating, conducting, or terminating a flight.

OSHA Occupational Safety and Health Administration.

PANS Procedures for Air Navigation Services (ICAO).

PAR Precision approach radar. Ground-based radar providing precision approach capability to aircraft, normally associated with military installations.

PBN Performance-based navigation (ICAO).

Personal aviation   On-demand air transportation used in support of personal needs and desires.

PF   Pilot flying. The pilot actually controlling the aircraft at a given time during flight.

PIC   Pilot in command. The person who (1) has final authority and responsibility for the operation and safety of the flight, (2) has been designated as pilot in command before or during the flight, and (3) holds the appropriate aircraft category, class, and type rating, if appropriate, for the conduct of the flight.

Pilot-in-command time   Flight time logged while the pilot is (1) the sole manipulator of the controls of an aircraft for which the pilot is rated or (2) the sole occupant of the aircraft or (3) acting as pilot-in-command of an aircraft on which more than one pilot is required under the type certification of the aircraft or the regulations under which the flight is conducted.

PIREPS   Pilot reports. En route weather conditions reported by pilots.

PNF   Pilot not flying. Sometimes referred to as the *monitoring pilot.*

Positioning flight   Relocating an aircraft to another airport, normally without passengers on board, to meet a specific mission or need. Also known as *deadheading* or *ferrying.*

Precision approach procedure   A standard instrument approach procedure in which an electronic glide slope is provided, such as ILS and MLS.

Preventive maintenance   Simple or minor preservation operations and the replacement of small standard parts not involving complex assembly operations.

RNAV   Area navigation. Generic acronym for any device capable of aircraft guidance between pilot-defined waypoints.

RNP   Required navigation performance. En route and terminal navigation performance standards imposed on aircraft operating in specified areas.

Rotable parts   Aircraft parts subject to reuse after overhaul or repair and recertification have taken place.

RTCA   RTCA, Inc., formerly Radio Technical Committee on Aeronautics. U.S. federal advisory committee that develops aircraft electronic equipment standards.

RVR   Runway visual range. Runway visibility measured in feet by precision instruments located on a runway.

RVSM   Reduced vertical separation minimums. Areas of the upper airspace in which aircraft vertical separation requirements have been reduced. Requires special aircraft modification and certification.

SATCOM   Satellite communications. Normally associated with commercially available nonaviation communications.

SATNAV   Satellite navigation. Usually associated with a global navigation satellite system (GNSS).

SIC   Second in command. Pilot who is designated to assist the pilot in command during flight.

SOP   Standard operating procedure.

Stabilized approach   Defined safe aircraft operating condition, not to be exceeded during approach and landing.

Stage 2 or 3 aircraft   Noise emission classifications for aircraft defined by ICAO and state regulatory standards. Stage 2 aircraft are several times more noisy than stage 3 aircraft, creating a number of airport restrictions on those aircraft.

Standard industry fare level (SIFL)   U.S. Internal Revenue Service's aircraft valuation formula used to compute the value of nonbusiness transportation aboard employer-provided aircraft.

Standards and recommended practices (SARPS)   ICAO guidance promulgated to member states via 18 separate annexes.

Sterile cockpit   An SOP designed to minimize flight crew distractions in which unnecessary conversation and paperwork are restricted when operating on the ground and at low altitudes (usually up to 10,000 feet AGL).

TAWS   Terrain awareness warning system. See also GPWS/EGPWS.

TCAS   Traffic alert and collision avoidance system. Electronic system installed in aircraft designed to detect nearby aircraft that may create a collision hazard. Some systems provide escape maneuver directions known as *resolution advisories*. Also known as *ACAS*.

TCDS   Type Certificate Data Sheet. Specifications for an aircraft, engine, or propeller issued by a State certifying authority. This serves as the basis for determination of the item's airworthiness.

Time in service   Time from the moment an aircraft leaves the surface of the earth until it touches it at the next point of landing. Principally used in determining maintenance time intervals.

Time sharing   A method of leasing an aircraft for discrete periods to a second party and recouping only a portion of the total operating costs of the aircraft.

Transponder   The airborne radar beacon receiver/transmitter portion of the Air Traffic Control Radar Beacon System (ATCRBS) that automatically receives radio signals from interrogators on the ground and selectively replies with a specific reply pulse or pulse group only to those interrogations being received on the mode to which it is set to respond.

Type rating   A special pilot rating required to act as pilot in command of specific types of large, turbojet, or special aircraft.

V_1   Takeoff decision speed. The computed speed beyond which an aircraft can no longer stop within the available runway but can sustain flight in the event of an engine failure.

V_2   Takeoff safety speed/takeoff climb velocity.

VFR   Visual flight rules. Rules that govern the procedures for conducting flight under visual meteorological conditions (VMC). The term *VFR* is also used in the United States to indicate weather conditions that are equal to or greater than minimum VFR requirements. In addition, it is used by pilots and controllers to indicate type of flight plan.

VMC   Visual meteorological conditions. Defined ceiling, visibility, and cloud clearance values that permit VFR flight.

VOR   Very high frequency omnidirectional range. Ground-based radio navigational aid.

V_R   Takeoff rotation velocity. Speed during the takeoff roll at which the airplane's pitch attitude is positioned for liftoff.

V_{REF}   Reference velocity. Computed speed used during landing approach.

Windshear   A change in wind speed and/or direction in a short distance resulting in a tearing or shearing effect. Constitutes a hazard for aircraft when windshear is severe.

Zulu   Designator for Coordinated Universal Time (UTC). Time at the prime meridian.

Index

Printed in the USA
CPSIA information can be obtained
at www.ICGtesting.com
JSHW050135270724
66940JS00002B/17